Library of
Davidson College

Great Currents of Mathematical Thought

Great Currents of Mathematical Thought

Edited by

F. Le Lionnais

in two volumes

VOLUME

I

Mathematics:
Concepts and Development

Translated by Dr. R. A. Hall and
Howard G. Bergmann

Dover Publications, Inc.
New York, N.Y.

Copyright © 1971 by Dover Publications, Inc.
All rights reserved under Pan American and International Copyright Conventions.

Published in Canada by General Publishing Company, Ltd., 30 Lesmill Road, Don Mills, Toronto, Ontario.
Published in the United Kingdom by Constable and Company, Ltd., 10 Orange Street, London WC 2.

Great Currents in Mathematical Thought is a new English translation of the 1962 enlarged edition of *Les Grands Courants de la Pensée Mathématique*. It is published by special arrangement with the original publisher, Librairie Scientifique et Technique, 9 rue de Médicis, Paris VI, France.

International Standard Book Number: 0-468-62723-3
Library of Congress Catalog Card Number: 74-151133

Manufactured in the United States of America
Dover Publications, Inc.
180 Varick Street
New York, N. Y. 10014

CONTENTS

The prefatory matter and Part I of the first volume were translated by Dr. R. A. Hall. Part II was translated by Howard G. Bergmann, except for paper No. 28, which was translated by Arnold Dresden.

	page
Preface to the Second Edition	ix
Foreword	xi
An Unpublished Letter of Paul Valéry	xv
Introduction	1

Part 1, *The Temple of Mathematics*

BOOK ONE, STRUCTURES, page 9

1. *Definitions in Mathematics* by Émile Borel . . . 12
2. *The Architecture of Mathematics* by Nicolas Bourbaki . 23
3. *Analogy in Mathematics* by Robert Deltheil . . . 37
4. *Symmetry and Dissymmetry in Mathematics and Physics* by Albert Lautman 44
5. *Intuitive Approaches Toward some Vital Organs of Mathematics* by Georges Bouligand 57

BOOK TWO, DISCIPLINES, page 67

A. NUMBER

6. *The Natural Number and its Generalizations* by Maurice Fréchet 70
7. *A Mathematical Enigma: Fermat's Last Theorem* by Théophile Got 81

		page
8.	*The History of the Mysterious Numbers* by Paul Dubreil	92
9.	*The Problem of the Infinite: Transfinite Numbers and Alephs* by Henri Eyraud	109

B. SPACE

10.	*From Three-Dimensional Space to the Abstract Spaces* by Maurice Fréchet	115
11.	*A Journey into the Fourth Dimension* by André Sainte-Lagüe	125
12.	*On the Curvature of Space and the Possibility of Forming a Conception of it by Elementary Means* by René Thiry	143

C. FUNCTION

13.	*The Origin and the Evolution of the Notion of an Analytic Function of One Variable* by Georges Valiron	156
14.	*The Role of Families of Functions in Mathematical Analysis* by Paul Montel	174
15.	*From Cauchy to Riemann, or the Birth of the Theory of Real Functions* by Jean T. Desanti	181
16.	*The Innateness of the Transfinite* by Arnoud Denjoy	191

D. GROUP

17.	*The Notion of Group its Power and its Limitations* by André Lentin	202

E. PROBABILITY

18.	*Modern Opinions on the Foundations of the Theory of Probability* by Robert Fortet	211
19.	*Chance and Mathematics* by Pius Servien	221

Part II, The Mathematical Epic

BOOK ONE, THE PAST, page 229

20.	*A General View of the Evolution of Mathematics* by Paul Germain	231
21.	*Views on Newton's Mathematical Thought* by Pierre Brunet	249
22.	*A Centenary: Sophus Lie* by Élie Cartan	262
23.	*Women Mathematicians* by Mme. Marie-Louise Dubreil-Jacotin	268

BOOK TWO, THE PRESENT, page 281

24. *Mathematics at the Beginning of the 20th Century* by Lucien Godeaux 284
25. *Henri Lebesgue: Renewer of Modern Analysis* by Louis Perrin 298
26. *David Hilbert (1862–1943)* by Jean Dieudonné . . 304
27. *The International Congresses of Mathematicians* by Rolin Wavre 312

BOOK THREE, THE FUTURE, page 319

28. *The Future of Mathematics* by André Weil . . . 321
29. *Modern Methods and the Future of Applied Mathematics* by Roger Godement 337

PREFACE TO THE SECOND EDITION

If there still exist pseudo-humanists who glory in their lack of comprehension of mathematics (a lack of comprehension that they share with all that is not human), the growing number of laymen who regret that they cannot fully participate in this banquet of the gods and who wish to profit at least from the leavings of the feast is rather reassuring. For evidence I need only note the success of this book—not, of course, with those knowledgeable in mathematics, for this success was guaranteed by the quality of the collaborators—but with the public at large.

Being unable to prepare the second collection of which I dreamed —and which I spoke of to friends in the deportation camp of Dora in 1944–1945—I am happy to include in this second edition two previously unpublished articles which consolidate and extend two chapters of Part I.

"Modern Axiomatic Methods and the Foundations of Mathematics," by Jean Dieudonné, elucidates and completes the article of Nicolas Bourbaki—whom Jean Dieudonné can take pride in counting among his oldest and most faithful friends—entitled "The Architecture of Mathematics." No connoisseur should be unacquainted with the simple and at the same time ingenious and conclusive way in which Hilbert and, following him, the contemporary mathematicians, give a theoretical basis for mathematical reasoning—in particular, reasoning about the infinite—by emptying it of all content. The reader will thus appreciate the high level of thought which Jean Dieudonné manifests in his manner of envisaging the possibility—an eventuality which seems scarcely likely at present—of a contradiction in mathematics.

As a direct extension of his article, "Intuitive Approaches Toward Some Vital Organs of Mathematics," the report of Georges Bouligand,

"Views on Mathematical Training" brings us a faithful reflection not only of the theses of the author, but also of his way of thinking. May I be permitted to see in some of his reflections on the heuristics of our science an illustration as valuable as it is original of the propositions I formulated several years ago in an article entitled "Toward a Mathematization of the Method?"[1]

F. LE LIONNAIS

Before sending the text of this new preface to the printer, it is of interest to note that the calculation of the initial decimal places of the number π has taken a giant step forward since the time when Professor Dubreil wrote his article, "The Mysterious Numbers." Giant electronic computers have been employed and we now have the first 100,000 decimal places of π. This is an indisputably remarkable performance inasmuch as it allows us to sense the extraordinary calculating power of the modern machines, but it is of rather slight interest to mathematicians. Actually, applied mathematics needs far fewer digits for the problems of technology and the physical sciences. As for pure mathematics, the least theorem about the number π would be of greater value, and there is scarcely any chance that the examination of the sequence of digits or of their statistics will lead to any discovery.

While we are on this topic—that of numbers which if not mysterious, are at least "electronic"—let us cite the largest prime number known at present: $2^{4423} - 1$.

---

[1] "Vers une mathématisation de la Méthode," published in *La méthode dans les sciences modernes*, Éditions Science et Industrie.

FOREWORD

This work, the idea for which came to us during the occupation, has undergone many vicissitudes. Let me be permitted to recount them here as my sole excuse for appearing in such learned company.

François Le Lionnais was in Marseilles in 1942. Fascinated by the extent and, above all, the lucidity of his knowledge, I suggested that he solicit expository articles from the best mathematicians of the time and assemble these into a book which would present a picture of the ideas and spirit of present-day mathematics.

We agreed that this work should attempt to show not an immobile panorama of the different disciplines belonging to this science, but, above all, the different directions in which they were moving. We hoped that this would inaugurate a series of works on scientific humanism. We have kept the title chosen at that time because the word "currents" [*courants*] seemed to define correctly the perspective in which we viewed our efforts.

Neither of us had any doubts about the difficulties confronting such an enterprise in 1942, but we were far from imagining that five long years would be necessary to bring it into being. No one could foresee all the obstacles which would retard the execution of our plans. One need only recall the period when the Frenchmen of the two zones could only correspond by those postcards so strictly rationed by our prudential occupiers! No room was left for the imagination and still less for intellectual exchange. Into those parsimoniously counted lines we had to compress immense questions, suggest a complex design and outline problems which permitted no trifling with their statements.

There was no hope of finding contributors outside of France. The exiles were lost to us, but we hope to invite them to participate in our second series [never published]. We were able to reach some of them since, but the same delays which allowed us to include André Weil, the brother of our sorely missed Simone Weil, caused us to lose the irreplaceable assistance of Paul Valéry, Léon Brunschvicg and Jean Cavaillès. During this period we had suffered the loss of two contributors who were unable to see the completion of the work, Albert Lautman, shot by the Germans at Toulouse, and Paul Mouy, whom we lost at the end of the year 1946.

Our work, despite so many obstacles and sorrows, went along slowly until a certain event stopped it cold. Le Lionnais, who belonged to a resistance group, was arrested at the end of April 1944 and taken to the concentration camp at Dora, where he was employed in the underground V2 factories during those endless months which were to pass until the advance of the Allies. Few enough returned from that notoriously dismal hell; our friend escaped the fate of the majority of the prisoners only because of his good health and high morale. Among the hopes which sustained him, the project of the *Main Currents of Mathematical Thought* so occupied his mind that one day it almost cost him his life. He had rewritten the summary from memory, when by an unfortunate accident he dropped in front of the guards a piece of wrapping paper on which certain names were written. He was called in to explain it to a police agent. The latter, as ignorant as only a prison guard can be, viewed everything in his own light. This man could see in the list of names Borel, Montel, de Broglie, Valéry, Brunschvicg, only a group of prisoners who wanted to give him the slip. It was impossible to make him change his mind; he demanded a complete confession: At what time, at what place were the accomplices to assemble? He finally let himself be convinced, but that did not prevent him from decreeing punishment for the crime of having written with a Nazi pencil on the paper of the Third Reich, and it was thus that for the love of mathematics our friend Le Lionnais received the mandatory lashes of the whip which might have killed him.

When he returned at the beginning of May 1945, long months were necessary in order for him to regain his health and readapt to normal life. His health had been greatly shaken and his papers had been destroyed, notably his article on beauty in mathematics. Thanks to the precautions he had taken, he was able to recover one by one copies of the texts already obtained; he sounded the rallying call again and the crew began anew to navigate toward their destination.

We cannot thank François Le Lionnais too much for his perseverance, praise him too much for his faith. In spite of many obstacles he has brought an important work into existence. All those who are devoted to works of the mind and especially to mathematical research will be grateful to him for having illuminated the problems which so arouse their interest with the best light the times can provide.

But let us take care not to forget the concern of the scholars who were not asked in vain, nor the collaboration of eminent minds, accorded with good will despite the many tasks which burdened them. We thank all the contributors to this work for their confidence and their devotion. We hope that they will find the justification of their efforts in the pleasure of seeing them assembled here.

JEAN BALLARD

AN UNPUBLISHED LETTER OF PAUL VALÉRY

THIS work had already gone to press when a friend, knowing of the gap left at first by the death of the man who was to write the preface, sent me a letter of Paul Valéry. In their correspondence my friend had doubtless questioned Valéry about the best introduction to the mathematical sciences. Here is Valéry's reply:

40, rue de Villejust XVIe.

                              Letter dated 29 February, 1942
                              (the date is erroneous, since the letter
                              arrived on the 28th)

Sir,

Your letter embarrasses me, which is not to say that it does not interest me. I was astonished first of all to be considered as a consultant in this matter, since you have a professional mathematician at hand. Then, since you are only pursuing a plan of intellectual cultivation, I cannot understand why you wish to commence with mechanics rather than some other branch of the science. If you have almost no mathematical training, you cannot hope for a long time to be able to reflect usefully on the works of Hamilton and Lagrange. Moreover, I can think of no book which entirely fulfills your requirements. In any case, if the mathematical aspect of thought, or rather the philosophical aspect of mathematics, interests you, read the works of Bertrand Russell, which are very remarkable, and combine the reading of them with that of the critical studies of Henri Poincaré. As for mechanics, you will find very valuable information on its origins in the two volumes of Jouguet, composed of original texts dating back to antiquity. Unfortunately, the war has interrupted the publication of this organized and annotated anthology, which stops (if I remember exactly) before the introduction of Mayer's principle.

But if, generally speaking, you do not intend to make mathematics your principal object of study, and if you only seek that characteristic fruit of it which attention and the analysis of arbitrarily defined concepts can offer to the mind, then I counsel you to reconsider the basic ideas of the science and to think through for yourself those problems which are seemingly most elementary. These foundations are, moreover, a perpetual source of reflection and discovery for the masters of the science. Even in number system you will find material for long reflection. Remember that Leibniz did not disdain to occupy himself with it. It is no less interesting to meditate on algebraic notation; all of its formal and symbolic part, which was abstracted from algebra little by little and has undergone immense development, is worthy of great interest. The same is true for the definitions and postulates of geometry, to which an infinitely subtle analysis has been applied in modern times, permitting physics to be conceived as a generalized geometry.

There, Sir, are some suggestions. I do not know if they accord with your desires, but I am not at all a specialist—at most an admirer and an unlucky lover of this most beautiful of the sciences.

With warm regards,
PAUL VALÉRY

These unpublished lines[1] of the poet who revered number and dreamed of algorithms for thought will not replace the preface that he had promised us and which would have been the perfect pediment for this edifice; but the absorbing interest that it indicates and the appropriateness of its counsel made it seem to us that it had a place here, especially since it would confirm the presence of its author among the mathematicians, his friends. It thus accomplishes the desire he expressed while he lived.

---

[1] This letter was communicated to us by Pierre Honnorat (Marseilles), to whom we are grateful for this gesture. Mme. Paul Valéry has kindly authorized us to publish it.

Great Currents of Mathematical Thought

INTRODUCTION

by F. Le Lionnais

> *Even today the true importance of mathematics as an element in the history of thought is not fully appreciated.*
> WHITEHEAD

> *All of our modern life is as if impregnated with mathematics. Man's daily acts and constructions bear its mark and even our artistic pleasures and moral life are subject to its influence.*
> PAUL MONTEL

> *Mathematics is one of those subjects which obliges one to make up his mind.*
> L. G. DES LAURIERS

The development of the sciences as the most significant factor in the liberation of human potentialities plays a decisive role in the evolution of civilizations.

It influences them first of all in a direct and material way by the improvement of the old and the creation of new techniques. These techniques in their turn transform the economy and thereby comes the metamorphosis of social relations, of political systems and soon even of the creative activities of thought and art.

But the influence of science is not limited to its action on and through technique. It is by a more intimate contact with the universe—and not by a defeatist withdrawal into himself—that man sees his dormant inner powers awaken and blossom. All these new directions in which science ceaselessly calls upon our mind to take part and to develop act upon its structure and mould it vigorously. How can there help but be a cascade of repercussions within all the sectors of everyday life, private and public, and in all the domains of civilization where this human thought, so radically remoulded by experience and knowledge, comes into play? The fruit of the tree of science has been stolen more than once since the dawn of history; every day it is plucked anew, bringing with it a train of changes in our conceptions, our mores and our history. In this sense we make the following thought of the mathematician Clifford our own: "Scientific thought is not an accompaniment or condition of human progress but human progress itself."

Already essential from the beginnings of humanity, the role of science ceaselessly grows larger with the march of history. In our time the pace of discovery has quickened, revolutionizing our manner of life at the same time as our knowledge of the world. The power which is the ineluctable corollary of science has not failed to awaken the liveliest and, in a certain sense, the most legitimate fears in the hearts of the many who are frightened by every novelty. The problems raised by these prospects have given rise to solutions, some of which scarcely appear unbiased and many of which appear to be chimerical or naive. Among these we count all those attempts to conjure away by a return to the past the problems thrown up by that explosion of modernism whose violence is unfurling above our heads. This would be equivalent to asking a single man to stop an avalanche solely with the strength of his body. The flow of the torrent of science, inscribed in the brain cells of Homo sapiens from the time that several mutations caused him to develop from the Neanderthal species, cannot be stopped; but might it not be possible not only to channel it, but even to tame it and make it serve the great and fascinating adventure in which all prior human history will have been only an episode?

We reply to this question in the affirmative, as do those for whom a new social and economic regime is at present being worked out amid painful convulsions, a regime which will be capable, when it is extended to the entire planet, of assimilating scientific progress without difficulty. Necessarily accompanying this regime will be a new type of man defined by a new kind of humanism, a rational humanism inheriting the fruits of the initial attempts made by the Renaissance and the Enlightenment.

The collection begun with this work has as its purpose the definition and the construction of that scientific humanism which will be the crowning achievement of the civilization of tomorrow, and it will attempt to avoid in so far as possible all useless or excessive technicality in the exposition of the facts and theses which confront it. We would like to be able to engrave on the pediment of our enterprise the sentence that Fontenelle placed in the preface of his Entretiens sur la pluralité des mondes:

"My wish has been to treat philosophy in a manner not at all philosophic; I have tried to bring it to a point where it would not be too dry for the common man, or too trifling for scholars."

If we have insisted in commencing our tour of the scientific horizons with an examination of the Main Currents of Mathematical Thought, *it is not only because it is generally agreed that mathematics has its place at the origin of the chain of sciences—this view is not at all false, though oversimplified—but* also because mathematics constitutes one of the most original, the most astonishing and, above all, the most revealing forms of human thought. Many attempts have been made to characterize man in different ways, no one of which, probably, is definitive. From certain points of view we are tempted to define man, without any paradox, as a mathematical animal. Doubtless this is not the only characteristic, for many elements of the principal human mental activities—art, science, morality, religious feeling, etc.—are found in the higher animals. Nevertheless, of all these activities, the aptitude for abstraction seems to us to mark most clearly the difference between man and his neighbors in the animal family. And is not mathematics the alembic in which the most quintessential abstractions are distilled?

Issuing from the real world (with which it may more or less lose contact, provided that it never loses it definitively), mathematics has become the most abstract of sciences. It is to this double nature that it owes that characteristic physiognomy which at all times has made such a great impression on mankind. In the first place, there is its extraordinary efficacity, which permits us to have recourse to its power each day and almost each second of our individual lives as well as in our collective endeavors, in order to insure the success of our most humble acts as well as our most ambitious enterprises. Next, the perfection and the infallibility which stupefy some, make others envious and fill all with admiration, and which cannot be seriously affected by controversies about details, those minor episodes which are effaced as easily as the ripples the wind forms on a lake. Then, its reputation for difficulty, which is justified in great part, but which causes some to deduce caluminous imputations of dryness, rigidity or monotony, which we hope this volume will help to dispel. And also its position, at once superior and seemingly ambiguous, which makes it like a bridge between philosophic speculation and scientific experimentation, and which causes the most ethereal of theories to be at the same time the most practical of techniques.

We cannot hope to cover the vast plain of mathematics in this work. We have undertaken, however, to sketch its configuration and to describe the principal streams which furrow it. The plan that we have adopted ought to make possible a preliminary idea which will be already clear and distinct and which we hope to be able to complete in a second collection[1]. We would like next to justify this plan.

Any examination of a subject can proceed in two ways. The first consists in

---

[1] The second collection to which Le Lionnais refers is the one mentioned in his Preface to the Second Edition. It was never published. This Dover edition in two volumes is an unabridged translation of the work originally published in one volume by Le Lionnais [publisher's note].

exploring the interior of its domain, first describing its general form, then proceeding step by step with a systematic inventory of the parts which compose the edifice. The second aims at delimiting the boundaries and next defining the relations of the subject under study with areas external to it. This constitutes the essential outline of our approach.

Mathematics forms a cyclopean structure, so vast and so full of life that one takes fright at the thought of embracing it and of commanding a view of this limitless architecture. We have devoted the nineteen articles of the first part to a sketch, if not to a description, of this "Temple of Mathematics." A daring enterprise, certainly, and we ask that it be judged only in its entirety, after the appearance of Collection II, in which the attempt to portray this difficult subject will be continued.

The second part, "The Mathematical Epic," will advantageously complement the first part by allowing us to be present at certain moments during the construction of this astonishing edifice in the past and present and by permitting us to participate in the projects of some of the present architects who dream of its future growth and modification; at certain points this second part will prepare the way for the third and final part, entitled "Influences." We shall then leave the "Mathematical Temple" in order to visit all the other monuments of the city and also the region in whose heart it is built, to which it owes its reason for being and with which it is constantly joined. We shall endeavor to take note of some of these interrelations in our inquiry into what is external to mathematics, and, notably, into science and technology, to which we shall devote the collection which is to follow the present one.

In undertaking this small mathematical crusade I did not conceal from myself the difficulties of every nature that would have to be surmounted. I would not have been able to succeed in this task without support and advice, of which I would at least like to indicate the most important.

In working with Paul Montel, whose jubilee was just celebrated, I benefited not only from that universal and profound comprehension of mathematics about which I shall say more later, but also from his personal acquaintance with all the mathematicians; and also from that witty conversation and inexhaustible affability appreciated so much by those who have come to know him. I owe to him my being able to surmount or to bypass many of the obstacles I encountered on this long road. I gained from Georges Bouligand much useful counsel and an inspiring sympathy for this work, whose development he facilitated in many ways. The friendship of Charles Ehresmann and his intimacy with the most advanced notions of mathematics largely contributed to the success of many parts of this volume. But if I am to parcel out my gratitude, then I must thank all the contributors to this work for their participation and for the good will they showed in adapting their personal views to my general directives. It is also a pleasure for me to tell how much this work owes to Jean Ballard, to his tenacious insistence which convinced me to undertake it, to the confidence he showed me through many vicissitudes and, finally, to the attention he gave to the form of the work.

I have tried to justify the structure of this volume and to show the connections between its parts by uniting them with a guideline. All those pages printed in italics between two articles form the continuation of the above remarks and appear as the fragments of a single article.

PART I

The Temple of Mathematics

"*The conceptions with which we work and those intimate connections we study are themselves the product of a prolonged effort of mathematical thought and are widely separated from the thoughts of everyday life.*"

FELIX KLEIN

"*It is difficult to give an idea of the vast scope of modern mathematics. The word 'scope' is not the best; I have in mind an expanse swarming with beautiful details, not the uniform expanse of a bare plain, but a region of a beautiful country, first seen from a distance, but worthy of being surveyed from one end to the other and studied even in its smallest details: its valleys, streams, rocks, woods and flowers.*"

ARTHUR CAYLEY

Book One

STRUCTURES

Before exploring in turn the different chapels of the Temple of Mathematics, it is appropriate to tell, in so far as possible, of its general plan, to specify the materials employed, to describe the architectural styles which are juxtaposed in it. This is the aim of the first book of Part I, entitled "Structures." Each of the articles which compose it has a general interest, and bears not only on this or that chapter of mathematics, but on its entirety, as well as on the connections and the relations of its parts.

We could not wish a better beginning to this work than that furnished us by one of the most eminent French mathematicians of the present day, Émile Borel. One cannot make an inventory in a few lines of the mathematical work of Professor Borel. His importance is due at the same time to the abundance and the variety of his works, to his intuitive penetration and to a certain stamp of perfection all his own. Professor Borel's research has taken three principal directions. In analysis, he has opened new paths in the theory of divergent series, in the examination of the notions of monogeneity, analyticity, quasi-analyticity, in the theory of growth, etc. He has contributed to the strengthening of the productive relationship between the theory of sets and the theory of functions, as much by his personal discoveries on the measure of sets as by the publication of the authoritative Collection de Monographies sur la Théorie des Fonctions, *which exerted a considerable influence on mathematics at the beginning of this century. Finally, he has left the mark of his genius on the theory of probability. A sort of realism, which inspires Professor Borel to discard all defective and gratuitous concepts from mathematics, marked him out as well-suited for writing this study on the role of definition in mathematics. Here he examines the foundation on which our entire edifice rests. All at once we find ourselves involved in one of the dramas of mathematics. This grandiose construction is, of course, collective, but* its architects sometimes belong to fundamentally different human types; their differences are reflected and echoed not only in the esthetics but also in the stability and solidity of the structure.[1]

The disputes among intuitionists, logicians and materialists, among objectivists and subjectivists, are in a certain sense the renewal of the old quarrel of Idealism in antiquity and the Middle Ages, which reappears as soon as one begins the examination of the foundations of mathematics. The position adopted by Professor Borel is well known, spelled out as it was in polemics which caused an uproar in their day. Whether the reader adopts or rejects it, he will in any case be grateful to the author for an exposition whose elegant clarity excludes all ambiguity.

After testing the materials, we next discuss the plan of the edifice, which was itself the object of heated controversy.

Having in mind the classical conception of "disciplines" (which we have very roughly adopted in the classification which follows in order to remain closer to the idea currently held of mathematics), we felt it necessary to give a place to one of the schools in the vanguard of present-day mathematics. In order to disentangle the meaning of the totality of mathematics, we addressed ourselves to Nicolas Bourbaki, that many-headed mathematician who has undertaken the reformulation of the exposition of mathematics from its origins—not historical but logical— and endeavors to reconstruct it in all its complexity with materials passed through the sieve of axiomatic critique. Understood thus, mathematics reduces to the study of very general laws that apply to collections of elements which are no longer necessarily numbers or points. Bourbaki's article could serve as a manifesto for an entire school which follows his teaching and as preface to a work of which we have as yet only six fascicules[2] and whose completion is awaited with impatience. It is interesting to note that, while proclaiming his adherence to the axiomatics of Hilbert and to the formalist movement, the author refrains from taking any fixed philosophical position and takes care to make precise the very technical sense in which he employs the term formalism.

The Bourbakist conception causes cleavages in the block of mathematics which are in some sense perpendicular to the stratifications corresponding to the traditional disciplines. Such cleavages cause relations which would never have been suspected to appear, often in a surprising light, between widely separated regions of mathematics. But just as there exist several traditional and more or less independent stratifications (arithmetic and geometry; then algebra and analysis; finally groups, sets, matrices and tensors, probability, etc.), so there

---

[1] Those who apply mathematics rarely notice this, since they have recourse only to the parts of mathematics which have been sheltered from these quarrels.
[2] Published by Hermann, Paris.

exist several cleavages (structures of algebra, of order, of topology; and their combinations), and thus we are faced with the problem of the unity of mathematics. While awaiting new works which will resolve this question in one way or another, cannot one remark that it is the intertwining of these two classifications (classical "disciplines" and modern "structures") which assures, as in a braid, the interdependence, cohesion and solidity of mathematics? This would assure that, even if lacking unity of content or method, it would have a sort of unity in its construction.

Bourbaki's message, so rich and so dense with meaning, has as its overall aim a systematic inventory of the analogies in mathematics and at the same time an elucidation of their validity and of their significance. The two articles which follow it can be related to this message from some quite different points of view.

That of Robert Deltheil, one of those most knowledgeable in geometric probability, harks back to the climate of classical mathematics. He converts the note of the previous article into small change and offers us a wittily plucked bouquet of analogies, the last of which lead us effortlessly to the heights of this subject.

The article of our late friend, Albert Lautman, on the other hand, takes up the modern conception again and, in a particular but most instructive case (a certain mixture of symmetry and dissymmetry), he extends it at one and the same time in both directions of the mathematical river: toward its physical source and toward its philosophical mouth.[3]

In contrast to these theses which adhere to the formalist or logical school, we next experience the sounding which one of the best defenders of intuition, Georges Bouligand, takes with agility and sagacity in the same directions. Besides his varied researches—notably on boundary value problems in potential theory—he is well known for his works on differential geometry. In his article we find notions uniquely his, such as those of causality in mathematics, the stability of propositions, extended intuition—notions which throw a new light on the foundations of mathematics.

F. LL.

---

[3] A. Lautman had suppressed, at our request, all those parts of his article (located especially at the beginning) referring to physics. As is well known, he was arrested in 1944 and shot for his participation in the Resistance during World War II.

On our return from deportation we insisted on publishing his article in its entirety, thus assuring the wide diffusion of ideas which are very often remarkable and which their author would no longer be able to express on other occasions.

We have since had the pleasure of learning of the separate publication of the same text by Éditions Hermann.

1

DEFINITIONS IN MATHEMATICS

by Émile Borel
THE ACADÉMIE DES SCIENCES

I

"MATHEMATICS is the science in which one never knows what one is talking about, nor whether what one is saying is true."

This sally of Bertrand Russell tends to emphasize the fundamental role that arbitrary definitions play in mathematics. But one could just as well say: "Mathematics is the only science in which one always knows exactly what one is talking about and in which one is certain that what one says is true." Of course, if we are discussing the properties of a line, the expression *straight line* does not have the same meaning in the geometries of Euclid, Riemann, or Lobachevsky; however, if a line is defined in conformity with the postulates of this or that geometry, then we will know exactly what we are talking about and we will be assured of the truth of this or that theorem: for example, the sum of the angles in a triangle is always equal to two right angles, or it is always greater than two right angles, or it is always less than two right angles. Although these theorems appear contradictory, each of them is true in one of the geometries, i.e., the one with the appropriate definition of a line.

Mathematics is coming to appear more and more as the science which studies the relations between certain abstract entities defined in an arbitrary manner, restricted only in that these definitions must not lead to a contradiction. However, to avoid the risk of confusing mathematics either with logic or with games such as chess, we must add that these definitions were suggested first of all by analogies to real objects; such is the case for the straight line, for the circle, for the rigid body of dynamics, etc. On the other hand, complex numbers, transfinite numbers and many other mathematical entities are pure creations of the human mind. They are justified by the fact that they permitted an easier solution of problems posed by mathematicians or physicists and clarified difficulties which they faced.

In addition, a distinction must be made between those definitions which can be called general, such as those of a function of a real or complex variable, and the particular definitions which allow one to single out classes of functions having a special interest, such as continuous functions of a real variable, discontinuous functions of the first Baire class, etc., and, among the functions of a complex variable, entire functions, meromorphic functions, Abelian functions, Fuchsian functions, etc. Finally, one can specialize still further and define individual functions, such as the exponential function, trigonometric functions, etc. The situation is similar in geometry and in all the other branches of mathematics. Of course, there is a certain degree of arbitrariness in the choice of the definitions considered to be interesting, but we cannot discuss this question here since it would entail a critical study of the entire field of mathematics, a task requiring numerous volumes, even supposing that an author could be found with sufficient competence to write it.

Therefore, we shall limit ourselves to these general statements on axiomatic definitions and go on to study in more detail some problems concerning those definitions which can be called individual and which define a specific mathematical entity rather than a class of such entities.

II

Let us start out with the simplest definitions of arithmetic, those of the integers, the rational numbers and the irrational numbers. Each of these defines a *set* of numbers and permits a precise, unambiguous definition of certain elements of these sets. For example, 3, 17, 10^{100}, $2^{10,000}$ are specific integers; their precise definition is obtained from the conventions relating to decimal representation and integral exponents. Similarly, 3/4, 5/7 or 0.3427 are well-defined rational numbers. We also know that $\sqrt{2}$, $\sqrt[3]{7}$, e and π are irrational numbers. The principal question which first arises is the following: which numbers can be *well defined* in this way? It is clear that a solution of this question would lead to a solution of the analogous questions that could be posed for all mathematical entities. For example, a circle is defined if its center and radius are given; moreover, two circles of the same radius are equal and are regarded as identical in Euclidean geometry; but to give the radius, a number must be given and we are back to the problem of the definition of number.

Let us first consider the integers.

By means of mathematical symbols it is easy to define arbitrarily large integers; if, for example, $\phi_1(n)$ denotes the number 10^n; $\phi_2(n)$ the number $\phi_1[\phi_1(n)]$; $\phi_k(n)$ the number $\phi_1[\phi_{k-1}(n)]$; etc.; the number $\phi_k(n)$ will be a very large number, if k and n are themselves very large. Then we could take $\psi_1(n) = \phi_n(n)$ as our point of departure and obtain still larger numbers, and so on. However, it seems that those numbers having an incalculable number of digits are exceptional among the numbers having a very large number of digits. It would be completely impossible for a man to write a number with several billion digits, if he were actually required to write down all these digits. But it is nevertheless conceivable, if the number n were defined as the billionth prime number, that the number n is well defined, for all mathematicians would be sure that they were talking of the same number when they talked of n. However, it is beyond the present capacity of the science to actually write down the number n, although it is known to have only 10 digits; the computation would take too long. This reasoning applies a fortiori to the 10^nth prime number N. But it is not absurd to think it will be possible someday to demonstrate certain specific properties of the numbers n and N.

If we limit ourselves to integers with a relatively small number of digits—for instance, not exceeding 50 or 60—then it would appear simple at first glance to define any one of them, since we could write down a sequence of 50 or 60 digits chosen arbitrarily. However, it could just as well be said that it is easy to write down a sequence of 40 to 50 letters of the French alphabet in such a way as to obtain an Alexandrine verse. And yet it is certain that the French language will have ceased to exist, at least in its present form, before all the French verses of twelve syllables have been written. It is as inconceivable that we could know all the numbers of 60 digits as it is to suppose that we know all the French verses of twelve syllables. For the man who knows all the properties of numbers, two integers of 60 digits would be as distinct from one another as a beautiful line by Victor Hugo and a random collection of 46 letters would be. However long mankind endures, men will not be able to define more than a very few of the 60-digit numbers, and will choose among them those which have special or interesting properties, just as they will not write more than a very tiny part of the prose and poetical works which possibly could be imagined.

Let us say a few words about artificial and meaningless combinations. If we apply a coding procedure to a French text, we will obtain an incomprehensible grouping of letters of no interest to anyone who does not have the key to the code. For example, two given texts could be summed letter by letter; that is, on encountering the third and fourth letters of the alphabet one above the other, the seventh letter would be written; if the total exceeded 26, then 26 would be subtracted, i.e., the 15th plus the 17th would give the 6th. One could proceed in the same way with numbers: after defining a number with a very large number of digits by a simple arithmetic formula, say 2^{100} or 3^{117}, another number could be obtained from one of these by replacing the figure 5 everywhere by the figure 7 and conversely, or an *artificial sum*[1] of two numbers could be formed by agreeing that if the two digits 2 and 7 occur one below the other, then the sum would be written down as the digit 4, the last digit of 14. Artificially defined numbers really seem to have as little interest for mathematicians as cryptograms have for readers of literature. However, for the sake of completeness it was necessary to mention artificial numbers, since they incontestably satisfy the first condition that we have required of a correct definition: the number that they define is certainly *the same* for all mathematicians; they can be discussed without the risk of any ambiguity. The trouble is that there is nothing to say about them.

It is interesting to try to classify the integers which can be defined in terms of the time required to define them, assuming that a word can be spoken or a digit written each second. If the decimal system is used, then about the same number of seconds as digits would be needed, if it is agreed purely and simply to say these digits one after the other. Of course, certain numbers with a very large number of digits can be defined more simply by arithmetic formulas, such as 2^{100} or 257! (which denotes the product of the first 257 natural numbers). One can also speak of the prime number of rank equal to one of the numbers previously defined. We shall call the minimum time necessary to define an integer its *height* (expressed in seconds in accordance with the preceding conventions).

---

[1] The *actual* sum of $2^{100} + 3^{117}$ would be a number defined naturally.

Let us try to give an idea of the actual extent of the set of integers which have been studied by mathematicians, i.e., for which they know some property. The construction of tables of prime numbers, more than anything else, has yielded knowledge of the arithmetic properties of a very large number of integers. Tables of primes have been calculated up to 10 million and these also give the divisors of numbers which are not prime. These tables would permit by means of laborious but not completely unmanageable calculations the decomposition into prime factors of a number having not more than 14 digits. But if a number is written with 30 digits chosen at random, then it would be impossible to decompose it into prime factors or to know if it were prime.[2] The same is true for integers defined by arithmetical procedures as soon as the height reaches 20 or 30. Thus, the problem (interesting from many points of view) of finding the divisors of numbers of the form $2^n + 1$, where n is a power of 2, has been solved only for relatively small values of n. The height of these numbers, however, exceeds that of n only by four; i.e., it is 7 if n has 3 digits. For instance, we would say: 2 to the exponent 512 plus 1.

To sum up, the number of integers that could be defined is unlimited in the sense that, having defined a certain number, we could always define one more, but in practice it is limited to a relatively small number because of the short duration of human life and the fact that the great majority of men have other things to do than to define integers other than the ones needed in their daily life. In fact, most of the integers which have been defined by mathematicians appear in the various numerical tables which have been published, notably in the tables of primes. To these can be added the ones which appear in the numerous statistical and financial publications, although the latter do not appear to have been studied for their arithmetical properties and, therefore, are not of interest to mathematicians. On the whole, although certain of these numbers exceed a billion and although mathematicians have frequently considered much larger numbers, defined by means of the exponential or other notations, the total number of integers actually defined is certainly much less than a billion. The creative imagination of man has been much more pro-

---

[2] If we consider a number N chosen at random, it is very likely that it will have a certain number of prime factors that are smaller than 1000 or 10,000. It is not very difficult to find these prime factors, and by dividing N by these prime factors, a considerably smaller number will be obtained. However, if N is prime, or has several sufficiently large prime factors, the calculations cannot be carried out in practice.

lific in the domain of language, since the number of lines of all the volumes that have been printed is of the order of magnitude of a thousand billions and, as we have already observed, each line is as complex as a number of several tens of digits.

III

The detail with which we discussed the integers will permit us to be much more brief in regard to the other numbers, rational or irrational, since it is only by taking the integers as the point of departure that one can precisely define these other numbers.

This is evident for the rational numbers, which are the quotients of two integers; this is also the case for the algebraic numbers, which are the roots of algebraic equations with integral coefficients. Such numbers are well defined when the integers on which they depend are themselves well defined.

Let us now move on to the irrational numbers. The simplest and most important of these are defined by series or by differential equations. Such is the case for the number e and the number π; such is also the case for the trigonometric ratios, sine or cosine. It must be noted, however, that for $\sin x$ and $\cos x$ to be well defined, it is necessary that the argument x be itself a well-defined number. The same is true for the values of the elliptic functions (provided the value of the modulus is itself well defined) and of other functions conceived by mathematicians.

More generally, if we consider an algebraic differential equation with coefficients which are well-defined numbers, and if equally well-defined initial conditions are given, then the value of the integral will be itself well defined for a well-defined value of the variable.

The idea of height established for the integers can be readily extended to all the numbers defined in terms of integers; their height will be the sum of the heights of the integers which are involved in their definition. In the case of numbers defined by a series it suffices, of course, to give the law for the series.

Finally, if it is considered worthwhile, artificial definitions analogous to those of cryptograms can be added to these natural definitions; for example, from a number such as $\sqrt{2}$ or π written approximately in the form of a finite decimal, another number can be derived by replacing everywhere the digit 3 by the digit 8, and conversely. However, such artificial numbers do not have any other known properties than the one expressed in their definition, and so they are without interest for mathematicians. It is thus hardly likely that they will ever cease to be anything but mere curiosities.

Is it now possible to go further and to *define* other irrational numbers? Naturally, one can imagine many artificial arithmetic definitions; for example, a decimal whose digits are the sequence of natural numbers:

$$x = 0\ 1\ 2\ 3\ 4\ 5\ 6\ 7\ 8\ 9\ 10\ 11\ 12\ 13\ 14\ 15\ 16\ 17\ 18\ 19,\ \text{etc.}$$

The definition of a number such as x can be varied in many ways, but it is scarcely likely that such a number would ever play a role in mathematics, since for that the discovery of a special property beyond that of its definition would be required.

It is equally easy to *define* certain sets of numbers and to demonstrate properties which belong to all the numbers of one of these sets. An example is the set of all real numbers between 0 and 1; such is also the case for many sets that can be obtained by means of a finite or infinite number of operations. For instance, such an operation might consist in removing from a given set the points[3] included in a given interval whose end points are well-defined numbers. The set is then well defined and can be the object of further study which might make possible statements of certain properties common to all of its points. However, if one desires to speak of a particular point of this set, is it sufficient to designate it by the letter x in order to consider it well defined? This is what certain mathematicians appear to believe; for my part, I cannot consider x as well defined if, when I discuss x with another person, I cannot be certain that we understand one another, i.e., that we are speaking of the same number.

Of course, just as in certain cases one can demonstrate properties common to all the points x of a set E, it would be possible to demonstrate properties common to all pairs of points x and y with the sole condition that x is different from y. However, this last condition does not at all entail the conclusion that x and y are well defined; neither of them has any real individuality if all that is known about them is that they belong to the same set and that they are distinct.

When the notion of a well-defined number is made precise, as we have done, we arrive at the somewhat paradoxical conclusion that

---

[3] I employ here geometric language equivalent to the arithmetic language; to define a point is the same thing as to define the number equal to the abscissa of the point.

the well-defined irrational numbers actually considered by mathematicians are, in fact, far fewer in number than the well-defined integers. For, if we take the example of the algebraic numbers, only a minute number of the innumerable equations of degree less than one thousand and with coefficients less than one million have been actually studied or even written down. As for the transcendental numbers, only a very small number have been considered or studied apart from the values of certain functions for which tables have been calculated.

I shall not insist upon an answer to the following question: What is the order of infinity of the transcendental numbers which could be well defined, if human life were of unlimited duration and, hence, arbitrarily large integers could be used? Actually, this question would be connected with the subject of trans-finite numbers and would require a lengthy development.

IV

Before concluding, let us say a word about a difficult question. We have made clear what is to be understood by a well-defined number and we have required that when two mathematicians discuss such a number a, they be sure that they are discussing the *same* number. Should it not be added that when two numbers a and b are well defined, it should be known with certainty whether the two numbers are equal or unequal? This problem can be posed in the following way. Let us consider the integer $N = 145! + 1$, that is, the product of the first 145 natural numbers, increased by one. We do not know whether this is prime, but there is no information to the contrary. If N is prime, then well-known formulas of analysis allow the approximate calculation of its rank in the sequence of primes; this rank is certainly between $a - b$ and $a + b$, if b is sufficiently small with respect to a. Now, if we define N' as the prime number of rank a, then it is possible that N' equals N, but it is also possible that they are not equal; in the present state of the science the calculations necessary for this decision cannot be carried out practically. Ought one then to regard N and N' as well defined?

The same question can be posed in a different form for the irrational numbers, even in the case in which their height is relatively small. If two such numbers a and b are defined in a simple manner permitting their methodical calculation, as is the case for e, π, π^2, ..., for example, it will be possible, if they are unequal, to determine this after a finite number of operations; but if they are equal, this can never be certainly known unless it can be demonstrated by an analytic method. Actually, agreement for any number of digits whatsoever would not suffice to demonstrate it.

For example:
$$\pi = 3.1415926\ldots$$
$$\tfrac{355}{113} = 3.1415929\ldots$$

If only six decimal places had been calculated, it might have been believed that these two numbers were equal. Similarly, certain people seeking to square the circle observed that

$$\pi = 3.14\ldots$$
$$\sqrt{2} + \sqrt{3} = 3.14\ldots$$

and concluded that these two numbers are equal.

In the two preceding cases anyone knowing that it is impossible to square the circle (since π cannot be the root of any algebraic equation with integral coefficients) could have asserted without any calculation that the two numbers whose first few decimal places were the same were certainly not equal. However, it is very rare that a demonstration of this type is available.

There are many ways to define the number π by means of a series or a definite integral. If we set

$$s = 1 - \frac{1}{3} + \frac{1}{5} - \frac{1}{7} + \frac{1}{9} - \frac{1}{11} + \cdots,$$

$$t = \int_0^1 \frac{dx}{\sqrt{1 - x^2}},$$

then we know that

$$s = \frac{\pi}{4}; \qquad t = \frac{\pi}{2}.$$

Let us suppose that some mathematicians, not aware of the preceding results, had sufficient curiosity to compute the values of $2s$ and t to a large number of decimal places. They would observe, for instance, that the first twenty decimal places of these values are the same. Could they then conclude without further calculation that $2s$ is exactly equal to t? In the present state of the science the answer to such a question must be in the negative. However, one cannot doubt that the mathematician who had obtained such a result would eagerly seek a rigorous demonstration of the equality $2s = t$, and, if he was not able to obtain it immediately, would have posed this problem to other mathematicians until such a demonstration was found. For, even lacking such a rigorous demonstration, he would nevertheless have been convinced that the agreement of the initial decimal places was not fortuitous and that the following decimal places would also coincide.

However, it would be absurd to claim that two irrational numbers with a large number of decimal places in common must necessarily be equal; in fact, nothing prevents us from defining an irrational number x as being obtained by increasing or decreasing by one the thousandth decimal place of π; the numbers x and π would thus have 999 decimal places in common but would still not be equal.

The great difference that exists between the comparison of $2s$ and t and that of x and π is that, in the first case, we are dealing with numbers defined in a simple manner by means of formulas which can be stated in a few words, whereas the number x is obtained from π in an artificial manner and its very definition implies that x is not equal to π.

It would, therefore, be very interesting to be able to demonstrate a general proposition which would permit us to state in interesting cases analogous to that of $2s$ and t that what everyone feels is true is actually true. For this it would be necessary to start from the following observations. After defining precisely the height of an irrational number, it would be observed that the number of numbers of height less than h is certainly finite. As a consequence, the minimum difference between two of these numbers is a quantity $\phi(h)$ which may be very small, but is not zero. It would now be a question of determining the function $\phi(h)$ or at least a function $f(h)$ such that $f(h)$ is definitely less than $\phi(h)$. Then, if two numbers of height less than h, which were defined by different methods had a difference less than $f(h)$, one could be sure that they were exactly equal.

Liouville's theorems on algebraic numbers and more recent work[4] on the number e are a first step toward the determination of the function $f(h)$. This determination would constitute great progress in our knowledge of the arithmetic properties of numbers.

---

[4] See Émile Borel, "Sur la nature arithmétique du nombre e," *Comptes rendus Acad. des Sc.*, CXXVIII, p. 1281.

See also Émile Borel, "Sur l'approximation les uns par les autres de nombres rationnels ou incommensurables appartenant à des ensembles énumerables donnés," *Comptes rendus Acad. des Sc.*, CLXXVI, p. 66.

V

I would like, in conclusion, to say a few words about those "definitions" that could be called enigmatic, i.e., those which suppose an enigma has been solved to which we do not have the key. Certain of these definitions have a mathematical appearance; for some of them there is hope that the progress of the science will allow the resolution of the enigma, although that appears quite unlikely.

For example, let us define the number x in the following manner: in the decimal expansion of π we replace the billionth digit by a zero. Thus, if this digit is a zero, x will equal π; if not x is different from π. Those adept in the logic of Brouwer would say that we have here a case in which the law of the excluded middle does not apply and that neither of the propositions "x equals π" and "x differs from π" is either true or false.

The case in which the enigma posed is unrelated to mathematics is still less interesting; it is not defining a number to say that it is equal to 0 or 1 according to whether the number of Frenchmen killed at the battle of Waterloo was even or odd, or according to whether Bacon was or was not the collaborator of Shakespeare.

One could even "define" a number by defining its successive digits, the value of each depending on the solution of a scientific or historical enigma. Knowledge of this number would thus give the solution of the entire denumerable infinity of enigmas that could be posed in the domains of science, history, metaphysics or popular imagination. All this is pure fantasy. The interesting domain of research which remains open to mathematicians in regard to the definition of numbers and other mathematical entities is sufficiently vast so that one may easily forego an artificial extension of it.

2

THE ARCHITECTURE OF MATHEMATICS

by Nicolas Bourbaki

MATHEMATICS, SINGULAR OR PLURAL?

To give a general idea of the science of mathematics at the present time seems at first glance to be an almost insurmountable task because of the breadth and variety of the subject. As in all the other sciences, there has been a considerable increase in the number of mathematicians and works devoted to mathematics since the end of the 19th century. The articles devoted to pure mathematics published throughout the world in the course of a normal year cover several thousands of pages. Doubtless, not all of them are of equal value; but after pouring off the inevitable waste, it still remains true that each year mathematics is enriched by a throng of new results, is constantly diversified and ramified in theories which are ceaselessly modified and remolded, confronting and combining with one another. No mathematician, even if he devoted all his time to it, would be capable today of following this development in all its detail. A number of them shut themselves up in one corner of mathematics, which they never seek to leave, and are not only almost completely ignorant of everything outside their subject, but would even be incapable of understanding the language and terminology employed by their colleagues in a specialty separated from their own. There is scarcely anyone, even among the most cultivated, who does not feel lost in certain regions of the immense mathematical world. As for those such as Poincaré or Hilbert who imprint the stamp of their genius in almost every domain, they constitute a very rare exception even among the greatest.

Thus there can be no question of giving to laymen a precise image of something that even mathematicians cannot conceive in its entirety. But it is possible to ask whether this luxuriant proliferation is the growth of a vigorously developing organism which gains more cohesion and unity from its daily growth, or whether on the contrary it is nothing but the external sign of a tendency toward more and more rapid crumbling due to the very nature of mathematics, and whether mathematics is not in the process of becoming a Tower of Babel of autonomous disciplines, isolated from one another both in their goals and in their methods, and even in their language. In a word, is the mathematics of today singular or plural?

Even if it is more real today, this question should not be thought to be new; it has been asked almost from the very beginning of mathematics. Actually, even disregarding applied mathematics, there exists an obvious duality of origin between geometry and arithmetic (at least in their elementary aspect); the latter is initially the science of the *discrete*, the former of the *continuous*, two aspects which have been radically opposed since the discovery of irrationals. Moreover, it was precisely this discovery which proved fatal to the first attempt to unify the science, the arithmetism of the Pythagoreans ("all things are numbers").

If we were to follow the vicissitudes of the unitary conception of mathematics from Pythagoreanism to the present day, we would be led too far from our task. It is, moreover, a task for which philosophers are better fitted than mathematicians, since a common trait of the different attempts to unify mathematics into a coherent whole— whether by Plato, Descartes or Leibniz, by arithmetization or the logicism of the 19th century—is that they have been made in combination with a more or less ambitious philosophic system, always starting out, however, with a priori ideas on the relations of mathematics with the double universe formed by the external world and the world of thought. We could not do better than to refer the reader on this point to the historical and critical study of M. Brunschvicg, *Les Étapes de la Philosophie Mathématique*.[1] Our task is more modest and more limited; we shall not undertake the examination of the relations of mathematics with reality or with the main categories of thought; we shall remain within the domain of mathematics and seek, in analyzing its characteristic steps, an answer to our question.

---

[1] Paris, Alcan, 1912.

Logical Formalism and Axiomatic Method

After the more or less obvious failures of the various systems to which we referred above, it seemed at the beginning of this century that there was little hope of seeing in mathematics a science characterized by a single goal and a unique method; there was rather a tendency to consider it as "a series of disciplines based on special notions and with precisely defined boundaries" connected by "a thousand channels of communication" which permitted the methods proper to one of these disciplines to fructify one or more of the others (Brunschvicg, *loc. cit.*, p. 447). Today we believe, on the contrary, that the internal evolution of mathematics has, in spite of appearances, tightened the unity of the various parts more than ever, and has created a sort of central kernel, more coherent than ever. The essence of this evolution has consisted in a systematization of the relations existing among the various mathematical theories, and is comprised in an approach generally known under the name of the "axiomatic method."

The terms "formalism" and "formalist method" are sometimes used; but one must be on guard from the outset against the risk of confusion caused by these poorly defined words, a confusion only too often exploited by the adversaries of axiomatics. Everyone knows that the external character of mathematics presents itself in the form of the "long chain of reasoning" spoken of by Descartes; every mathematical theory is a series of propositions, each derived from preceding ones in conformity with the rules of a logic which is essentially that codified since Aristotle under the name of "formal logic" and appropriately adapted to the particular needs of the mathematician. Thus, it is a banal truism to say that this "deductive reasoning" is a principle of unity for mathematics. Such a superficial observation could certainly not take into account the obvious complexity of the various mathematical theories any more than one could unify physics and biology into a single science by saying that they both applied the experimental method. Reasoning by series of syllogisms is nothing but a transforming *mechanism* indifferently applicable to all sorts of premises and hence not capable of characterizing their nature. In other words, it is the external *form* that the mathematician gives to his thought, the vehicle which makes it possible for others to assimilate it[2] and, in a word, the

---

[2] Moreover, every mathematician knows that a proof is not really "understood" as long as he has only verified the correctness of the deductions involved step by step, without trying to understand clearly the ideas which led to the construction of this chain of deductions in preference to all others.

language appropriate to mathematics; nothing else should be sought in it. To codify this language, to order its vocabulary and to clarify its syntax is to do very useful work, constituting one real facet of the axiomatic method, that which can be properly called the *logical formalism* (or, as one also says, the "logistic"). However—and we insist on this point—*it is only one facet*, and the least interesting.

What axiomatics sets as its essential goal is precisely what logical formalism alone cannot supply: the deep-lying intelligibility of mathematics. Just as the experimental method starts out from the a priori belief in the constancy of natural laws, the axiomatic method finds its base of support in the conviction that, if mathematics is not a series of syllogisms developing randomly, neither is it a collection of more or less "crafty" artifices consisting of fortuitous juxtapositions which are a triumph of mere technical astuteness. Where the superficial observer sees only two or more apparently very distinct theories lending one another "unexpected aid" (Brunschvicg, *loc. cit.*, p. 446), through the agency of a mathematician of genius the axiomatic method teaches one to seek out the deeper reasons for this discovery, to find the common ideas buried under the external apparatus of detail appropriate to each of the theories considered, to single out these ideas and to exhibit them.

The Notion of "Structure"

How is this operation to be carried out? It is here that axiomatics approximates most closely the experimental method. Likewise drawing from the Cartesian source, it "will divide up the difficulties, the better to solve them"; it will seek to *dissociate* the principal lines of reasoning which figure in the demonstrations of a theory; then, taking each of them in *isolation* and considering it as an abstract principle, it will unfold the consequences properly belonging to it; finally, returning to the theory being studied, it will *recombine* its constituent elements which had previously been separated out and will study their interactions with one another. Of course, there is nothing new in this classical balance of analysis and synthesis; all the originality of the method is in the manner in which it is applied.

In order to illustrate by an example the procedure just described schematically, we will take one of the oldest (and one of the simplest) of the axiomatic theories, that of "abstract groups." For example, let us consider the three following operations: (1) the addition of real numbers, where the sum of two real numbers (positive, negative or zero) is defined in the ordinary way; (2) the multiplication of integers "modulo a prime number p," where the elements considered are the integers $1, 2, \ldots, p - 1$, the "product" of two of these numbers being taken as the *remainder* after division by p of their ordinary product; (3) the "composition" of translations in the Euclidean space of three dimensions, the "composition" (or "product") of two translations S, T (taken in that order) being by definition the translation obtained by applying first the translation T and then the translation S. In each of these three theories there corresponds to two elements x and y (taken in that order) of the set of elements considered (in the first case the set of real numbers; in the second the set of numbers $1, 2, \ldots, p - 1$; in the third the set of all translations) a uniquely determined third element which we shall agree to denote in the three cases by the symbol $x \tau y$ (it will be the sum of x and y if x and y are real numbers; their product "modulo p" if they are positive integers $\leq p - 1$; their "composition" if they are translations). Now, if the properties of this "operation" are examined in each of the three theories, a remarkable parallel will be observed. However, these properties are mutually dependent within each of the three theories, and an analysis of their logical connections leads to the isolation of a small number of independent properties (i.e., no one of them is a logical consequence of the remaining properties). We can, for instance,[3] take the three which follow and which we shall express in the symbolic notation common to the three theories, but which may be easily translated into the special language of each one:

(*a*) For any three elements x, y, z, we have $x \tau (y \tau z) = (x \tau y) \tau z$ ("associativity" of the operation $x \tau y$);

(*b*) There exists an element e such that, for any element x, we have $e \tau x = x \tau e = x$ (for addition of real numbers it is the number 0; for multiplication "modulo p" it is the number 1; for the composition of translations it is the "identity" translation which leaves every point of the space fixed);

(*c*) For every element x, there exists an element x' such that $x \tau x' = x' \tau x = e$ (for the addition of real numbers, x' is the negative $-x$; for the composition of translations, x' is the "inverse" translation

---

[3] There is nothing absolute about this choice, and numerous systems of axioms are known which are "equivalent" to the ones which we are expressing. The statements of the axioms of each of these systems are logical consequences of the axioms of any one of the other systems.

of x, i.e., the one which takes every point displaced by x back to its original position; for the multiplication "modulo p," the existence of x' follows from elementary arithmetic arguments.[4]

It will then be observed that the properties capable of being expressed in the same manner in all three theories with the aid of the common notation are consequences of the three preceding ones. For example, if we propose to show that the relation $x \tau y = x \tau z$ implies $y = z$, then it could be done in each of the three theories in terms of special arguments for each; but it is also possible to proceed as follows in a manner applicable to all cases: from the relation $x \tau y = x \tau z$ it follows that $x' \tau (x \tau y) = x' \tau (x \tau z)$ (x' is defined in the above sense); then, applying (a), $(x' \tau x) \tau y = (x' \tau x) \tau z$; using ($c$), this relation can be written $e \tau y = e \tau z$, and finally, applying (b), $y = z$, which was to be proved. The *nature* of the elements, x, y, z was completely irrelevant in this argument, i.e., we were not interested in knowing if they were real numbers, or positive integers $\leq p - 1$, or translations; the sole hypothesis which came into play was that the operation $x \tau y$ on these elements satisfied properties (a), (b) and (c). If only to avoid tedious repetition, it is obviously useful to develop *once and for all* the logical consequences of just the three properties (a), (b), (c). Naturally, for convenience of language it is necessary to adopt a common terminology; it will be said that a set on which an operation $x \tau y$ is defined satisfying the three properties (a), (b), (c) is furnished with a *group structure* (or, more briefly, that it is a group); the properties (a), (b), (c) will be called the *axioms*[5] of the group structure, and to develop their logical consequences is to pursue *axiomatic group theory*.

It is now possible to indicate in a general way what should be understood by a *mathematical structure*. The common trait of the various notions designated by this generic name is that they apply to sets of elements whose nature[6] *is not specified*; in order to define a structure

---

[4] One may note that the remainders under division by p of the numbers $x, x^2, \ldots, x^n, \ldots$ could not all be distinct; next, by equating two of these remainders, it is readily shown that a power x^m of x has a remainder equal to *one*; if x' is the remainder under division by p of x^{m-1}, then it may be concluded that the product "modulo p" of x and x' equals 1.

[5] It goes without saying that there is nothing in common between this sense of the word "axiom" and the traditional meaning of "self-evident truth."

[6] Here we take the "naive" viewpoint and do not approach the troublesome questions, half philosophic, half mathematical, raised by the "nature" of mathetmrical "entities" or "objects." Let it suffice to say that for the initial

one or several relations involving these elements are given[7] (in the case of groups it is the relation $z = x \tau y$ involving any three elements); it is then postulated that the given relation or relations satisfy certain conditions (which are enumerated) and which are the *axioms* of the structure envisaged.[8] To study the axiomatic theory of a given structure is to deduce the logical consequences of the axioms of the structure, *while excluding all other hypotheses* about the elements considered (in particular, any hypothesis concerning their special "nature").

The Main Types of Structures

The relations involved in the beginning of the definition of a structure may be quite varied in nature. The one which comes into play in the structure of a group is what is called a "law of composition," that is, a relation among three elements uniquely determining the third as

pluralism of the mental conception of these "entities"—imagined at first as ideal "abstractions" from sense experience and conserving all the heterogeneity of the latter—the axiomatic research of the 19th and 20th centuries has there, too, substituted little by little a unitary conception progressively leading all mathematical notions back, first to the notion of number, then, as a second step, to that of set. The latter, long considered as "primitive" and "undefinable," has been the object of endless polemics because of its extremely general character and the very vague nature of the mental conceptions that it evokes. These difficulties did not disappear until the notion of set itself disappeared (and with it all the metaphysical pseudo-problems about mathematical "entities") in the light of recent research on logical formalism. In this new conception mathematical structures become, properly speaking, the *sole "objects"* of mathematics.

The reader will find a more ample development of this point in the following articles:

J. Dieudonné, "Les méthodes axiomatiques modernes et les fondements des mathématiques," *Revue Scientifique*, LXXVII (1939), pp. 224–232. [This article now appears in the present book, Vol II, p. 251; see Preface to the Second Edition, p. ix.]

H. Cartan, "Sur le fondement logique des mathématiques," *Revue Scientifique*, LXXXI (1943), pp. 3–11.

[7] In reality, this definition of structure is not sufficiently general for the needs of mathematics; we must also consider the case in which the relations defining a structure would hold, not between *elements* of the set considered, but also between *parts* of this set, and even more generally, between elements of sets of still higher "degree" in what is called the "scale of types." For more details on this point, see our *Éléments de Mathématique*, Book I (results), Actualités Scientifiques et Industrielles, No. 846.

[8] A completely rigorous treatment in the case of groups would require us to consider as axioms not only (*a*), (*b*), (*c*) as stated, but also the fact that $z = x \tau y$ determines a unique z when x and y are given; ordinarily, this property is considered as tacitly implied in the writing of the relation.

a function of the first two. When the defining relations of a structure are "laws of composition," the corresponding structure is called an *algebraic structure* (for example, the structure of a *field* is defined by two laws of composition with appropriate axioms: addition and multiplication of real numbers define a field structure on the set of these numbers).

Another important type is that given by structures defined by an *order* relation; this time it is a relation between two elements x, y which is usually stated "x is *at most* equal to y" and for which we shall use the general notation $x \, R \, y$. Here we do not at all suppose that it determines one of the elements x, y uniquely as a function of the other; the axioms which the relation must satisfy are the following: (a) for all x, we have $x \, R \, x$; (b) the relations $x \, R \, y$ and $y \, R \, x$ imply $x = y$; (c) the relations $x \, R \, y$ and $y \, R \, z$ imply $x \, R \, z$. The set of integers provides an obvious example of a set supplied with such a structure (also the set of real numbers), if we replace the symbol R by the symbol $\leq$. However, it will be observed that we do not include in the axioms the following property which seems to be inseparable from the common notion of "order": "Given any x and y, we have $x \, R \, y$ or $y \, R \, x$." In other words, we do not exclude the possibility that two elements are *incomparable*. This may seem at first to be paradoxical, but it is easy to give very important examples of orderings in which this phenomenon occurs. This occurs when X and Y signify subsets of a given set and the relation $X \, R \, Y$ signifies "X is contained in Y"; or again if, when x and y are positive integers, $x \, R \, y$ signifies "x divides y"; or finally if, when $f(x)$ and $g(x)$ are real functions defined on the interval $a \leq x \leq b$, $f(x) \, R \, g(x)$ signifies "for every x, $f(x) \leq g(x)$." At the same time these examples show the great variety of domains in which order structures are met with and give an indication of the interest accorded to their study.

We shall say a few words about a third main type of structure, *topological structures* (or *topologies*): they supply an abstract mathematical formulation of the intuitive notions of *neighborhood*, *limit* and *continuity* to which we are led by our conception of space. The statement of the axioms of such a structure requires much greater effort of abstraction than for the preceding examples, and the scope of this treatment requires us to refer readers desirous of obtaining further details to specialized treatises.[9]

[9] See for example our *Éléments*, Book III, Introduction and Chapter I, Actualités Scientifiques et Industrielles, No. 858.

The Standardization of the Mathematical Machinery

We believe that we have said enough to give the reader a sufficiently precise idea of the axiomatic method. Its most salient feature, according to what has been said, is that it allows a considerable *economy of thought*. "Structures" are *tools* of the mathematician; as soon as he has discovered relations among the elements that he studies satisfying the axioms of a structure of a known type, he immediately has at his disposal the entire arsenal of general theorems relating to structures of this type, where before he had to laboriously devise his own means of attack, the power of which depended on his personal talent and which were frequently encumbered by overly restrictive hypotheses arising from the special nature of the problem studied. It can thus be said that the axiomatic method is nothing but the "Taylor System"—the "scientific management"—of mathematics.

However, this comparison is not sufficiently close; the mathematician does not work mechanically as does the worker on the assembly line; the fundamental role that a special *intuition*[10] plays in his research cannot be overestimated. This is not the intuition of common sense, but rather a sort of direct divination (prior to all reasoning) of the normal behaviour he has a right to expect from the mathematical entities which a long association has rendered as familiar to him as the objects of the real world. Now, each structure has its appropriate language charged with special intuitive overtones which the above-described axiomatic analysis has selected from the particular theories; and for the researcher who suddenly discovers this structure in the phenomena he is studying it is like a sudden modulation which all at once reorients the intuitive current of his thought in an unexpected direction and throws a new light on the mathematical terrain on which he moves. Consider—to take a classic example—the progress initiated at the beginning of the 19th century by the geometric representation of the complex numbers; from our point of view this was the discovery of a well-known topological structure, that of the Euclidean plane, in the set of complex numbers. The possibilities of application suggested by this discovery, when placed in the hands of Gauss, Abel, Cauchy and Riemann, caused analysis to be completely recast in less than a century.

---

[10] An intuition, moreover, which frequently errs, as does all intuition.

Such examples have multiplied in the last 50 years: Hilbert space and, more generally, function spaces, which introduced topological structures into sets of elements which were no longer *points*, but *functions*; the p-adic numbers of Hensel, where, even more astonishingly, topology invaded what up till then was the domain of the *discrete*, of discontinuity par excellence, the set of integers; Haar measure, which immeasurably enlarged the field of application of the notion of integral and permitted a profound analysis of the properties of continuous groups—these represent decisive moments of mathematical progress, sudden turns where a flash of genius decided the new orientation of a theory by revealing in it a structure which did not appear a priori to play a role there.

Thus, less than ever is mathematics reduced to a purely mechanical game of isolated formulas; more than ever does intuition reign in the genesis of discoveries; but it now has at its disposal powerful levers furnished it by the theory of the main types of structure, and it takes in at a single glance the immense regions unified by axiomatics, where the most shapeless chaos formerly seemed to reign.

An Overall View

Guided by the axiomatic conception, let us try to present the entirety of the mathematical universe. Certainly, we shall scarcely recognize there the traditional arrangement which, like the first attempts at classifying animal species, was limited to arranging side by side theories presenting the greatest external resemblance. In place of the well-defined compartments of algebra, analysis, number theory and geometry, we shall see, for example, the theory of prime numbers in a neighboring position to that of algebraic curves, or Euclidean geometry alongside integral equations; and the ordering principle will be the conception of a *hierarchy of structure*, going from the simple to the complex, from the general to the particular.

At the center will be the main types of structures, the *mother-structures*, one might say, the principal ones of which we have just now enumerated. Within each of these structures a considerable diversity already reigns, for one must distinguish between the most general structure of the type considered, with the fewest axioms, and the ones obtained by enriching it with additional axioms, each of which carries with it its harvest of new consequences. Thus, the theory of groups, besides general statements valid for all groups and depending only on the axioms stated above, includes a special theory of *finite* groups (where one adds the axiom that the number of elements of the group is finite) and a special theory of *Abelian* groups (where $x \tau y = y \tau x$ for all x, y), as well as a theory of *finite Abelian* groups (where both these axioms hold simultaneously). In the same way, among the *ordered* sets, we can single out those sets in which (as in the order of the integers or of the real numbers) any two elements are comparable, and which are called *totally ordered*. Among the latter, specializing further, one may study the sets called *well-ordered* (in which, as for the positive integers, every subset has a "least element"). There is an analogous gradation in the topological structures.

Besides this primary core, structures appear which could be called *multiple*, where two or more mother-structures come into play at the same time, not simply juxtaposed (which would yield nothing new), but organically *combined* by one or more axioms which relate them. Thus one has *topological algebra*, the study of structures in which one or more laws of composition and a topology figure, related by the condition that the algebraic operations are *continuous* functions (under the topology considered) of the elements to which they apply. Not less important is *algebraic topology*, in which certain sets of points of a space defined by some topological properties (simplexes, cycles, etc.) are themselves taken as elements on which laws of composition operate. The combination of the structures of order and algebra is also a fertile source of results, leading, on the one hand, to the theory of divisibility and of ideals, and on the other, to integration and the "spectral theory" of operators, in which topology also plays a role.

Further on, begin the specialized theories, properly speaking, in which the elements of the sets under consideration, completely undetermined in the general structures up until now, receive a more fully realized individuality. Here one rejoins the theories of classical mathematics; analysis of functions of a real or complex variable, differential geometry, algebraic geometry, theory of numbers; but now they have lost their former autonomy and are the crossroads where numerous more general mathematical structures meet and act upon one another.

In order to maintain a proper perspective, we must immediately add that this rapid sketch is to be considered only a very rough approximation of the present state of mathematics as it actually exists; it is at once *schematic, idealized and rigid*.

Schematic—because in their detail, things do not happen as simply or as regularly as we seem to imply; among other things there are unexpected turnabouts, in which a thoroughly specialized theory such as that of the real numbers lends indispensable aid to the construction of a general theory such as topology or integration.

Idealized—because the exact role of each of the main structures is far from perfectly recognized in all the areas of mathematics; in certain theories (for example, in the theory of numbers) there exist many isolated results that up till now no one has been able to classify or relate in a satisfactory manner to known structures.

Lastly, rigid—because nothing is more foreign to the axiomatic method than a static conception of the science, and we do not want the reader to think that we claim to give a definitive account of the latter. The structures are not immutable, either in their number or in their form; it is quite possible that the further development of mathematics will augment the number of fundamental structures by revealing the fecundity of new axioms, or of new combinations of axioms. We may take it for granted that there will be decisive progress in the *inventions* of structures, to judge by those which have produced the structures presently known. Of course, the latter are by no means completed edifices, and it would be very surprising if the full essence of their source had been already drawn out.

Thus with the aid of these indispensable qualifications one can better understand the internal vitality of mathematics, that which gives it at the same time both unity and diversity; like a great city whose suburbs never cease to grow in a somewhat chaotic fashion on the surrounding lands, while its center is periodically reconstructed, each time following a clearer plan and a more majestic arrangement, demolishing the old sections with their labyrinthine alleys in order to launch new avenues toward the periphery, always more direct, wider and more convenient.

A Glimpse at the Past and Conclusion

The point of view that we have tried to expound above was not developed all at once; it is only the end product of a half-century's evolution, during which time it has met with serious resistance, as much from philosophers as from mathematicians themselves. For a long time many of the latter could see in axiomatics only the empty subtleties of logicians, something incapable of giving birth to any theory whatsoever. This criticism can doubtless be explained in terms of pure historical accident: the first, and also the most celebrated axiomatizations (those of arithmetic by Dedekind and Peano and that of Euclidean geometry by Hilbert) dealt with *univalent* theories, i.e., theories which were uniquely determined by the totality of the axioms, which were thus incapable of application to any theory other than the ones from which they had been drawn (contrary to what we have seen for group theory, for example). If such had been the case for all the structures, then the reproach of sterility made against the axiomatic method would have been fully justified.[11] However, the method has shown its worth as it has developed; and the aversion still displayed toward it here and there is only to be explained by the difficulty that the mind naturally experiences in admitting that a form of intuition other than that directly suggested by the facts of a concrete problem (and one that is often obtained only through a higher and sometimes difficult abstraction) can show itself to be equally fruitful.

As for the objections of the philosophers, they deal with a region in which, for lack of competence, we would rather not venture: the large problem of the relations between the experimental and the mathematical worlds.[12] That there is a close connection between experimental phenomena and mathematical structures seems to be confirmed in a most unexpected manner by the recent discoveries of contemporary physics; but we do not know at all the deep-lying reasons for this (in so far as these terms have any meaning), and we may never know them. In any case, this observation ought to inspire the philosophers with more prudence in the future: before the revolutionary developments of modern physics much effort was expended in an all-out attempt to deduce mathematics from experimental facts, notably from immediate spatial intuitions; but, on the one hand, quantum physics showed that this "macroscopic" intuition of reality covered up "microscopic" phenomena of an entirely different nature, dependent on

---

[11] There also took place, especially at the beginning of axiomatics, a flowering of teratological structures totally devoid of applications, whose only merit was to show the exact role of each axiom by observing the effect of its suppression or modification. Clearly, one would be tempted to conclude from this that these were the sole results that could be expected from this method!

[12] We shall not confront here the objections raised by the application of the rules of formal logic to the reasoning in axiomatic theories; these are connected with the logical difficulties encountered by the theory of sets. We simply note that these difficulties can be surmounted in a way which obviates all the objections and allows no doubt about the correctness of the reasoning. In this regard the reader is invited to consult the articles of H. Cartan and J. Dieudonné cited above.

branches of mathematics which had certainly never been imagined capable of application to experimental science; and, on the other hand, the axiomatic method showed that the "truths" which were supposed to be the pivot of mathematics were only very special aspects of general conceptions whose bearing was by no means limited to these. So that at last, this intimate fusion whose harmonious necessity we were supposed to admire appeared more as a chance contact between two disciplines whose relations were much more deeply hidden than could have been supposed a priori.

From the axiomatic point of view mathematics appears on the whole as a reservoir of abstract *forms*—the mathematical structures; and it sometimes happens, without anyone really knowing why, that certain aspects of experimental reality model themselves after certain of these forms, as if by a sort of preadaptation. It cannot be denied, of course, that the majority of these forms had a well-determined intuitive content at the beginning; but it is precisely by voluntarily emptying them of this content that it has been possible to employ them with all their potential efficacy and to render them capable of new interpretation and a complete fulfillment of their elaborative role.

It is only in this sense of the word "form" that the axiomatic method can be said to be a "formalism"; the unity that it confers on mathematics is not the supporting framework of formal logic, the unity of a lifeless skeleton, but the nourishing sap of an organism in full development, the supple and fruitful tool of research consciously worked on by all the great thinkers of mathematics since Gauss, all those who, according to the formula of Lejeune-Dirichlet, have tried always "to substitute *ideas* for *calculations*."

3

ANALOGY IN MATHEMATICS

by Robert Deltheil
HONORARY RECTOR AT THE UNIVERSITY OF TOULOUSE,
PROFESSOR OF DIFFERENTIAL AND INTEGRAL CALCULUS
AT THE FACULTY OF SCIENCES AT TOULOUSE

DOES reasoning by analogy, which consists in concluding from a resemblance of certain objects in some respects that they resemble each other in other respects, play a role in the methods of the mathematical sciences? It is superfluous to point out that this is certainly not the case as far as proofs are concerned. However, its role as an instrument of discovery is immense, orienting the researcher as he tries to find out the full extent of the resemblance and the lack of resemblance in the analogy observed.

The history of mathematics furnishes numerous examples of important progress due to the observation of suggestive analogies between facts which up till then had appeared to be completely independent. Thus, by comparing the observations of Tycho Brahe with the geometric properties of the ellipse studied by the ancient Greeks, Kepler was able to deduce the laws of planetary motion from these observations. And our readers will surely be familiar with the celebrated account of Henri Poincaré in which the great geometer connects his discovery of the Fuchsian functions with a striking observation which entered his mind while on a geological excursion, the fact that the transformations which his research into this subject had just led him to consider reproduced exactly those of non-Euclidean geometry.

The establishment of such comparisons is the very essence of the genius for discovery and the goal of a mode of intuition of the first importance, analogical intuition.

Mathematical notions present examples of all degrees of analogy, from the literal identity of two polynomials with integral coefficients where the analogy is complete, or the equality of two geometric figures which differ only in their position in the plane or in space, to the correspondence between two figures as distinct as the line in space and the sphere associated with this line by the transformation of Sophus Lie.

We shall not stress the role of analogy in the classifying pure and simple of mathematical objects and theories; it is known in this respect that the catalogue whose origin goes back to the International Congress of Bibliography presided over by Henri Poincaré in Paris in 1889 presents a variety of classes, subclasses, divisions, sections and subsections which has nothing to lose by comparison with the catalogues of the naturalists.

We shall base our presentation on purely classical examples and limit ourselves to a summary sketch of the results (positive or negative, but rarely without interest) to which logical verification of the suggestions of analogy can lead.

An important class of mathematical propositions has as its object to show that certain kinds of resemblance between two elements have as a necessary consequence their resemblance on all points. The simplest of these propositions is the case of equality of triangles, which may be compared, for example, with the identity of two integral polynomials of degree n which agree for $n + 1$ different values of the variable. On a more elevated level, let us cite the conclusion that two analytic functions of the same complex variable are identical on the whole of the connected domain on which they are defined whenever they are identical on an arbitrarily small subdomain. The notion of "characteristic property," whose importance is rightly stressed in all basic educational programs, is connected with this category of propositions.

A much larger collection of propositions establish only that the resemblance of two mathematical objects in certain properties can be extended to a certain number, but not all, of the remaining properties. Such is the case with similarity of triangles, or with the theorem of Poncelet relating to polygons of n sides inscribed in one conic and circumscribed by another.

Analogical intuition, whose role is to suggest possible extensions of the resemblance between two mathematical notions, problems or theories, can frequently lead to conclusions contrary to reality if all its suggestions are not subjected to verification.

This is the case with the analogy between triangles and trihedral angles, one of the most striking in elementary geometry and completely deceptive in certain respects. Two similar trihedral angles are necessarily equal, which is not the case for two similar triangles. The existence of similar figures is peculiar to Euclidean geometry, while the geometry of trihedral angles (or of spherical triangles) is not Euclidean, considered as a geometry of two dimensions.

The lack of resemblance between the properties of spherical triangles and planar triangles stems from the curvature of the sphere, which is the basic reason for the difference between the very notions of spherical and planar triangle. In other notable cases the dissimilarity results from the fact that the analogy considered involves a certain amount of generalization.

Numerous problems in the geometry of the tetrahedron, for example, are completely analogous to corresponding problems in the geometry of the triangle. Essential differences appear, however, as soon as the three-dimensional space plays a more profound role. Thus, the four altitudes of the tetrahedron are not generally concurrent as are the three altitudes of the triangle; to the inscribed circle and the three escribed circles for the triangle there correspond for the tetrahedron not only the inscribed sphere and the four escribed spheres, but also the three escribed spheres; in space the faces of the tetrahedron actually bound regions which have no analogues in the geometry of the triangle.

An example of much wider import is furnished by the determination of the point transformations which preserve angle in the geometries of two and three dimensions. If it is observed that in the plane, transformations thus defined form an extremely large set with the same cardinality as the analytic functions of a complex variable, then it would be expected that in space an even more extensive set of solutions in the set of transformations would be found. However, according to the result expressed by the celebrated theorem of Liouville, the conformal mappings of space form a ten-parameter group obtained by arbitrary combinations of translations, similarity transformations and inversions.

Finally, the history of mathematics has recorded the vain efforts of the algebraists of the 18th century and of Lagrange, who, inspired by analogical intuition, attempted to solve polynomial equations of arbitrary degree by means of algebraic formulas generalizing the classic solutions of equations of second and third degree (these permit the solution of the fourth-degree equation in the most general case). Abel's explanation, which created such a stir at the beginning of the 19th century, demonstrated the impossibility of solving by radicals the general equation of degree greater than or equal to five.

Thus, when the mathematician extrapolates by analogy, he exposes himself to the risk of contradiction by the facts; but this risk cannot bar him from this form of exploration as long as logic continues to retain its validity. Moreover, investigation by analogy leads in certain domains to general perspectives whose harmony constitutes an essential element of the beauty of mathematics.

One of the most remarkable of these perspectives is that of the geometry of transformations. The very mechanism of the transformation gives rise to a profound analogy between a figure F and its transform F'. If, for example, it is a question of a similarity transformation, the analogy may include a resemblance in the external aspect of the two figures; if it is an inversion or a transformation by reciprocal polars, the resemblance may become much less obvious; we have cited an example above in which it disappears completely, that of the transformation of Sophus Lie.

In any case, however, to each property P of figure F there corresponds a property P' of figure F'. This property P may consist of a special feature presented by figure F, either given or obtained by construction; two circles of this figure may be orthogonal or three lines may be concurrent. It may be the object of a proposition; for example, "all the circles of F which intersect a given circle orthogonally in two points and pass through a fixed point of the space necessarily pass through a second fixed point."

From this there results a transformed proposition concerning the figure F'. The two propositions P and P' are logically equivalent, and the transformation can act as a tool of discovery if P is considered as known and P' as unknown, or as an instrument of demonstration if P is considered as a consequence of P', the latter being easier to establish directly.

In addition, propositions P and P' may both have been already established and considered as independent up to this point; then the transformation creates a more or less unforeseen connection between them which is often a sinigficant one. Thus the Lie transformation associates properties of spheres which are tangent to three given spheres to properties of lines in space based on three given lines.

Thus, to every class of transformations there corresponds for a given property P an entire class of properties P' equivalent to P and among themselves. It may happen that all these properties P' reduce to the property P itself, which is left invariant under all the transformations of the class considered. This is the case with the proposition cited above concerning the circles in space which pass through a fixed point and intersect a given circle orthogonally in two points; this proposition is left invariant under all the transformations of the conformal group of the space.

Thus there appear in connection with the group of translations (to which the similarities may be adjoined), or with the projective group, or with the conformal group of the space, "domains of causality" (according to the expression of Bouligand) described by Felix Klein in his celebrated "Erlangen Program" and of fundamental importance. Let us note at this point that the "continuity principle" of Poncelet, according to which geometrical reasoning is valid whether the elements involved in a figure are all real or whether some are imaginary, may be placed in the domain of causality of the projective group as an example of successful analogical exploration.

We shall close this exposition by recalling to mind possibly the most extensive domain of causality which can be defined in mathematics. One reads in the geometry of Rouché and Comberousse the following statement of conditions that two quantities should vary proportionally: "Two quantities are proportional to one another if to two arbitrary but equal values of the first quantity there correspond two equal values of the second, and if, in addition, to the sum of any two values of the first there corresponds a value which is the sum of the two corresponding values of the second."

This statement defines in ordinary language the role of the additive property

$$f(x_1) + f(x_2) = f(x_1 + x_2)$$

as the characteristic property of the homogeneous linear function $y = Cx$. It is known that this additive property, extended to functions of several variables, characterizes the linear forms $f = Ax + By$ (A and B are constants). In the case of two variables, for example, this is written

$$f(x_1, y_1) + f(x_2, y_2) = f(x_1 + x_2, y_1 + y_2).$$

This proposition is fundamental to the entire theory of systems of linear equations.

But the linear additive property has an import much more considerable still; the reading of the *Principes d'Algèbre et d'Analyse* by Émile Borel (published by Albin Michel) will give a very suggestive general perspective on this point. For example, the linear property permits one to predict the form of the most general solution of a linear differential equation and to establish the ordinary properties of the fundamental functions of analysis defined as solutions of differential equations such as $y' - y = 0$, or $y'' + y = 0$, etc.

The usefulness of this property extends to linear finite difference equations, to linear partial differential equations and to many other mathematical theories. Its field of application is so extensive that analogical intuition finds broad possibilities in it, even beyond linear properties.

For instance, it is known how the integration of the differential equation

$$y'' + ay' + by = 0,$$

in which a, b designate constant coefficients, is related to the solution of the second-degree equation

$$r^2 + ar + b = 0;$$

it is also known how this same second-degree equation enters into the determination of the most general solution of the finite difference equation

$$af(x+2) + bf(x+1) + cf(x) = 0,$$

or into the integration of the second-order partial differential equation

$$a\frac{d^2z}{dy^2} + b\frac{d^2z}{dx\,dy} + c\frac{d^2z}{dx^2} = 0.$$

If the second-degree equation in r has two real, distinct roots r_1 and r_2, then the general solutions of the three equations which we have associated with it are respectively

$$Y = C_1 e^{r_1 x} + C_2 e^{r_2 x},$$
$$f(x) = C_1 r_1 x + C_2 r_2 x,$$
$$Z = F(x + r_1 y) + G(x + r_2 y),$$

in which C_1 and C_2 denote two arbitrary constants, F and G two arbitrary functions of one variable.

Thus, these three problems in the linear domain depend on the same second-degree equation; if r_1 and r_2 are imaginary, or if the equation in r has a repeated root, the analogy can be carried on and we leave the details to the reader. It clarifies many an observation arising from mechanics, or physics, or the theory of probability. The development of these observations cannot find a place in this modest survey.

4

SYMMETRY AND DISSYMMETRY IN MATHEMATICS AND PHYSICS

by Albert Lautman

THE important role that the so-called paradox of symmetric objects plays in the philosophy of Kant is well known. The difference in orientation of symmetric figures with respect to a plane in ordinary space appeared to Kant as a sense intuition not susceptible of conceptual resolution. This necessary intervention of sense perception in the knowledge of left and right is at the origin of the Kantian distinction between sense perception and understanding. The specific nature of sense perception is thus made evident in the incongruence of symmetric figures. It is certainly a remarkable achievement on Kant's part to have characterized sense perception as early as 1770 by means of a property that contemporary science has found at the center of all the investigations into the structure of phenomena.

In the language of crystallography, two isomeric crystals symmetric to one another with respect to a plane, and not superimposable, are said to be enantiomorphs. In order that two symmetric crystals be enantiomorphs, it is necessary that each present in isolation a certain internal dissymmetry—for example, the absence of a center of symmetry. The importance of dissymmetry by enantiomorphy appeared in the early work of Pasteur, when he recognized the relation that exists between the difference of orientation of two enantiomorphic (or hemihedral) crystals and their opposite effects on the plane of polarization of light. Following these discoveries, Pasteur conceived the theory of molecular dissymmetry, manifesting itself essentially in the hemihedry of two isomers and characteristic of living phenomena: "Synthetic substances thus have no molecular dissymmetry, and I cannot indicate a more profound separation than this between the substances formed under the influence of life and the others" (L. Pasteur, *Recherches sur la dissymétrie moléculaire des produits organiques naturels*, *Œuvres complètes*, Vol. I, p. 333). Even though the conceptions of Pasteur on the relation between life and dissymmetry by enantiomorphy are no longer defensible at the present time, they nonetheless gave birth to the theory of asymmetric carbon, which is at the base of all the structural theories of modern stereochemistry.

The idea of enantiomorphy presents itself on analysis, as we have seen, as an intimate union of symmetry and dissymmetry. This same necessity for a mixture of symmetry and dissymmetry is found in the work of Pierre Curie on symmetry in physical phenomena. Curie no longer intends this to characterize only biological phenomena in opposition to physical phenomena; the mixture of symmetry and dissymmetry becomes for him a necessary condition of physical phenomena in general. The determination of the elements of symmetry of a physical phenomenon is carried out, as in crystallography, by searching for the center, the axes and planes of internal symmetry that the phenomenon possesses. To each physical phenomenon is linked the idea of a saturation of symmetry, of a maximal symmetry compatible with the existence of the phenomenon and characterizing it. A phenomenon can only exist in an environment possessing its characteristic symmetry or a lesser symmetry. Thus, if the absence of a certain element of symmetry is called an element of dissymmetry, it becomes possible to understand how Pierre Curie could write: "Certain elements of symmetry can coexist in certain phenomena, but they are not necessary. What is necessary is that certain elements of symmetry do not exist. It is the dissymmtery that creates the phenomenon." (P. Curie, *Sur la symétrie dans les phénomènes physiques*, *Œuvres*, p. 126). Thus the presence of an electric field is incompatible with the existence of a center of symmetry and a plane of symmetry normal to the axis of the field, and the presence of a magnetic field excludes the existence of planes of symmetry passing through the axis of the field. The dissymmetric constituent of physical phenomena is thus defined by Curie by means of the idea of a limited symmetry, the presence of certain elements of symmetry in conjunction with the necessary absence of other elements, and the enantiomorphy of Pasteur is nothing but one of these dissymmetries within symmetry which give rise to the world of the senses.

In seeing the sensory world thus defined as a mixture of symmetry and dissymmetry, of identity and difference, it is impossible not to recall the *Timaeus* of Plato. The existence of bodies is there based on the existence of that vessel which Plato calls place and whose function consists, as Rivaud has shown in the preface to his edition of the *Timaeus*, in making possible the multiplicity of bodies and their alternation in a single place in the world of sensation, in the same way that the role of the Idea of the Other in the world of intellect is to assure, in its mixture with the Same, the connection and at the same time the separation of types. This reference to Plato enables one to understand that the materials of which the universe is formed are not so much the atoms and molecules of physical theory as these major pairs of ideal opposites such as the Same and the Other, Symmetry and Dissymmetry, related to one another by the laws of a harmonious mixture. At the same time Plato suggests more. The properties of place and substance, according to him, are not purely those of sense perception; they are, as Rivaud goes on to say, the geometric and physical transposition of a dialectical theory. In the same way, perhaps, the distinction between left and right as observed in the world of sense is nothing but the transposition to the plane of experience of a dissymmetric symmetry, which is equally a constituent of the abstract reality of mathematics. This common participation in the same dialectic structure would thus make clear an analogy between the structure of the world of sensation and that of mathematics, and would permit a better understanding of how these two realities accord with one another.

In this respect the development of contemporary mathematical physics offers an extremely suggestive lesson. If we consider the theories developed to account for the facts of perception, in which the distinction between right and left plays a fundamental role, i.e., essentially the phenomena of rotation, of the electromagnetic field, of the polarization of light, we see that they bring into play abstract mathematical theories developed independently of any need for physical application, but which nevertheless possess the following property: we find there either a duality of opposite elements capable of being permuted with one another as left and right are permuted in a symmetry; or, what is even more characteristic of the mixture of symmetry and dissymmetry, a division of mathematical entities into two classes, a rigorously symmetric class comparable to an ambidextrous being and a class that mathematicians call antisymmetric, i.e., a class of objects which change orientation under symmetry, as do the left hand and the right hand, or change sign under permutation of two variables, as a line AB in space changes orientation when it is traversed from B to A.

Our first example will be taken from the theory of spinors, from its relations with the spin of the electron. The physical necessity which gave rise to the theory of electronic spin was the experimental necessity of endowing the electron with a moment of rotation, or spin, which could at any time take on two opposite values with two different probabilities. The elaboration of this idea led, in Dirac's theory, to the assignment to an electron of a four-component wave, or rather a wave with two groups of two components each, constituting what is called a spinor. These two groups of two components, these two semi-spinors, play a role in relation to one another which could be considered as left and right in the space-time of restricted relativity. To each symmetry of space in space-time which preserves the sense of time and changes the sign of the directions in space there can be made to correspond an algebraic operation which permutes the two semi-spinors with one another and thus represents in the abstract space of components of the spinor the physical operation of a change of orientation in physical space. The spinors, these abstract entities split in two internally and of such importance in quantum mechanics, were not, however, invented by physicists; they were discovered by É. Cartan in 1913 in the course of his research on the linear representations of the rotation group of a space of an arbitrary number of dimensions. This is a typical case in which the physics of dissymmetric symmetry leads back to an algebra in which opposite terms exchange roles.

The example of the unitary theories of the electromagnetic and gravitational fields offers an analogous case of a physical theory based on the dissymmetry of certain mathematical entities discovered by É. Cartan long before their utilization by Einstein. In his theory of 1928, Einstein associates the phenomena of electromagnetism and those of gravitation by considering spaces endowed with *torsion*, which were introduced by É. Cartan in 1922. Here is a single example due to Cartan which may illustrate the nature of torsion. Let us consider two systems of curves on a surface, such as the meridians and the parallels on a sphere. Suppose that a ship moves a distance d along one meridian, going from point A to point A′, then a distance δ along a parallel, going from A′ to A″. Now suppose that the ship moves in the opposite way, first traveling δ along the parallel passing through A, which leads it to A‴, then d along the meridian passing through A‴, which leads it to a point $\bar{A}″$ different from A″. If we denote these respective operations by $d\delta$ and δd, then we see that $d\delta \neq \delta d$. The existence on the sphere of a vector of torsion $A''\bar{A}''$ (or of the oppositely directed vector $\bar{A}''A''$) thus realizes the non-commutativity of the operations d and δ. In a more general way, the distinction between left and right in this case is nothing other than a concrete expression of a non-commutative algebra in which the product $A \cdot B$ differs from $B \cdot A$.

We must stop a moment to lay stress upon the fashion in which the distinction between left and right in the real world can symbolize the non-commutativity of certain abstract operations of algebra. The fundamental property of symmetry with respect to a plane is that, applied once, it gives a figure distinct in orientation from the original, and that applied a second time, it gives the original figure again. It is for this reason that this symmetry is called an *involution*. Let us next consider an algebraic operation defined on two quantities X and Y and which we shall write (X, Y); the parenthesis may denote the ordinary product or any other operation defined for the two variables. It is a non-commutative operation if $(X, Y) \neq (Y, X)$ and the most fecund type of non-commutativity in mathematics is that in which $(X, Y) = -(Y, X)$. The parenthesis (X, Y) is dissymmetric in X and Y, but it may be easily verified that it defines an involution, as does ordinary symmetry. The expressions (X, Y) and (Y, X) are called antisymmetric, and this word translates well the mixture of symmetry and dissymmetry which is thus seen to be deeply imbedded in the heart of modern algebra. All of the theory of continuous Lie groups is based on the non-commutativity of the product of two infinitesimal operations of the group. This theory, which is closely associated with the theory of Pfaffian forms, expressions with an antisymmetric multiplication, permitted Cartan to discover a profound analogy between the generalized Riemann spaces which appear in the physico-geometric theories of relativity and the space of Lie groups.

The examples from mathematical physics that we have cited up to now show us how the physical fact of dissymmetric symmetry can be conceived as the equivalent in the world of phenomena of the antisymmetry inherent in certain mathematical entities.

Recent developments in the wave mechanics of systems of particles have shown that the distinction between symmetry and antisymmetry serves as a foundation for observable properties of matter, possibly more important from the philosophical point of view than the properties of orientation. Here it is a question of the constitution and stability of molecular structures, of the very notion of system, of the "whole," in the sense in which a "whole" possesses global properties which characterize it qualitatively and make of it something other, and more, than the sum of its parts.

The wave mechanics of systems of particles considers a system composed of identical particles and assigns to this system a wave function Ψ (1, 2, ... n), which is a function of the n particles of the system. When the trajectories of these identical particles encroach upon one another, the particles become indistinguishable, and the system must thus be equally well described by any function obtained by permuting in any way the n variables of the function Ψ as by Ψ itself. It results from this theorem that for systems containing only two particles with the same physical nature, the functions Ψ which describe their evolution must necessarily be either symmetric or antisymmetric with respect to these two particles, i.e., we have either $\Psi(1, 2) = \Psi(2, 1)$ in the symmetric case, or $\Psi(1, 2) = -\Psi(2, 1)$ in the antisymmetric case.

As far as systems with an arbitrary number of particles are concerned, the experimental facts give rise to an analogous result. According to de Broglie, it is certain that for each type of particle the wave functions are either symmetric or antisymmetric, and antisymmetry seems to play a much more fundamental role in nature than symmetry. In fact, if the elementary particles are distinguished from the composite particles, it is observed that the systems of elementary particles such as the electron, the proton, the neutron, the neutrino (if it exists) are in the antisymmetric state; in contrast, the complex particles formed by the union of several elementary particles such as photons or α particles are in the symmetric or the antisymmetric state according to whether the number of their components is even or odd.[1] These results have allowed the notion of chemical valence and, consequently, the constitution of molecules to be linked to the antisymmetry of the spin of two electrons belonging to two distinct atoms. Antisymmetry thus plays a fundamental role in the study of the chemical bond.

If we study the mathematical basis for this differentiation of wave functions into symmetric functions and antisymmetric functions, we find that it lies in the internal dissymmetry of the permutation group of n objects.

---

[1] See Louis de Broglie, *La Mécanique ondulatoire des Systèmes de corpuscules*, Gauthier-Villars, Paris, 1939.

This group decomposes into two classes: the subgroup of even permutations and the subset of odd permutations, which does not form a subgroup since it does not contain the identity permutation. As an example, let us consider the permutation group of three variables 1, 2, 3. The subgroup of the three even permutations is obtained by cyclic permutation: 1, 2, 3; 2, 3, 1; 3, 1, 2; the three odd permutations are 1, 3, 2; 3, 2, 1; 2, 1, 3. Here we have the second type of dissymmetry discussed above, that of an entity which divides itself into a symmetric and a dissymmetric part. The example of the wave mechanics of systems of particles, in which we see the saturation of electronic levels in an atom and the formation of chemical molecules related to a mathematical dissymmetry as essentially simple as that of the permutation group of n objects, shows in the clearest possible fashion in what sense it is possible to speak of the common participation of the real world and certain mathematical theories in the same dialectical structure, composed of a mixture of symmetry and dissymmetry.

We should like to go further and show the importance of this structure not only for those mathematical theories which apply to the real world, but in a general way, for the most abstract mathematical domains. In order to do this let us reconsider for a moment the analysis of the distinction between left and right. We found two ideas there: (1) the division of a complete entity into two distinct parts, at least by inversion of their orientation; and (2) the existence of an involute relation between the two parts, such that A is to B as B is to A, the symmetry reapplied yielding again the original element. Transposed into more abstract language, this situation is equivalent to the possibility of distinguishing within a single entity two distinct entities X and X', which will be said to be in a dual relation, if an orientation or an ordering can be found for each of them such that they are inverse to one another, and if, in addition, we can find an involution relating them, i.e., if X is to X' as X' is to X, or $(X')' = X$.

A certain number of mathematical theories basing themselves on this kind of structure of duality have been known for a long time. The most celebrated is the calculus of propositions established by Boole in 1847, which is the foundation of modern mathematical logic. Let Σ be the set of all possible propositions in a theory. This set can be subdivided into complementary subsets S and S' which have no elements in common, that is, whose logical product (in symbols: $S \cap S'$, the set of elements common to S and S') is empty, and whose logical sum (in symbols: $S \cup S'$, the set of elements belonging either to S or to S')

equals the whole set Σ. If we wish, we may consider these two subsets to be the set of true propositions and the set of false propositions, but this special interpretation is not at all necessary; it is sufficient to consider them to be two complementary subsets. The involutive character of this complementation is made evident by the fact that the complement of the complement of S is equal to S. This duality embracing S and S' enables us to establish a duality between a formula P in S and a formula P' in S' which will be called the negation, or the contrary, or the complement of P, and is such that the logical sum $P \cup P'$ is always true and the logical product $P \cap P'$ always false. The fundamental property of this duality is to interchange the symbols of logical sum and logical product: If we are given a formula constructed with elementary propositions p, q, etc., and the logical symbols of sum, product and negation $\cup$, $\cap$, $'$, we can obtain the negation (also called the contrary or complement) of this formula by replacing all the elementary propositions in the formula by their negations and by interchanging the symbols $\cap$ and $\cup$. Thus we have

$$(p \cup q)' = p' \cap q' \quad \text{and} \quad (p \cap q)' = p' \cup q'.$$

If, by using the notion of implication, we introduce an ordering between two propositions, then we can easily demonstrate that duality changes the direction of the implication,

$$(p \supset q)' \equiv q' \supset p',$$

which helps to identify the duality of geometric symmetry with the duality of logical negation.

There is another mathematical theory in which the notion of duality has played a fundamental role since its discovery by Poncelet in 1822; this is projective geometry. It is well known that in projective plane geometry, given a true proposition constructed from the notions of point, line and the relation of inclusion, we can obtain another true proposition by interchanging the words point and line, and by changing the direction of the inclusion relation. To the points situated on a line there correspond the lines passing through a point. Thus, we can define a curve by means of the points which compose it or by the lines tangent to it at each point.

We now examine the algebraic expression of this duality. Consider the equation $u_1 x_1 + u_2 x_2 + u_3 x_3 = 0$. This equation can be interpreted in two ways: If we consider the three quantities u_1, u_2, u_3 as defining a line in the space of the projective plane, the coordinates of all the points of this line are determined by the values of x_1, x_2, x_3 satisfying the equation; conversely, if we take the three coordinates x_1, x_2, x_3 of a fixed point, the equation is that of all the lines u_1, u_2, u_3 passing through this point. The quantities u_1, u_2, u_3 and x_1, x_2, x_3 can thus interchange their roles of coefficients and variables in the proposed equation; as a result of this, we can consider the projective plane either as a set of points or as a set of lines, and these two sets are said to be in *correlation* with one another. More generally, in a projective space S_n of n dimensions the points (elements of dimension 0) and the hyperplanes (elements of dimension $n - 1$) of this space are in correlation, and this correlation constitutes a true duality in the above sense. Modern axiomatic research has even permitted us to rigorously define the sum and the intersection, or product, of two projective spaces S_p and S_q of respective dimensions p and q, and in this way to associate with each subspace S_m of a space S_n a dual space S_{n-m-1} such that

$$S_m \cup S_{n-m-1} = S_n \quad \text{and} \quad S_m \cap S_{n-m-1} = 0.$$

The sum of a subspace and its dual is the whole space, and their intersection is empty. We can easily demonstrate that this duality is involute and that it reverses the inclusion relation

$$\text{if} \quad S_p \subset S_q, \; S_{n-q-1} \supset S_{n-p-1}.$$

There is thus a projective duality as well as a logical negation or a geometric symmetry.

At first sight, it appears that the goal of the calculus of propositions and that of projective geometry are different, and yet the logical structures of the two disciplines, as we have shown, present many analogies. The reason behind this analogy did not appear until very recently, in the light of research into the domain called abstract algebra. Inspired by Dedekind, a large number of contemporary mathematicians, among them Birkhoff, von Neumann, Glivenko,[2] Ore and others, have constructed a general theory of lattices which embraces the theory of sets, the theory of numbers, projective geometry, combinatorial topology, the theory of probability, mathematical logic, the theory of function spaces, etc. The basic notions of this theory follow. In each case we begin with a set S and we select not individual elements but subsets of the set as its parts. We next define

---

[2] Cf. Valère Glivenko, *Théorie générale des structures*, Librairie Hermann, Paris, 1938.

between any two parts either an order (in the case of a set composed of a finite number of parts), or at least a partial order, making use of relations such as that of size, of dimension, of inclusion, of implication, of boundary, and the two operations of sum and product. Following this, we consider the set S^* obtained by inverting the order between the parts and interchanging the symbols for sum and product. The new set thus obtained is called the dual of the first, and in the most interesting cases the dual set is nothing other than the original set with all the order relations reversed. Duality thus establishes an *antiisomorphism* (an inverse isomorphism) between S and itself; moreover, it is an involution, since the dual of a dual set is the original set. The general theory of lattices thus is based on the possibility of ordering the same set in two mutually inverse ways. To us it is a result of fundamental philosophical importance to see an internal duality of two distinguishable antisymmetric entities embedded in one entity and to see this duality become the generating principle behind an immense harvest of mathematical results; the theories which we have cited above, according to Glivenko, allow us to consider as lattices the totality of rings, the set of convex fields, the set of subgroups of any group, the set of all positive integers, the set of all the elements of a projective geometry, the set of all simplexes contained within a given topological simplex, the set of events of the theory of probability, the set of propositions of the calculus of propositions, etc.

This development of the theory of lattices has led naturally to the establishment of distinctions between all the structures satisfying the duality law. Thus, for example, in the case of mathematical logic, the complementary formula of a given formula is uniquely determined, whereas, given $m + 1$ points which determine a subspace S_m of a projective space S_n, there exist an infinity of ways of choosing $n - m$ other points in order to determine the complementary subspace S_{n-m-1} of S_m. The determination of the complement is therefore possible in both cases (this is not the case for all lattices), but the uniqueness of this element, established in the calculus of propositions (or Boolean algebra), does not hold for projective geometry.

These represent two distinct realizations of the same dialectical structure of a simultaneously complemented and antisymmetric duality, and it is of interest to stress for a moment the basis of this difference.

Boolean algebra satisfies the following distribution laws for sum and product:

$$x \cap (y \cup z) = (x \cap y) \cup (x \cap z),$$
$$x \cup (y \cap z) = (x \cup y) \cap (x \cup z),$$

while projective geometry satisfies only one weaker distribution law called the modular law and discovered by Dedekind:

$$\text{if } x \leq z, \quad x \cup (y \cap z) = (x \cup y) \cap z.$$

If we replace the symbols $\cap$ and $\cup$ by the usual symbols of sum and product, we see that in distributive lattices

$$x(y + z) = xy + xz,$$

while in modular lattices we have only

$$\text{for } x \leq z, \quad (x + y)z = x + yz.$$

In a celebrated article, "The Logic of Quantum Mechanics" (*Annals of Mathematics*, 1936, p. 823), Garrett Birkhoff and von Neumann, basing their work on this difference, established that the calculus of propositions relative to certain observations of classical mechanics has the structure of a Boolean algebra, while the calculus of propositions relative to the observable facts of quantum mechanics satisfies the modular but not the distributive law, and thus has the structure of a projective geometry. Birkhoff and von Neumann support this thesis with the following example: Let proposition a correspond to the observation of a train of waves Ψ on one side of a plane, proposition a' to the observation of Ψ on the other side of the plane and b to the observation of Ψ in a symmetric state with respect to the plane; neither $b \cap a$ nor $b \cap a'$ can be observed simultaneously, therefore $b \cap a = b \cap a' = 0$ (0 represents the identically false proposition). Thus $(b \cap a) \cup (b \cap a') = 0$. On the other hand, since $a \cup a'$ is an identically true proposition, the conjunction $b \cap (a \cup a')$ is equivalent to b; since clearly $b > 0$, we conclude that

$$b \cap (a \cup a') > (b \cap a) \cup (b \cap a').$$

Thus, the distributive law is not satisfied.

The logical difference between two lattices both satisfying the duality law is thus translated in the real world into the difference between classical and quantum mechanics. These applications of the theory of lattices to physics cannot substitute at present for mathematical physics properly speaking. They nevertheless appear to us to justify the hypothesis of a like importance of symmetric dissymmetry in the real world and of antisymmetric duality in the mathematical world. Moreover, it is remarkable that even in those mathematical theories which do not at the present time seem to be related to lattice theory, we find reciprocity laws comparable to the duality that we have just studied. In certain cases this reciprocity presents itself in the form of a possible exchange of roles between arguments in the same relation or function, accompanied by a change in sign of the relation. This reciprocity is therefore comparable in every way to the antisymmetry of non-commutative products. In the same way that we had $(X, Y) = -(Y, X)$, we have $f(x, y) = -f(y, x)$. In other cases the reciprocity appears as a total symmetry not accompanied by any change in sign, i.e., any dissymmetry. Nevertheless, it seems that the symmetry that we shall call symmetric is nothing but a limiting case of an antisymmetric symmetry which remains the general case. Just as when $(X, Y) = -(Y, X)$ we still have $(X, Y) = (Y, X)$ when (X, Y) has the special value 0, so an antisymmetric relation between two elements related to numbers p and $n - p$ with constant sum n can become symmetric in the special case in which $n - p = p$, i.e., if $n = 2p$.

We find something analogous in the extremely important quadratic reciprocity theorems of number theory. Legendre introduced into number theory the symbol m/p, which equals $+1$ if m is a quadratic residue modulo p, that is, if there exists an integer x such that $m - x^2$ is a multiple of p, and -1 in the opposite case. Now let us consider two positive odd integers a and b; they satisfy the fundamental reciprocity law

$$\left(\frac{a}{b}\right) = \left(\frac{b}{a}\right)(-1)^{[(a-1)/2][(b-1)/2]}.$$

This law thus includes at once all the cases of reciprocity in the proper sense of the word, i.e., the cases in which a is to b as b is to a, as well as the cases in which reciprocity does not hold. The general law contains an element of dissymmetry (the factor -1) which drops out in the special cases in which reciprocity actually holds.

While seeking to determine the nature of mathematical reality, we showed in a previous work[3] that the theories of mathematics could be viewed as material ideally designed to give substance to a dialectical ideal. This dialectic seems to be principally constituted from pairs of

---

[3] A. Lautman, *Essai sur les notions de structure et d'existence en mathématiques*, Librairie Hermann, Paris, 1938.

opposites, and the ideas of this dialectic present themselves in each case as opposed notions between which relations must be established. These relations can only be determined in the heart of the domain in which the dialectic takes concrete form; that is why we have been able to discern the concrete design of the edifices in a large number of mathematical theories. The effective existence of this concrete design may be looked upon as an answer to the problems that the ideas of this dialectic pose. In this respect, it appears certain to us that the idea of a mixture of symmetry and dissymmetry plays a dominant role not only in physics but also, as we have tried to show, in mathematics. The two realities are thus presented in accord with one another as two distinct realizations of the same dialectic which gives birth to them in comparable acts of genesis.

5

INTUITIVE APPROACHES TOWARD SOME VITAL ORGANS OF MATHEMATICS

by Georges Bouligand
PROFESSOR AT THE SORBONNE

1. As mathematics progresses, chains of ideas are built into it which gain universal acceptance. Little by little, this constitutes the constantly enlarged patrimony of collective intuitions, many times renewed; a patrimony which will always be fruitfully exploited in the first introduction to the deductive sciences.

Not all these assembled intuitions are of the same nature. This will be shown first of all by an example capable of revealing short cuts taken by thought that are not accessible by direct sampling of reality.

If we next call to mind the first appearance of non-Euclidean geometry, we shall understand that the integration of collective intuitions into the patrimony does not always occur without overcoming some obstacles.

An adaptation therefore takes place. As the geometer becomes progressively more familiar with the abstract objects that he studies, he comes to have a concrete idea of them almost as real as his memories of real objects. In regions of mathematics already sufficiently well explored, he will be able to perceive highly abstract relationships: equivalence between problems a priori quite distinct, categories of objects far apart in their nature, but in which the same system of relations holds.

Perceptions of this type have very wide ramifications and are capable of great variety. They lose nothing of their effectiveness when we approach the fundamentals of mathematics. Thus the natural desire to embed a theorem within a deductive system that is as restricted as possible causes the *general notion of group* to come into being;

the latter is the common element in the structure of many theories. If we interpret conclusions as effects resulting from the choice of the premises and the hypotheses, we shall understand that there is need to analyze the causal mechanism; something like this is done in dynamics, where the idea of cause occurs at the primary level. Viewing this situation in its entirety, it is not difficult to understand that this causal mechanism will be sometimes continuous and sometimes discontinuous.

I shall not go beyond these remarks. Certain results relating to the principles of mathematics are still too recent to be integrated into the generally accepted fund of intuitions. We should, however, take note of the work recently done in regard to the great classical problems of analysis, namely, to derive them directly from finite problems by means of limit processes. Each existence theorem concerning a transcendental case of one of these problems could then be expressed in terms of a class liberally supplied with neighboring finite cases. I believe that I have already shown the significance of this observation.[1] It will suffice simply to take note of it.

2. In approaching the development of the announced program, I shall first give an example which will permit us to follow an intuitive chain of reasoning, which begins with concrete pictures and is extended by a uniform process in the direction of the Cartesian arithmetization of geometry.

When analysts consider the double integral over the planar region A of a function assumed to be positive at any point P of this region, they find it difficult to avoid evoking two equivalent problems which are faithful representations of this integral. The first consists in finding the *mass* of a lamina occupying this region, whose planar density at each point P of this region equals the value at P of the function mentioned. The second consists in finding the *volume* of a region of space on one side of the plane of A composed of the points M, projected above points P of the region A, such that each length PM does not exceed the value of the function at P. In this case, the intuition associates two concrete images with the fundamental operation for which Riemann worked out a purely numerical definition in the case of continuous functions.

If, instead of a planar region A, we consider a volume V and there take the triple integral of a positive function of the point P, the first

---

[1] On this point see Chap. V of my book *Structure des théories, problèmes infinis*, Act. Scient. et Ind., Hermann, Fasc. 548.

of the above representations still permits us to see in this triple integral the computation of the *mass* of a three-dimensional body which occupies the volume V and has volume density at each point P equal to the value of the function there. However, the second representation is lost; it does not give rise to any concrete interpretation, since we lack the space of four dimensions needed for this purpose.

In spite of this, even if the geometry of such a space does not arise in the real world, we can have a clear conception of it thanks to the Cartesian principle of the equivalence of algebra and geometry, a principle which allows us to arithmetize the Euclidean theorems concerning the plane and three-dimensional space. Our conception of four-dimensional space (or of space of a higher dimension) crystallizes in this arithmetic form thanks to a process which allows us to add a dimension by adjoining a coordinate to those already present, while leaving intact the symmetry possessed by the totality of coordinates.

Finally, it would be a profound disappointment to the intellect not to find the two representations that were available in the case of the double integral present also in the case of the triple integral. An attempt at unification seems to be necessary here, even at the cost of considerable effort: we have to establish the theory of a four-dimensional space without the aid of any concrete representation. This project is based on the recollection of analogous deductive constructions already carried out by Cartesian arithmetization for the plane and for ordinary space. Thus we only prolong a stairway whose first steps are embedded in reality, a stairway of fixed architecture throughout its extent. In the course of the work to be carried out in building up the four-dimensional theory, the geometer will not find himself constrained to follow the unbroken thread of computations called for in the arithmetization; he will make use of short cuts which will enable him, often with a stroke of the pen, to dispense with tedious formal operations. If the image which guides him in such a case does not derive purely from concrete reality, it does not lack the essential aspect of an intuitive approach of a special type. It is what I have already proposed calling *extended intuition*.

3. Throughout a long period in the history of mathematics, this kind of idea or result has been glimpsed under very different conditions. Regarded for several centuries, especially, as essentially fictitious quantities, the complex numbers have since been given a very simple geometric interpretation which today makes them indispensable to us in questions of the most concrete type—for example,

in aerodynamics. In this area the development of intuitive insight required considerable time. The same observations could be made about the non-Euclidean geometries, even if we consider their history only from the point at which Legendre, making use of the premise (explicitly assumed by Euclid) that a line can be indefinitely extended, proved that the sum of the angles of a triangle is less than or equal to two right angles.[2] The question of the legitimacy of the deductive constructions based on the rejection of the parallel postulate was not cleared up until various concrete models were available, appropriated from Euclidean geometry itself and furnishing a representation of the theorems of Lobachevskian geometry. It was at this point that its adoption into the body of mathematics was achieved, to the great advantage not only of geometric theory taken as a whole, but also of the theory of functions.

4. We live in a period in which the attention of mathematicians is focused on the foundations of mathematics, for the latter has recently been the subject of many a polemic. One may ask whether there is a place for precise intuitive conceptions in regard to these questions. If a study of this nature can even be *begun* at present, this is due, as in analogous cases (examples of which we have just cited), to contingencies of a historical nature. It is because of the superabundance of theories since the beginning of the present century that we are able today to add some ideas on the structure of these theories to the patrimony of collective intuitions.

This superabundance, first of all, has caused an essential idea to become widespread: mathematical truths are not absolute. There are only true propositions in this or that system of postulates. The sum of the angles in a triangle, which equals two right angles in the system of Euclid, is less than two right angles in the system of Lobachevsky.

With that in mind, when a theorem is of interest in a field in which certain postulates are given, it is the conclusion stated by the theorem that creates the interest. It is therefore natural to seek the deductive system with the least number of hypotheses in which the theorem is true. For example, if we are interested in the theorem of Thales (that is, in the proportionality relation brought about by drawing a parallel to one side of a triangle), we observe that, by omitting the notions of perpendicularity and angle, we can construct a system possessing logical autonomy in which, in addition to the theorem of Thales, all the propositions of plane geometry invariant under cylindrical pro-

[2] See Lucien Godeaux, *Les géométries* (Chap. IV), Collection A. Colin.

jection hold. On the other hand, the Pythagorean theorem has no meaning in this system.

5. In regard to this matter, the idea of invariance gives rise to a most surprising short cut. Moreover, this idea plays an essential role in every case, in the most diverse areas, in which there is a question of unearthing a *cause and effect relation*.

In the preceding point of view, postulates and hypotheses play a causal role; and in order to clarify this role, we are led to eliminate *excess postulates and hypotheses*.

Let us consider a proposition which leads to a previously given conclusion about the objects of a well-defined class C. We will designate as a *modification* any operation which transforms one of these objects (antecedent object) into a second (consequent object). The modifications to be discussed form a family which will occupy our attention for a time. It is a family F which possesses the following two characteristics:

(1) For each modification which occurs in F there exists an inverse modification;

(2) Given two modifications such that the consequent of the first is the antecedent of the second, the family F contains their product (composition).

If we limit our attention to those modifications which apply to the objects of class C, but now without reference to this or that antecedent O_1 and its consequent O_2, between which a one-to-one correspondence is set up (which makes the said modification equivalent to the set of ordered pairs O_1, O_2 related by this correspondence),[3] then it is possible to consider the composition of these modifications without requiring a special pair of objects that any one of them relates. Thus we obtain the *general notion of group*, at least in the frequently occurring case[4] that the family F possesses the characteristics (1) and (2), in which the second has the simplified form: If two modifications of the class C belong to F, then so does their product. If we add to this property that of possessing an inverse for any modification in the family, then we get the specific attributes of a group.[5]

---

[3] In linear geometry of three dimensions, for example, if the objects of class C are *tetrahedra* taking one into another, then the type of one-to-one transformation to be preferred is that of *linear transformation*.

[4] However, the majority of the cases cannot be considered without restrictions. On this point, see a recent article: "Sur une catégorie de propositions," *Comptes rendus Acad. des Sc.*, CCXXIII (1946), p. 495.

[5] An objection must be presented. Given the variety of ways in which the effective determination of one-to-one correspondences within the class C can be carried out, it is not surprising that by starting out from different points of view

The notion of group affords a most precious organizing principle for the seeming labyrinth of deductive structures. For example, geometric propositions will be found to be organized in terms of their invariance properties. Thus, within the Euclidean structure we single out:

(1) *The metric propositions* (example: the Pythagorean theorem), which are invariant under translation, or, what is a little more specialized in this geometry, even under similarity transformations which preserve rectilinearity and the measure of angles;

(2) *The linear propositions* (example: The theorem of Thales), invariant under those transformations which simultaneously preserve rectilinearity and parallellism;

(3) *The projective properties* (example: Pascal's theorem on the hexagon), invariant under those transformations which preserve rectilinearity only.

The gradation of these properties is clear: the first are possessed by the translation group, the second by the linear group, the third by the projective group, groups of which each is a subgroup of the following.

From this appropriate intuitive point of view we are able to encompass in a single glance the structural ideas developed by Felix Klein in his celebrated Erlangen Program, ideas which have bearing outside the Euclidean geometry. To each autonomous geometric system there corresponds a group.[6] In three-dimensional Euclidean

---

it would be possible to assign several groups to the same proposition. Thus the theorem on the square of the hypotenuse, invariant under similarity transformations, can be related to other groups; more precisely, if we specify a right triangle by giving the sides adjacent to the right angle, we can go from one right triangle to any other by means of a one-to-one continuous transformation of a pair of positive numbers to another pair of positive numbers (which gives a group with an infinity of parameters). But for us it is not so much a question of arriving at this or that specified group, as of making clear *what is the essence of the group structure*. And if the intuitive approach which permitted us to obtain this result seems commonplace, it would be appropriate to take up the question of the role of groups again in the light of the observations made by Élie Cartan in a recent study (Chap. III of Section A in the mathematical part of the first volume of *L'Encyclopédie française*, 1937). In particular, the fact that a collection of transformations leaves invariant a certain fixed property P does not permit us, in the most general cases possible, to assert that these transformations form a group. The question of deciding which properties P, said to be of the *première catégorie* in the text cited, will give rise to a group in this way seems to be a rather difficult one, if we are concerned with general cases. (Cf. *Revue Philosophique*, CXXXIV (1944), note on p. 202.)

[6] At least, with the reservations already made. What is called today the *geometry of distance spaces* is not based on any group.

space, the group of rotations about a fixed point, i.e., translations on a sphere, becomes the group of a two-dimensional abstract geometry in which there are no notions of translation or of parallels.

It is clear, of course, that two rigorous proofs of the same theorem may have different characteristics which cause us to prefer one of them. For instance, if it is a question of projective geometry, and if one of the proofs uses only projective considerations, we would certainly give our preference to that mode of reasoning which excludes any element not within the domain of invariance of the theorem that it claims to establish.

What has been said could be greatly enlarged; metric propositions and linear propositions could be given a common base, provided by the topological propositions left invariant by all one-to-one bicontinuous transformations. This is the root of the mathematical idea of number of dimensions.

5. The intuitions which revealed to us this very fundamental organ, the group, derive from *the idea of cause*. They were able to take on greater force only in the light of research into the principles of dynamics, considered as much in the Newtonian as in the relativistic form. Certain mathematical theories had attested to the efficaciousness of the concept of group, whose applications were developed in the course of the 19th century. But its universality had not yet been discovered. Mechanics, closely linked to the idea of cause, has certainly contributed to disengaging the concept of group from its special applications, even to making it clear that the unification of important fields of physics required the unification of their groups (the group of distance and time invariance for Newtonian dynamics and the Lorentz group for electromagnetism) under whose aegis the rational coordination of these sectors was carried out.

These clarifications taken from mechanics carry us in a new direction, which leads us to the heart of the deductive sciences, toward other essential aspects.

6. Now that we have introduced the idea of a causal mechanism into the various branches of mathematics, it is appropriate to study the nature of the continuity or the discontinuity of this causal mechanism.

Suppose that a certain conclusion has been established on the basis of an irreducible system of hypotheses within the context of given axioms; what will happen if we modify this system of hypotheses by substituting another for it in such a way as to provide a system arbitrarily close to it?

If we pose the question thus, it will be easily seen that it may or may not be resolved in the continuous sense, depending on the situation at hand. Let us illustrate this by an example.

If the slope of the graph of a function of x is everywhere zero over an interval I, then the function is a constant, and conversely. Let us vary slightly the conditions of the theorem and its converse. A very small slope over the length of an interval I would cause the function to be nearly constant, but it is not true that if the function is almost constant over the interval, then the slope is always very small. In other words, the continuity of the causal mechanism which exists in the theorem does not exist in its converse.

Such contrasts may also be expressed by distinguishing between *stable* and *unstable* propositions. It is again dynamics, in its most familiar aspect, which provides the best illustration. A simple pendulum, capable of oscillating in a given vertical plane, remains in equilibrium if its initial position is vertical (downward) and its initial velocity is zero. If its initial position is almost vertical and its initial velocity is very small, then the pendulum will diverge but little from the vertical, which may be expressed by saying that the pendulum is in stable equilibrium. In the terminology of the beginning of this paragraph, it is just this somewhat special example which affords us a prototype of the logical stability which Pierre Duhem saw as early as 1907 (in almost these terms)[7] as a test of effectiveness to be applied to a mathematical deduction that one wishes to employ in the solution of physical problems.[8]

7. The preceding approach should not cause us to forget that other approach which considers the *hypothesis* as an independent variable and the *conclusion* as a function thereof, and proceeds to examine any resulting discontinuities. Thus, after having considered the notion of correspondence in its most diverse aspects and having made it the special object of its speculations, mathematics, in reflecting on its own

---

[7] See P. Duhem, *La Théorie Physique, son objet et sa structure*, Part II, Chap. III.

[8] It should be kept in mind that the stability or instability of a proposition in each case depends on the type of definition given for the neighborhood of the hypothesis and, of course, on the type of definition given for the neighborhood of the conclusion. For example, the property of a line segment of being the minimum path between its extremities is unstable if the change in the hypothesis is limited in such a way that the new path to be followed must remain within the union of the spheres of radius ϵ centered on all the points of the segment. On the other hand, this proposition becomes stable if we limit the change in the hypothesis by the maximum angle made with the segment AB by a chord of the varied path (supposing that it is traversed in the same direction as AB).

nature, recognizes this theme, this structure, to be entirely adapted to its own organism.

This indicates that we will not be able to disregard the logic. In fact, we become deeply involved with it the moment we consider for classes properties which are independent of the nature of their elements.[9]

But the accord between mathematics and logic considered in terms of operators is much less perfect than it is often thought to be. It is impaired as soon as the concept of infinity enters in some way: this is already the case for mathematical induction and for those arguments of classical analysis which make use of the linear continuum and its generic element, number, the latter defined in terms of an infinite decimal expansion.

Adapting to our conceptions certain expressive terms used by Cournot in the last century, we may say that there is no *logical order* which can be completely separated out; what we actually have is a *rational order* inspired by our need to know the true state of affairs. This order is progressively revealed to us in the form of intuitions, even though we would prefer to operate solely within the framework of logic. Let us again give an example.

8. The conception of statements overloaded with excess hypotheses, which occur frequently in the area of classical differential geometry, cannot fail to call forth its opposite, that of statements whose hypothesis is insufficient to imply the desired conclusion. Relative to this conclusion, there will be a need to distinguish between ordinary and

---

[9] These properties can in fact be deduced from the following relations: *inclusion*, as a consequence of which one set is included in another; the fact that a set is either the *union* or the *intersection* of some others (in the latter case it is composed of their common elements); the fact that two sets without common elements are *complementary*—that is, their union reproduces the universal set from which they were formed.

Let us make correspond to a characteristic property the set of all elements in the universal set which possess the property. If property *A implies* property *B*, then *A* will correspond to a set which contains the set corresponding to *B*. If several properties occur simultaneously in a theory, then we shall necessarily encounter the set corresponding either to the truth of at least one of the properties (the *logical sum*) or to the truth of all of the properties (the *logical product*). Finally, if a property corresponds to a set *A*, then the *negation* of this property will correspond to the complement of this set in the universal set.

In this classical conception, established in the framework of the law of the *excluded middle*, the wheels of the logical mechanism are thus the formal properties of sets. This conception is valid for all those cases in which the usual scheme of true and false is applicable. This gives rise to the possibility that other collections of formal properties may represent other logics.

exceptional cases and to study if possible (if extended intuition can provide an appropriate abstract space as a quasi-concrete representation) the degree of rarity of the exceptional cases with respect to the ordinary cases. This degree of rarity may be expressed sometimes in topological terms, as in theorems relating to the determination by the appropriate number of points of the algebraic curves or sufaces of a given degree;[10] sometimes by using the idea of measure, as in the theorem of Lebesgue which asserts that every curve of finite length will have a tangent at each point with the possible exception of a set of measure zero, i.e., that it can be approximated by a sequence of segments of arbitrarily small length. At this point we approach the idea of a statistically exact conclusion.

9. This latest example shows that the facts of the rational order, in the sense of Cournot (we might also say: the *economic order* of mathematics), may be perceived by us intuitively. Certainly, logic has created symbols and used them successfully in formalizing arithmetic and extensive parts of the theory of sets. But it was not able to extract all this exclusively from within itself, any more than algebra was able to come into existence without the benefit of long geometric experience. It is for this reason that it is possible to believe today that the formalism of Hilbert has created an especially favorable climate for a new approach to mathematics of a more refined and technical nature. For all that, it has not yet forged tools which would allow us to resolve certain of the remaining difficulties, inherent in the foundations of mathematics itself.[11] Doubtless, intuition will be the guide to future progress in this domain as well.

[10] Statements of this type have the characteristic of being stable in the ordinary cases, unstable in the exceptional.

[11] Above all, it would have to be shown that by constructing an *enlarged* (or at least: *not reduced*) *mathematical system*, i.e., a system permitting of potential elements (numbers, functions) which will never be actually introduced, and by causing certain paradoxes to arise by this abuse of the power of extension, we do not encounter them except in a purely verbal context in which they lose their potency. But by limiting what is verbal in this way, we shall have a tendency to assign a *lesser or greater degree of existence* to mathematical entities! And then, would not any criterion invoked for this purpose require the most subtle intuition? At this point one would be subject to mental decisions difficult to separate from sense experience. Compared with corpuscular physics, mathematics is an incomparable fruit which has ripened thanks to the illusions of our senses.

Book Two

DISCIPLINES

After viewing mathematics in this somewhat panoramic way, we are now going to follow the various paths which lead to the chapels of the mathematical temple. It was not possible to present a series of articles in the overly scholarly order of the table of contents in a mathematical encyclopedia. We had to resign ourselves to heavy sacrifices and deny a place to certain aspects of mathematics in spite of their great importance; they will find a place in a second volume. However, we have managed to describe within this first volume the main lines and principal axes of this architecture. Following the traditional order, we shall investigate in turn number, space, function, group *and* probability.[1]

Of course, these concepts, along with that of equation,[1] *are still very concrete notions, by reason of their relative familiarity, especially the first two, whose connections with the infinitely complex history of the relations between man and the universe are evident. These precious and moving "impurities" allow the double countenance of thought and nature to appear as through a veil. This is why we have given preference to these traditional divisions (somewhat modernized, of course) over a classification more satisfying from the point of view of reason alone, like that which is so clearly and validly discussed in the article of Bourbaki, but which is also less suggestive for the public at large.*

[1] While also reconsidering these fundamental notions in order to complete their study and to make them more precise, we shall make use of a second series of articles to introduce the ideas of *determination* or *law* (especially in its algebraic and analytic aspects) and of *metamathematics*, which are not given their proper importance in a sufficiently direct manner in this first series.

A. NUMBER

"Number resides within all that is known. Without it, it would be impossible to reason or to know," wrote the Pythagorean philosopher Philolaus. We cannot share this categorical certainty without serious reservations, but in spite of that we still feel that it is appropriate to begin our tour with the notion of number. Number, at least in classical mathematics, plays the role of the cell in a living organism. It was appropriate to devote an all-encompassing study to this fundamental notion. Maurice Fréchet has brushed a beautiful fresco celebrating that prodigious ascent which, moving from peak to peak, from the concrete, natural numbers, to the negative numbers, fractions, irrational numbers, transcendental numbers, complex numbers, . . . , terminates in the transfinite numbers and the theory of sets. Fréchet's efforts have been divided between the theory of sets, which he has treated in a spirit of very great generality, and the theory of probability, which he has greatly enriched by his studies on stochastic convergence and on the probability associated with a sequence of events, as well as on random variables of an arbitrary nature. Thus, as much by his aptitude for abstraction as by his care in respecting the concrete origins and practical application of general ideas, he is very qualified to trace the course of this evolution.

Fréchet was able to discuss only briefly each category of numbers as it passed in review. We have singled out for closer examination three categories which seem to be of greatest significance: the integers, transcendental numbers and transfinite numbers.

Although directed toward a very special problem of Diophantine analysis, that of Fermat's last theorem, the article by Théophile Got, who has been principally interested in the theory of Fuchsian groups and the theory of numbers, underscores the profoundly basic character of the integers and of the prime numbers, as well as the immense difficulties that arise in their study.

Paul Dubreil, one of the foremost members of the French school of modern algebra, has spent much of his time in the study of ideal theory and its geometric applications, as well as in the study of the very general notion of equivalence. He invites us to a grand performance in which appear the greatest actors among the transcendental numbers: the numbers e *and* π, *mathematical stars without which the opera of analysis could never be performed. "They say," the Platonist philosopher Proclus assures us, "that the people who first divulged the secret of the irrational numbers perished in a shipwreck to the last man, because the inexpressible, the formless, must be kept absolutely secret; those who divulged it and depicted this image of life perished instantaneously and will be tossed about by the waves for all eternity." By good luck, the ancient gods have become gentle and we shall not have to praise M. Dubreil for his courage but only thank him for this expression of his knowledge.*

"The infinite! Never has a question so profoundly troubled the mind of man," exclaims Hilbert. The article by Henri Eyraud, who has conducted research in domains as varied as tensor theory and statistics, and who has an equal interest in the foundations of mathematics, commences the study of the theory of sets, a subject which inaugurated the computability of the infinite at the end of the 19th century in an atmosphere of violent controversy which is still far from settled.

Eyraud stresses the arithmetic—or, if we prefer, the numerical—aspect of the theory of sets. However—and this is made clear within the article itself—the theory of sets should not be too closely linked to the notion of number alone. It is to be found at the base of all the other notions of mathematics,[1] especially those of space and function, and that is why it will be encountered in other articles within this work, in particular that of Denjoy.

F. LL.

---

[1] It is in this sense a unifying principle for mathematics.

6

THE NATURAL NUMBER AND ITS GENERALIZATIONS

by Maurice Fréchet
PROFESSOR AT THE SORBONNE

It is an immense program that Le Lionnais has asked me to cover in a few pages, and it is conceived in a lyrical spirit in no way disproportionate with its aim, but one which is difficult to sustain in the course of somewhat abstract exposition.

We shall try, however, to give as clearly as possible a brief idea of the evolution which is one goal, and, since the facts themselves are well known to the majority of our readers, the reasons which motivated it.

The Natural Numbers. For the sake of brevity we shall omit any discussion of the nature, the origin and the theory[1] of the positive integers 1, 2, 3, ..., also called the natural numbers, as well as of the four fundamental operations (addition, etc.) relating them.

Let us observe, however, that while two of these operations, addition and multiplication, can be carried out for any pair of natural numbers, this is not the case for subtraction and division. In order that the difference $a - b$ be defined, it is necessary that a be greater than b, and in order that the quotient a/b be defined, it is necessary that a be a multiple of b. These two restrictions are troublesome. When we have to solve a problem in which the given quantities depend on the particular case of the problem, then in place of obtaining a solution which can be expressed by a general formula, we have to consider several cases, to carry out what mathematicians call a complete treatment.

---

[1] See, for example, the treatment of this nature intended for the general public on p. 1602 of the article by C. Chevalley in the volume "Mathématiques" of the *Encyclopédie française*.

Negative and Rational Numbers. This necessity can be avoided by the introduction of zero and of the negative and the rational (also called commensurable) numbers.[2] The introduction of zero allows us to give a meaning to the expression $a - a$; the introduction of the negative numbers allows the operation $3 - 4$, for example; that of the rational numbers or fractions confers a meaning on the division of 3 by 4, for example.

It should be observed that it is not sufficient to decree that what was impossible has now become possible by the introduction of a new language. This extension cannot be carried out without some thought, and certain conditions must be satisfied. The four operations with their restrictions had been defined for the natural numbers and gave natural numbers. When we introduce the negative numbers (after zero), for example, in order for this to be a true extension of the notion of natural numbers it is necessary that the four operations, which now relate pairs of numbers taken not only from among the positive natural numbers (the category already studied), but also from among the negative numbers and zero, agree with the previously defined operations when applied to two positive natural numbers, and that no restriction remain on subtraction, i.e., that $a - b$ be defined for any a and b (positive, negative or zero). The classical definitions of the operations on the integers (of whatever sign) are therefore not arbitrary, but the result of a carefully thought-out choice. The same is true of the notation. Other less convenient but legitimate notations could have been adopted; in fact, some of them remain in use. For example, for latitude we still employ the notation 32° north latitude, 46° south latitude. If we used the notation $+32°$ and $-46°$, we could say that the angle formed by two points a and b on the same meridian of latitude with the center of the earth is always equal to $a - b$, while for the above points it is not $32° - 46°$, but $32° + 46°$.

This introduction of the negative and the rational numbers was justified above in terms of the tendency to eliminate troublesome restrictions. It is a mathematical stratagem; but in both cases the introduction could also have been motivated by the need to solve concrete problems. For example, fractions were introduced to represent certain modes of sharing. A will gives to two heirs a piece of land of which one heir will have a part twice as large as the other: they will

---

[2] We shall comment on the introduction of the negative, rational and irrational numbers without recalling their definitions, which we suppose to be familiar to the reader.

have $\frac{2}{3}$ and $\frac{1}{3}$ of the land, respectively. Similarly, any time that we have to deal with oriented quantities, we have to adopt a special notation to distinguish those oriented in one direction from those oriented in the other. This is the case with latitudes; in the same way, we distinguish between the dates before and after Christ, the temperatures above and below zero, etc. The adoption of the minus sign preceding numbers of the second orientation supplies first of all a general, convenient notation in place of special notations for each case. But, above all, it has the advantage of eliminating the restrictions on certain formulas.

Irrational Numbers. The same advantage was also sought in the introduction of the irrational numbers such as $\sqrt{2}$ or π.

It would be possible to set forth all of the known mathematical results without making use of these new numbers, but this would cause the statements and proofs to become much longer and less intuitive. For example, $\sqrt{2}$ would be replaced by the sequence of fractions $1, \frac{2}{2}, \frac{4}{3}, \frac{5}{4}, \ldots$, with denominators $1, 2, 3, 4, \ldots$, consisting of the largest among fractions with square less than 2. But then it would be necessary to replace the statements in which simple operations on $\sqrt{2}$ appear by others which would involve an infinity of rather complex operations on the terms of the above sequence.

Thanks to this introduction, it was possible to give many propositions a simpler form. For example, it used to be said that a rational number has a cube root if, when it is reduced to its simplest form, its terms are perfect cubes. Now that we have extended this class by adjoining the irrational numbers, we can say that every number, rational or irrational, has a cube root; the previously necessary restriction has disappeared. Thus, we were not able to assign a measure to the diagonal of a square with sides of unit length using rational numbers, but after the introduction of the irrational numbers we were able to say that this measure is $\sqrt{2}$ and to make use of it in further calculations.

We have just indicated one of the ways of introducing the irrational numbers, that based on the notion of the natural order of the numbers (a way which leads to the concept of "Dedekind cut"). They can also be introduced by making use of the notion of distance. Let us denote by the distance between two numbers a and b the absolute value $a - b$, i.e., the difference between the two numbers a and b, apart from the sign of this difference. Given this, we then determine a real number in terms of the class of all sequences of rational numbers

$r_1, r_2, \ldots, r_s, \ldots$ such that the distance $r_s - r_{s+q}$ tends to zero (that is, the distance can be made less than any given positive number—for example, a millionth—by taking the integer s sufficiently large, the integer q being arbitrary and independent of s).

In what follows, we shall denote by the term real number (in contrast to the imaginary numbers to be defined later) any rational or irrational number (positive, negative or zero).

Classification of Real Numbers. The simplest real numbers are the rational numbers (of any sign). Within the category of irrational numbers, we can distinguish a still simpler category (containing the rational numbers), that of the real *algebraic* numbers; by this we mean any real number X which is a root of a polynomial $a_0 x^n + \cdots + a_n$ whose coefficients $a_0, \ldots, a_n$ are integers (of any sign). Moreover, if we set $x = X/a_0$, substitute this expression into the polynomial, and then multiply the polynomial by a_0^{n-1}, we see that X will be a root of the polynomial

$$X^n + a_1 X^{n-1} + a_0 a_2 X^{n-2} + \cdots + a_0^{n-1} a_n$$

with integral coefficients and leading coefficient 1 (monic polynomial). X is called a *real algebraic integer*. We see, on the one hand, that we can discern within the set of real algebraic numbers a still simpler category (which contains the rational numbers) and, on the other hand, that every real algebraic number is the quotient of a real algebraic integer divided by an ordinary integer.

The Question of Existence. We have defined a series of families of numbers, each more general than its predecessor and containing it. However, a question remains to be cleared up. Is each of these defined extensions really an extension? Is a new family containing the preceding families distinct from them? We already know that there are rational numbers ($\frac{1}{2}$, for instance) which are not integers, that there are algebraic numbers (like $\sqrt{2}$, the solution of $x^2 = 2$) which are not rational. We can easily prove that there are algebraic numbers (such as $x = \sqrt{\frac{2}{3}}$) which are not algebraic integers. But are there *transcendental* numbers, i.e., irrational numbers which are not algebraic?

Transcendental Numbers. This is a question which remained unsolved for centuries. In relatively recent times we have been able to give three kinds of answers. We can content ourselves with proving (as is done further on, p. 78) that the set of algebraic numbers is only a part of the set of rational and irrational numbers. Liouville in 1851 gave a better answer to this question by showing us a way in which we can determine as many transcendental numbers as we wish. Let a be one of these special transcendental numbers, called *Liouville numbers*; by definition, a is such that, given any integer n, there exists an infinite number of irreducible fractions p/q for which

$$\left| a - \frac{p}{q} \right| < \frac{1}{q^n}.$$

We can easily prove that such a number could not be the root of a polynomial with integral coefficients; and in order to prove that such numbers exist, it suffices to give an example. This is furnished by the sum of the series

$$\sum_{n=1}^{n=\infty} \left(\frac{1}{2}\right)^{n^n}.$$

Finally, the third and most satisfying answer consists in showing that the numbers that naturally arise in analysis are actually transcendental. This was done for the first time by Hermite when he considered the number $e = 2.718\ldots$, the base of natural logarithms and equal to the limit, as $m \to \infty$, of $(1 + 1/m)^m$, a number of fundamental importance in analysis. Later, Lindemann showed that $\pi = 3.14159\ldots$ is also a transcendental number. After these, still other categories of real numbers have been shown to be transcendental. On the other hand, in spite of its importance, it is still not known whether "Euler's constant"

$$C = \lim_{n \to \infty} \left(\left(1 + \frac{1}{2} + \cdots + \frac{1}{n}\right) - \log n \right) = 0.577216$$

is transcendental or not.

p-adic Numbers. The p-adic numbers of Hensel are in an interesting creation and possess many of the properties of real numbers. However, we can only give a brief sketch of their definition here. We can base this on the notion of distance, as was done on page 72 for the irrational numbers. But here it is a rather odd notion of distance. If we are given two rational numbers (of any sign) r', r'', and a prime natural number p, then the difference $r' - r''$ can, as any rational number, be put into the form $(u/v)p^n$, in which u, v and n are integers of either sign, v is not zero and u, v, p are relatively prime in pairs. We then agree to take as the distance (r', r'') of r' and r'' the quotient $1/p^n$ if r' is distinct from r''; otherwise, of course, zero. We next consider as defining a "p-adic" number any sequence of rational numbers $a_1, a_2, \ldots, a_s, \ldots$ such that the distance (a_s, a_{s+q}) defined as above tends to zero as s increases without bound and for any integer q. We thus obtain results such as: 7 is a 3-adic number, for in the 3-adic sense:

$$7 = 1 + 1 \cdot 3 + 1 \cdot 3^2 + 0 \cdot 3^3 + 2 \cdot 3^4 + \cdots.$$

Ideals. It has been possible to build a theory of the algebraic integers by extending to these numbers many of the properties of the ordinary integers. However, other properties are lost and the reason for this is usually that the notion of greatest common divisor loses its meaning in the class of algebraic integers. Kummer surmounted this difficulty when he created a new class of numbers, the ideal numbers, which include the algebraic integers, but in which the notion of g.c.d. again has its classical meaning.

Complex Numbers. Before terminating this list of various numbers of interest especially in the theory of numbers, it will be sufficient to mention that their definitions have been extended to the complex domain, which will be discussed shortly, and even to the domain of more abstract numbers such as those which are considered at the end of this article.

The creation of algebraic integers, algebraic numbers and ideals is due rather to scientific curiosity and esthetic interests than to a practical need. At present, we would search in vain for any mention of these numbers in works written for engineers; this does not exclude the possibility, of course, that they may eventually occur in such publications.

This is not the case for the introduction of the complex numbers, also called imaginary numbers.

Let us designate the numbers considered in what has preceded as the real numbers and let us consider a trinomial of the second degree (that is, a sum of terms

$$ax^2 + bx + c$$

in which the variable x enters at most to the second degree). We know that if the coefficients a, b, c are real and we assume that x is real, then this trinomial can be decomposed into two factors of first degree with real terms

$$a(x - x')(x - x'')$$

when we have the inequality $b^2 - 4ac \geq 0$. It is this troublesome restriction that we succeed in avoiding by the introduction of the complex numbers. If we appropriately define the extension to these numbers of the four elementary operations,[3] the decomposition of the trinomial into two factors of first degree (with real or complex terms) becomes possible in every case, even when a, b, c are complex. It even happens that the simplification goes much farther than we had sought. For example, the preceding theorem generalizes (under the name of the theorem of d'Alembert) to a polynomial

$$ax^n + bx^{n-1} + \cdots + f$$

of any degree n; it can always (for $a \neq 0$) be decomposed into a product of n factors of first degree in x (with real or complex terms). Likewise, we have the remarkably simple theorem of De Moivre:

$$(\cos x + i \sin x)^n = \cos nx + i \sin nx.$$

At the time when the complex numbers were conceived there were no concrete objects whose study would lead naturally to this introduction. Later, however, the complex numbers were found to be very useful in representing certain electrical circuits, for example. It is characteristic of great mathematical discoveries that they prove to be useful later on in many areas other than the ones considered when they were made.

Quaternions and Hypercomplex Numbers. The ordinary complex numbers have been still further generalized in the form of quaternions and, later, of hypercomplex numbers.

---

[3] In brief, we still define subtraction and division here as in the preceding cases, i.e., as inverse operations of addition and multiplication. One of the ways of defining the latter two is the following (perhaps not the best, but the one which seems most convenient here). Suppose first of all that we have represented every complex number in the form $a + bi$, where a and b are real numbers and i is a new symbol; the result $a'' + b''i$ of taking the sum or product of two complex numbers $a + bi$, $a' + bi$ is obtained by performing the same operation on the polynomials in x: $(a + bx)$, $(a' + b'x)$ and by denoting by $a'' + b''x$ the polynomial obtained by this operation, replacing x^2, if it occurs, by -1. Thus

$$(3 + \pi i) + (\sqrt{2} - i) = (3 + \sqrt{2}) + (\pi - 1)i$$
$$(3 + \pi i)(\sqrt{2} - i) = (3\sqrt{2} + \pi) + (\pi\sqrt{2} - 3)i.$$

We can identify any real number a with the complex number $a + 0i$, for it is clear that if we apply the rules of the operations on the complex numbers to the real numbers put into this form, the results will coincide with the results of the corresponding operations previously defined for the real numbers.

The introduction of the symbol i was considered mysterious for a long time and explains why the numbers $a + bi$ were first called imaginary numbers. In reality, any complex number is nothing but an ordered pair (a, b) of real numbers, for which precise rules of operation are given and which, though perhaps novel, have nothing paradoxical about them.

By the term hypercomplex number, or more precisely *complex number of order* n, we mean an ordered sequence, a, of real numbers a_0, $a_1, \ldots, a_{n-1}$, which satisfies the following operational rules. We make correspond to a a "linear form"

$$a_0 x_0 + \cdots + a_{n-1} x_{n-1}$$

—which we may abbreviate by $a(x)$—then we say that two complex numbers of order n, a and a', are equal if the corresponding forms

$$a(x) \text{ and } a'(x) = a'_0 x_0 + \cdots + a'_{n-1} x_{n-1}$$

are identical, i.e., if

$$a_0 = a'_0, \ldots a_{n-1} = a'_{n-1};$$

the sum or product a'' of a and a' has the corresponding form

$$a''(x) = a''_0 x_0 + \cdots + a''_{n-1} x_{n-1},$$

that obtained by performing the same operation (sum or product) on the corresponding forms $a(x)$, $a'(x)$, of a and a', replacing, however (in the case of the product), the products $x_j x_k$ by the linear expression

$$x_j x_k = \sum_{s=0}^{n-1} \gamma_{jks} x_s,$$

where the γ_{jks} are given constants.

From these definitions it results that a complex number a of order n can be put into the form

$$a = b_0 e_0 + \cdots + b_{n-1} e_{n-1},$$

in which the b's are still real numbers and the e's are certain complex numbers of order n called units of the system of numbers considered and such that

$$e_j e_k = \sum_{s=0}^{n-1} \gamma_{jks} e_s$$

(it is not necessarily true that

$$e_j e_k = e_k e_j).$$

According to the set of values given to the γ_{jks}, we obtain one or another system of complex numbers of order n.

The ordinary complex numbers are of order 2 and correspond to

$$e_0 e_0 = e_0, \quad e_0 e_1 = e_1 e_0 = e_1, e_1 e_1 = -e_0.$$

The most celebrated system of hypercomplex numbers is that of the *quaternions*, discovered by Hamilton. It is a system of order 4 in which

$$e_j e_j = -e_0; \qquad e_j e_h = \pm e_k$$

(where j, h, k is a permutation of the numbers 1, 2, 3 and where the sign is positive for the cyclic permutations of 1, 2, 3; 2, 3, 1; 3, 1, 2 and negative for the others).

These various complex numbers have numerous applications in geometry, mechanics and physics, and they have also been studied for their intrinsic interest. An excellent résumé of all these works can be found in an article by Study and Cartan.[4]

Non-Denumerable Sets, Transfinite Numbers. An entirely different extension of the notion of number presented itself when Cantor pointed out the existence of two kinds of infinites.

It is easy to show that we can number the set of rational numbers of any sign with integral indices; this property is expressed by saying that the set of these numbers is *denumerable*. Let S_n be the set of irreducible fractions $\pm p/q$ such that $p + q = n$. There is only a finite number of such fractions. We can then number them a, $a + 1$, ..., $a + s$, starting from some suitable number a, and by successive ordering of $S_1, S_2, \ldots, S_n, \ldots$ we are able to number the sequence of all rational numbers.

In an analogous way, we can number all the algebraic integers, this time taking for S_n the set of roots of all the polynomials with integral coefficients (the first coefficient is taken equal to 1), in which the sum of the absolute values of the coefficients added to the degree of the polynomial is equal to n.

It is clear that every subset of a denumerable set is itself denumerable and that every denumerable set of denumerable sets is denumerable.

Before Cantor, it was understandably thought that in order to represent an infinite set, i.e., one comprising an infinite number of elements, it sufficed to take one of its elements, denote it by a_1, take one of the remaining elements, denote it by a_2, etc., and that the set could thus be represented by the ordered sequence: $a_1, a_2, \ldots$. Mathematicians must certainly have noticed that if the set were, for example, the sequence of integers $1, 2, \ldots, n, \ldots$, and if the preceding operation had been to number by n the even number $2n$, then

---

[4] Pages 329–428 of the first volume, "Arithmétique," of the French edition of the *Encyclopédie des Sciences mathématiques*, Gauthier-Villars, 1904.

the sequence of the a_n would not exhaust the set. But they were apparently convinced that it sufficed to choose an appropriate ordering in order to exhaust an infinite set by numbering it with integral indices.

But then Cantor proved that no matter what attempt was made to number the set of all real numbers, some will always remain: *the set of all real numbers is non-denumerable*. This was a noteworthy discovery, at the base of an entire theory, the theory of sets, whose influence has made itself felt throughout all of mathematics.

For example, we can immediately deduce from this the proof of the existence of transcendental numbers mentioned above, for if the set of real numbers coincided with the set of algebraic numbers, the former would be denumerable.

As for the subject at hand, this discovery had as an immediate consequence the creation of the transfinite numbers.

It is sufficient for this to extend to the case of infinite sets the notion of the number of elements of the set such as can be formulated for finite sets. This can be carried out in two ways.

Of course, it may be assumed or not in the definition of the number of elements in a set that these elements are arranged in a given order; the number will then be ordinal or cardinal, respectively.

But here a new phenomenon arises. Whereas, in the case of finite sets, if we start out from different definitions the cardinal and ordinal numbers coincide for practical purposes, this is not the case for infinite sets. We shall therefore treat these two notions separately.

In both cases, we shall be wise to avoid the philosophical difficulty involved in giving a precise clarification of the nature of the notion of number. It will be sufficient in order to make use of this notion to specify, as we are about to do, what we mean when we say of two sets A and B that A and B have the same number of elements, or that A has a greater number of elements than B. We can even go further and point out two sets C and D such that the number of their elements, respectively, will be considered as the sum and product of the numbers of elements of A and B.

Cardinal Numbers. How do we compare the numbers of eggs and apples in two piles placed on a table? The practical method would be to count the eggs and the apples separately and compare the numbers found for each. But this method is indirect; it consists in replacing the sets of eggs and apples by two sets of numerical symbols and comparing the latter among themselves. The direct method consists in placing an egg beside each apple in so far as possible and observing if eggs remain, or apples, or nothing. We shall agree to say in each of these cases that the number of eggs is respectively greater than, less than or equal to the number of apples.

This leads to the following general definition: Let there be given two sets E and F; if there exists a one-to-one correspondence, i.e., element to element, between the elements of E and a subset of the elements of F, then we shall say that the cardinal number of the elements of E is less than that of the elements of F; if the correspondence extends to all the elements of F, then these two cardinal numbers are said to be equal.

Now, this definition—which is applicable to the case in which E and F have only a finite number of elements and is then equivalent to the ordinary definitions—retains its meaning when at least one of the sets E or F is infinite.

Thus we see that: (1) the cardinal number of any finite set is less than that of any infinite set; (2) the smallest cardinal number of any infinite set is equal to that of the infinite sequence of natural numbers; (3) the cardinal numbers of infinite sets are not all equal.

Ordinal Numbers. A set of elements E is said to be *ordered* if there is given a rule by which it is possible to determine, when given any distinct pair of elements of E, which one precedes the other. (However, this rule must satisfy the condition that if a precedes b and b precedes c, then a precedes c.)

We note that in any finite ordered set E every subset of E has a first element (we say that such a set is well ordered). This is not necessarily the case for an infinite set. For example, if we arrange the natural numbers in decreasing order $\ldots, n, n - 1, \ldots, 3, 2, 1$, then we have an ordered set which is not well ordered.

The ordinal numbers have been defined only for the well-ordered sets (which include the finite sets). We say that the ordinal number of a well-ordered set E is less than or equal to that of a well-ordered set F, if it is possible to establish a one-to-one, order-preserving correspondence between the elements of E and those of a subset of F.

We limit ourselves to these remarks and refer the reader to the exposition, still rather brief, but more detailed and intended for the general public, in our pamphlet: *L'Arithmétique de l'infini*.[5]

---

[5] *L'Arithmétique de l'infini*, 37 pages, Hermann, Paris, 1935.

7

A MATHEMATICAL ENIGMA: FERMAT'S LAST THEOREM

by Théophile Got

FORMER MARINE ENGINEER, HONORARY PROFESSOR IN THE FACULTY OF SCIENCE AT POITIERS

God created the natural numbers, all the rest is the work of man.
KRONECKER

There are no solved problems; there are only problems that are more or less solved.
H. POINCARÉ

GAUSS—called by his peers the prince of mathematicians—said that mathematics is the queen of the sciences, but that the theory of numbers is the queen of the mathematical sciences. A few pages could never suffice to give an idea of the difficulty, but also of the beauty, of the problems which arise in its domain. A sketch, half theoretic, half historic, on the most celebrated of these—Fermat's last theorem—will, however, allow us to form an idea of their interest.

This unsolved problem of Diophantine analysis remains the only one among the numerous propositions of Fermat which has not yet been proved after three centuries of effort by a host of mathematicians both small and great. It is for this reason that it is called Fermat's "last" theorem, and that Édouard Lucas could say in 1891 that "this problem seems to be thrown out as a perpetual challenge to the human mind." Can we be assured that the human mind can meet this challenge? This is the question that we shall attempt to answer in the light of the research to date.

Diophantine analysis has as its goal to decide whether certain equations in several unknowns admit *integral* solutions and to find a general expression for these solutions. Diophantus of Alexandria, a mathematician of the 4th century, was the first to become interested in it. The problem of Diophantine analysis apropos of which Fermat stated his theorem is: *Find all those right triangles whose three sides have integral length*, i.e., find the integral solutions of the indeterminate equation

$$x^2 = y^2 + z^2.$$

We can suppose that x, y, z are relatively prime, for otherwise it would be sufficient to divide throughout by the square of their greatest common divisor. Not all three can be odd; therefore, at least one is even, and it cannot be x, for then the left side would be divisible by 4 but the right would not be. Let us suppose therefore that y is even and write the equation in the form

$$x^2 - y^2 = z^2$$

or

$$(x + y)(x - y) = z^2.$$

The two factors of the left-hand side are odd and relatively prime, for a common factor would divide their sum $2x$ and their difference $2y$, and thus x and y, which is impossible, since x and y are assumed to be relatively prime. Therefore, in order that the product of these two factors be a square, it is necessary that they be squares individually:

$$x + y = a^2, \qquad x - y = b^2,$$

from which we obtain

$$x = \frac{a^2 + b^2}{2} \qquad y = \frac{a^2 - b^2}{2} \qquad z = ab,$$

where a and b are any odd integers.[1] This is the general solution.

It is natural to ask whether a cube can also be the sum of two cubes, and more generally, whether an arbitrary power can be the sum of two powers, all the same degree. Fermat replied to this question in the negative. He did this in 1637 in the following terms, which appeared in a marginal note to the *Œuvres de Diophante* which had just been reedited and supplemented with a commentary by Bachet de Méziriac:

---

[1] If we set $(a + b)/2 = a'$, $(a - b)/2 = b'$, the solution takes the form
$$x = a'^2 + b'^2 \qquad y = 2a'b' \qquad z = a'^2 - b'^2;$$
here a' and b' are integers of different parity.

Cubum in duos cubos aut quadrato-quadratum in duos quadrato-quadratos et nullam in infinitum, ultra quadratum, potestatem in generaliter duas ejusdem nominis fas est dividere.

Cujus rei demonstrationem, mirabilem sane, detexi; hanc marginis exiguitas non caperet.

Which means:

"It is not possible to divide a cube into two cubes, a fourth power into two fourth powers and, in general, a power of any exponent greater than two into two such powers.

"I have discovered a rather remarkable proof of this proposition, but there is not room for it in the margin."

Unfortunately, the complete proof that Fermat said he possessed has not come down to us; he left to us only the principle of the one that he used for fourth powers, the method of *infinite descent*. Before we speak of the main works in which an attempt was made to affirm or deny the assertion of Fermat, perhaps it would not be out of place to devote a few words to his biography and discoveries.

Fermat

His Works. Pierre Fermat was born in 1601 in Beaumont de Lomagne, a small town on the border between Languedoc and Gascony. After giving him a good basic education at home, his father, a leather merchant, sent him to study law at Toulouse. Fermat later became a member of the Parlement there. He died at Castres in 1665.

"While his career proceeded in obscurity, he acquired the reputation of a geometer without peer on the basis of his handwritten treatises composed in Latin and his correspondence in French with several savants which was exclusively devoted to mathematical questions.

"In addition to his mathematical talents, Fermat possessed very great erudition.

"According to his correspondence, his character appears to be affable, relaxed, humble, but with a small touch of vanity that Descartes, who, his opposite in every respect, mixed bitterness into his polemics, characterized by saying: 'M. de Fermat is Gascon; I am not.'

"It was not by his publications that his name became widely known in the intellectual world; he himself published but one geometric dissertation, and that anonymously. This pamphlet appeared in 1660.

"He left the greater part of his theorems without proofs. It was the fashion of the time for mathematicians to propose problems to one another. The method of solution was most often kept secret in order to be used again for new triumphs for oneself and for one's nation, because there was rivalry especially between the English and the French geometers. For this reason the greater part of Fermat's proofs were lost" (from the preface to Legendre's *Théorie des Nombres*).

All of Fermat's works are of the highest quality.

His *Introduction aux lieux plans*, precisely contemporary with the geometry of Descartes, not only attempts to restore presumably the lost work of Apollonius on plane loci, but constitutes in addition a concise treatise of analytic geometry, more complete in certain respects than that of Descartes.

According to d'Alembert, Lagrange, Laplace, Fournier and Émile Picard, the origin of differential calculus should be traced back to Fermat's two works, *Mémoire sur la théorie des Maximas* and *Mémoire sur les Tangentes et les Quadratures*; the paper of Leibniz on differential calculus, *Nova methodus pro maximis et minimis*, followed by five years the—posthumous—publication of Fermat's *Mémoires*, and in one of his letters to Wallis Leibniz admitted how much he owed to Fermat.

The principle of least time which allowed Fermat to prove the law of refraction and to find the exact expression for the index of refraction as the ration of the velocities of light in the two media is already an ingenious anticipation of the methods of the calculus of variations created by Euler and Lagrange more than a century later, and of the principle of least action of Maupertuis and Hamilton, which plays such an important role in analytical mechanics.

Fermat shares with Pascal the merit of creating the *theory of probability*; his ideas on the fundamental principles of this theory were even more exact than those of Pascal.

But it was especially in the *theory of numbers* begun by Diophantus in antiquity that Fermat showed himself to be without rivals; Pascal, who called him "the greatest man in the world," wrote him: "Look elsewhere for someone who can follow your numerical inventions; as for myself, I confess that they are far beyond me; I am only capable of admiring them."

Let it suffice to cite the theorem—which bears his name—according to which for any integer a the difference $a^p - a$ is divisible by p if p is prime, and his results on expressions and indeterminate equations of the second degree, for example on the equation $x^2 - Dy^2 = 1$, which also bears his name. Fermat had never been found to be in error; mathematicians have succeeded in proving all the propositions that he left without proof with the exception of the theorem in which we are interested, and this they have not succeeded in refuting.

Fermat's Equation. We have now to consider the present state of affairs in the study of the question

$$x^n = y^n + z^n,$$

which Fermat declared to be insoluble in integers for any integer n greater than 2.

Let us first show that it is sufficient to prove the theorem for the case in which the exponent either is equal to 4 or is an odd prime number.

Of course, if the exponent n is not divisible by any odd prime number, then it is a power of two; and since it is assumed to be greater than 2, n must be equal to or divisible by four—$n = 4m$—and the equation may be written

$$(x^m)^4 = (y^m)^4 + (z^m)^4,$$

which proves that the equation with exponent 4 would have a solution.

On the other hand, if n is divisible by an odd prime p, let $n = pm$, and we have in the same way

$$(x^m)^p = (y^m)^p + (z^m)^p,$$

so that the equation would have a solution for the odd prime exponent p.

Case 1. Unsolvability of the Equation for n = 4; *Infinite descent.* Fermat's proof in the case of fourth powers is a proof by contradiction, which consists in showing that if the equation had a solution, then another solution of the same form could be deduced from it in which the numbers x, y, z are less, *and so on indefinitely*; this is impossible, since the numbers solving the equation are assumed to be non-zero. Without supplying the details, Fermat gave the principle of his method, which he called *infinite descent,* in a letter to his friend Carcavi which is now to be found in the Library of Leyden. Frenicle de Bessy, a friend of Fermat, followed his indications and completed the proof in his *Traité des triangles rectangles en nombres,* published in 1676.

Case 2. n *is an odd prime* p.

$$x^p = y^p + z^p$$

The steps to be taken in an attempt to solve this equation are analogous to those which are successful in the case $p = 2$: the difference $x^p - y^p$ is equal to the product $(x - y)(x^{p-1} + x^{p-2}y^1 + y^{p-1})$, and if the equation has a solution, then we have

$$(x - y)(x^{p-1} + x^{p-2}y + \cdots + xy^{p-2} + y^{p-1}) = z^p,$$

or, in abbreviated notation,

$$(x - y)\phi(x, y) = z^p.$$

The prime factors q common to $x - y$ and to ϕ are the prime factors common to $x - y$ and to p, for if $x - y$ is a multiple of q, $\phi(x, y) = py^{p-1} + $ mult. q, and q must therefore divide py^{p-1}. Now it cannot divide y, otherwise it would divide x as well, in contradiction to the assumption that x, y, z are relatively prime in pairs; thus q must divide p. As a consequence either $x - y$ and q have the greatest common divisor 1 or else they have the greatest common divisor p. The first case occurs if z is not divisible by p, the second if z is divisible by p. Thus we have in the first case:

$$x - y = a^p, \phi(x, y) = b^p, z = ab;$$

in the second case,

$$x - y = p^{p-1}a^p, \qquad \phi(x, y) = pb^p, z = pab.$$

In the first case, a and b are not divisible by p; in the second case, b is not divisible by p, but we can show that a is, so that z is divisible not only by p but by p^2.

The difficulty of the question now resides in the fact that, except in the case $p = 3$, we are unable to solve the preceding equations algebraically for x and y. A very brief historical résumé will now show how much time was needed to overcome the problem and in just the simplest cases, although it will not be possible to give an idea here of the magnitude of the efforts that the progress achieved has cost.

Cubes; p = *3.* It was only in 1774, *more than a century after Fermat*, that Euler succeeded in proving the theorem for cubes (or third powers), although his proof was not complete and could not be made completely rigorous until after the results obtained by Lagrange, Legendre and especially Gauss—twenty-five years later—on the divisors of binary quadratic forms were available.

p = *5.* It was Legendre who succeeded in 1823 in solving this case.

His very original proof is based on the identities

$$4(x,y) = 5(x^2 + y^2)^2 - (x - y)^4 = (2x^2 + xy + 2y^2)^2 - 5(xy)^2$$

and on the form of the divisors of these expressions, from which we find by a method analogous to that of infinite descent (here it is an *infinite ascent*) that the equation cannot be solved, since x, y and z would have to be greater than any given number.

In his great paper of 1823 on Fermat's theorem, Legendre made known the following very important proposition, communicated to him by a woman mathematician of great talent, Sophie Germain: Let p be a prime; if $2p + 1$ is a prime, or, more generally, if $2kp + 1$ is a prime for which there do not exist two consecutive residues of pth powers, then the equation of Fermat has no solution if one of the numbers x, y or z is not divisible by p. The contrary case still remains, but the work has been reduced almost by half. Thanks to the extensive tables of primes available today, Léon Pomey, an engineer of the Manufactures de l'État, was able to find (1925) prime numbers p such that $2p + 1$ is also prime up to $p = 5,003,249$.

p > 5. From the work of Legendre up to that of Kummer, only the proofs of the unsolvability of the equation by Lejeune-Dirichlet in 1832 for $n = 14$, by Lamé in 1837 for $p = 7$ and some attempts by Lebesgue, Liouville and Cauchy need be mentioned.

But it is to Kummer that we owe an advance which, if not decisive, is at least of fundamental importance.

Kummer

Ideals. Kummer, born in 1810 in Silesia and a pupil of Gauss and Dirichlet, was first professor at the Gymnasium in Liegnitz, later at the Universities of Breslau and Berlin. He came to Paris after his retirement in 1884 and resided there until his death in 1893.

Among his works stand out his beautiful discoveries in geometry based on studies in optics, but above all his inventions in the theory of numbers. He always had a predilection for this part of mathematics— "its only pure branch," he said, "unsullied by applications." In a series of papers dating from 1844 and published in Latin or German in Crelle's journal and in French in Liouville's journal, Kummer generalized the notion of integer and created the *ideals*. In the first, "De numeris complexis qui radicibus unitatis et numeris integris realibus constant," he generalized the notion of a complex integer which was due to Gauss. Today, as a consequence of the work of Cauchy, the complex or imaginary quantities and their functions are familiar not only to analysts, but even to electricians for their use in studying alternating currents and to engineers in designing airplane wings. But, as we shall see, a geometric interpretation permits anyone to form a conception of the complex integers of Kummer.

Let us mark on a half-line starting from its origin O an infinite series of equidistant points, which will be the images of the integers 0, 1, 2, etc. To the operations there will correspond very simple geometric relations between these points; for example, the multiplication $a \times b = c$ corresponds to the fact that A, B, C, the images of a, b, c, and U, that of the number 1, form two figures OAC and OUB, which are directly *similar triangles*.

Next let us imagine a regular convex polygon of p sides with its center at O and one of its vertices at U, and translate it in its plane parallel to itself in such a way as to cause its center to coincide with each of the original vertices successively. We then carry on this operation with the new polygons thus obtained and repeat this indefinitely. This yields a regular network of points in the plane which are the images of the complex integers of Kummer. These integers form what is called a *field*. We shall continue to represent them by a single number, even though two numbers are required to define them, and we shall *define* their multiplication by saying that $a \times b = c$ *if the triangles* OAC *and* OUB *are directly similar*. When the multiplication is thus defined, the theory of divisibility follows as a consequence.

The significance of the Kummer numbers for the problem of Fermat is due to the fact that $\phi(x, y)$ is a product of such numbers:

$$\phi(x, y) = (x + y\theta)(x + y\theta^2)\ldots(x + y\theta^{p-1}),$$

where θ is an imaginary pth root of unity.

This led Kummer to generalize the problem and try to find out whether the equation would be solvable using the new complex integers that he had defined. The answer would have been easy to find if the decomposition of the complex integers had been *unique*, as is the case with the ordinary integers. But, as Dirichlet pointed out to Kummer, this is not the case; the latter, however, was able to remove the difficulty by means of a deeper analysis which led him to single out among the complex prime numbers certain elements which he called *ideal numbers*, because they are algebraic integers of a more complicated nature not belonging to the set of integers previously defined. He isolated them on the basis of certain congruence properties; he compared these properties to the precipitates, obtained by the use of certain reagents, whose nature allows us to verify the presence of certain non-isolatable chemical radicals in a substance.

Later on, Dedekind was able to avoid the introduction of external elements into the field under study by extending the notions of divisibility to sets that he called *modules* and *ideals*; just as the set of all multiples of an ordinary integer has the characteristic property of containing the sums and differences and more generally all the multiples of its elements, so every *ideal* of Dedekind is the set of integers of a field which contains the sums and differences and more generally all the multiples of its elements by any integer of the field.

It is easy to define the product of two ideals. Prime ideals exist and —this is the main point—*every ideal is uniquely decomposable into a product of prime ideals.*

We say that two ideals L_1 and L_2 are *equivalent* or belong to the same *class*, if we can find two of Kummer's integers x_1 and x_2 such that the ideals $x_2 L_1$ and $x_1 L_2$ are identical. The number of classes h is finite, and Kummer succeeded in finding an expression for it after very arduous calculations on the basis of a principle due to Dirichlet.[2]

Kummer's Final Result. On the basis of these generalizations— familiar to us today, but quite bold and original at the time—Kummer was able to arrive at the fundamental result that the theorem of Fermat holds not only for the ordinary integers but also for the complex integers that he had defined *whenever the number of classes* h *is not divisible by* p. Fermat's theorem was thus rigorously proved for a host of prime numbers; within the first hundred integers, only 37, 59 and 67 are exceptions. In a paper of 1857, Kummer tried without complete success to extend his proof to the exceptional primes. In any case, by perfecting the method of Kummer, Mirimanoff, a professor at the University of Geneva, succeeded in 1893 in rigorously solving the case $p = 37$.

Since then only improvements in the details have been made, due especially to Wieferich, who established the validity of Kummer's proof for an infinity of primes, to Mirimanoff, Edmond Maillet, a civil engineer, and Léon Pomey. These latter authors used the

---

[2] See my note in our translation (with the late A. Lévy) of D. Hilbert's "Théorie des Corps de Nombres algébriques" (*Annales de la Faculté des Sciences de Toulouse*).

methods of Kummer and those—much simpler—of Legendre and Sophie Germain concurrently.[3]

It is none the less true that we are far from a definitive solution. Can we even be assured of success, given the necessary time? Not at all, for if the solution of exceptional cases always gave rise to new exceptional cases, then this would mean that the work would never be terminated; thus we still cannot know if the "perpetual challenge thrown out to the human mind" will ever be met.

Conclusion

We may wonder whether these efforts put into the solution of a question without any practical interest are justified and would not be better employed in other research capable of application.

It is easy to reply to this that some of the purest research has later had entirely unexpected applications. For example, if the geometers had not—without any idea of application—advanced the study of the conic sections to a high degree of perfection, then the discoveries of Kepler and Newton in the 17th century would not have been possible and the progress of mechanics could have been retarded by several centuries. As far as the works inspired by Fermat's problem are concerned, considerable progress in the theory of numbers, notably a theory of *algebraic number fields* of remarkable beauty, is due to them. The fertility of the notion of ideal introduced by Kummer has not, moreover, been restricted to the realm of arithmetic, but it has also manifested itself in the theory of algebraic functions; it will enable great progress to be made in the study of algebraic curves and surfaces and their fundamental topological invariants, such as the genus.

Finally, if it is true that the basic source of rigor in all the domains of mathematics is to be found in arithmetic, then we may be permitted to assume that the integers will come to play a more and more important role in the physical sciences at the time when a need for greater rigor makes itself felt.

Let us, therefore, continue to pursue the study of Fermat's problem without losing heart. There is much to be hoped for in the work of the young French school of mathematicians, who, in returning to Fermat's problem for the ordinary integers—a problem less general than that of Kummer—are seeking a solution in the study of rational

---

[3] We should also mention the important work of Dickson and of Mordell.

points on algebraic curves and thus are in a position to profit by all the progress achieved in the study of algebraic functions since Riemann. Let us hope, then, that one day an emulator of Fermat, our *Divus Arithmeticus*, will be found among them who will finally succeed in snatching the secret of his enigma from the Gascon Sphinx.

8

THE HISTORY OF THE MYSTERIOUS NUMBERS

by Paul Dubreil

DEPUTY LECTURER AT THE SORBONNE

$$e \quad \overset{\pi}{\underset{C}{}} \quad i$$

"Also he made a molten sea of ten cubits from brim to brim, round in compass, and five cubits the height thereof; and a line of thirty cubits did compass it round about."

(I KINGS 7:23; II CHRONICLES 4:2)

$\pi =$ 3.14159	26535	89793	23846	26433	83279
50288	41971	69399	37510	58209	74944
59230	78164	06286	20899	86280	34825
34211	70679	82148	08651	32823	06647
09384	46095	50582	23172	53594	08128
48111	74502	84102	70193	85211	05559
64462	29489	54930	38196	44288	10975
66593	34461	28475	64823	37867	83165
27120	19091	45648	56692	34603	48610
45432	66482	13393	60726	02491	41273
72458	70066	06315	58817	48815	20920
96282	92540	91715	36436	78925	90360
01133	05305	48820	46652	13841	46951
94151	16094	33057	27036	57595	91953
09218	61173	81932	61179	31051	18548
07446	23799	62749	56735	18857	52724
89122	79381	83011	94912	98336	73362
44065	66430	86021	39501	60924	48077
23094	36285	53096	62027	55693	97986
95022	24749	96206	07497	03041	23668
86199	51100	89202	38377	02131	41694
11902	98858	25446	81639	79990	46597
00081	70029	63123	77381	34208	41307
91451	18398	05709	85...		

(Shanks, *Proc. Roy. Soc. London*, 23, 1874, p. 45.)

The gap separating the rough approximation given by the Bible from the calculation of π to 707 decimal places carried out by Shanks in 1874 is so great that it seems to be difficult to give an idea of it. Four decimal places for π are sufficiently precise for practical needs; 16 decimal places allow us to give the length of the circumference of a circle whose radius is the mean distance from the earth to the sun to within the thickness of a hair; if we replace the sun by the most distant nebula and the hair by the tiniest particle known to physicists, we need only 40 decimal places in order to obtain a fantastic precision.

Do not say that "this is beyond imagining"; on the contrary, by reading the history of the number π one reaches to the center of the mathematical world, that marvelous world in which the imagination plays its most beautiful role.

The length of the circumference is proportional to the diameter; the area of a circle is proportional to the square of the radius; the coefficient of proportionality is the same number, π, *approximately equal to* 3.14. What a child knows today took the Greeks two centuries to establish. A contemporary of Socrates (469–399 B.C.), Antiphon, inscribed a square in a circle, then an octagon, and supposed that the number of sides continued to double until the polygon thus obtained practically coincided with the circle. At about the same time, Bryson made use of circumscribed polygons.

Following the work of Hippocrates and Eudoxus, Euclid (300 B.C.) made the necessary passage to the limit precise in his *Elements*; he stated the axiom which would later be taken up by Archimedes, whose name it bears: "Every length is surpassed by an appropriate multiple of a given length." He developed the method of exhaustion used by Antiphon, which consisted in doubling the number of sides of regular circumscribed and inscribed polygons and showing, as we would say today, that the process was convergent.

Archimedes (287–212 B.C.) assembled and developed these results. He showed that the area of a circle is equal to half the product of its radius and its circumference and that the ratio of the circumference to the diameter is contained between

$$3\tfrac{10}{71} = 3.14084 \quad \text{and} \quad 3\tfrac{10}{70} = \tfrac{22}{7} = 3.14285.$$

He established and utilized the fundamental theorem that we state as: An arc of length $x(0 < x\,\pi/2)$ is greater than its sine and less than its tangent. He obtained recurrence relations of remarkable form for the areas and perimeters of inscribed and circumscribed regular polygons of n and $2n$ sides, which make use of geometric and harmonic means, and allow us, at least in theory, to calculate π to any given accuracy; this method of computation bears the name of Archimedes' algorithm.

In order to appreciate properly the work of Archimedes, which includes the determination of both the surface areas and volumes of the principal solids of circular cross section, we must recall the considerable difficulties presented by the arithmetic operations at that time, because of the clumsiness of the notation and the complicated form of writing numbers. We do not know the procedure that Archimedes used to obtain square roots, but it appears certain that he had a perfected process of approximation available to him. Take account also of the rigor of his proofs. As Hobson remarked, the absence of certain definitions is the only criticism we have to make of the work of the Greeks; for them the length of a circumference, the area of a circle are as self-evident notions as the length of a line segment or the area of a square. In fact, precise definitions of the length of the arc of a curve and of the area of a region did not come until after the creation of the calculus, and it was Jordan (1838–1922) who first gave definitive results on rectifiable curves and measurable regions.

In antiquity, we also find several remarkable determinations of π outside of Greece. In Egypt, the Rhind papyrus, copied by Ahmes (1700 B.C.), gives

$$\pi = (\tfrac{16}{9})^2 = \tfrac{256}{81} = 3.1604,$$

which differs only a little from the value

$$\pi = \sqrt{10} = 3.1622$$

which was pointed out by Brahmagupta in India (born c. 598 A.D.) and by Ch'ang Hêng (died 139 A.D.) in China, and was later the delight of the "squarers of the circle."

In India, Āryabhaṭa (500 A.D.) considered the value 22/7 as insufficiently accurate and gave the "exact" value 3.1416. But the Chinese astronomer Tsu Ch'ung-chih (born 430 A.D.) did even better: he showed that π is contained between 3.1415926 and 3.1415927, and proposed as a close approximation $\tfrac{355}{113} = 3.1415929$. The Orient was in advance of the Occident during this epoch.

Starting in the 13th century, the calculation of the number again began to interest the curiosity of mathematicians in Europe. In his *Practica Geometriae*, written in 1220, Leonardo of Pisa proposed the value 3.1418 on the basis of bounds obtained with polygons of 96 sides:

$$3.1410 < \pi < 3.1427.$$

Works in cyclometry began to multiply with the Renaissance. Purbach (1423–1461) constructed a table of sines for angles with 10′ intervals and adopted the value $\frac{377}{120} = 3.14666\ldots$ for π; he even cast doubt on the existence of an "exact," i.e., a rational value. The 15th and 16th centuries were characterized by the development of trigonometry under the impetus of Copernicus and Kepler. Rhaeticus constructed a table of sines, not published until 1613, which implicitly gives π exact to 8 decimal places. Adrien Romain (Adriaen van Roomen; 1561–1615) obtained 15 decimals and Ludolph van Ceulen (1539–1610) went up to 32, not without prodigious labor, since his methods were not any real improvement over those of Archimedes. According to his wish, the 32 decimal places were engraved upon his tombstone, but, within his own country, posterity rewarded him much more by calling π the "Ludolph number."

The tour de force of Ludolph was soon eclipsed thanks to the remarkable improvements upon the method of Archimedes due to Snell (1580–1626) and Huygens (1629–1695); this method was based upon the fact that the arc is contained between its sine and its tangent, while the new idea consisted in replacing these limits by more complicated ones in the form of combinations of trigonometric lines, which were much better approximations.

Snell found that the arc x is contained between

$$\frac{3 \sin x}{2 + \cos x} \quad \text{and} \quad \frac{2 \sin x + \tan x}{3}.$$

Huygens' work *De circuli magnitudine inventa* has been styled a model of geometric reasoning. Like those of Snell, the approximate expressions given by Huygens,

$$\sqrt[3]{\sin^2 x \tan x}, \quad \frac{8 \sin x/2 - \sin x}{3},$$

differ from the arc x by only *fifth*-degree terms. And can one withhold admiration for this theorem given by Huygens: *Every circle is greater than the inscribed polygon, greater by more than a third of the excess of the latter over the inscribed polygon with half as many sides?* (Here it is a question, of course, of regular convex polygons, and the statement is as valid for the area of the circle as for the length of the circumference.)

With these methods, Snell obtained 34 decimal places by starting with a square and doubling the number of sides 28 times, operations from which Ludolph had been able to extract only 14 decimal places. Grünberger went up to 39 decimal places. Huygens calculated π to 9 decimal places without using any other polygon than—the hexagon. Here we see what progress can be achieved in mathematics by the use of the simplest methods of research.

Starting with the 15th century, the number of "squarers" continued to grow. The problem of squaring the circle consists in constructing a square of area equal to that of a given circle by means of straightedge and compass. It is equivalent to that of the rectification of the circle; to construct, still with straightedge and compass, a line segment with the same length as a given circumference. If the solution of these problems existed, then π would satisfy an algebraic equation with integral coefficients and of degree equal to a power of 2; thus it would be an *algebraic number*. The same would be true if, for example, π were a rational number such as $\frac{355}{113}$, or the square root of a natural number, such as $\sqrt{10}$.

The problem of squaring the circle haunted the minds of mathematicians for centuries, and even today, in spite of the fact that its impossibility has been solidly established, as we shall see later on, it continues to attract certain persons characterized by a great ignorance of geometry, contempt for "official science" and an unshakable confidence in themselves, accompanied by a total incapacity for exact reasoning. Some do not even understand the nature of the problem and propose approximate or even "mechanical" solutions: measure the circumference with a wire, weigh a circular disk, etc. Among those who seek an exact solution, many are completely incomprehensible.

If we may believe Plutarch, the ancestor of the squarers was Anaxagoras of Clazomenae (500–428 B.C.) who while imprisoned devoted his enforced leisure to the study of geometry and especially to squaring the circle. No doubt, he obtained an approximate construction which he considered to be exact. Shortly after that, the problem of squaring the circle gained considerable momentum from the remarkable results of Hippocrates of Chios; the latter, who lived in Athens in the second half of the 5th century B.C., discovered that the area of the line ADBC contained between the quarter-circumference ACB with center O and the half-circumference ADB with diameter

AB is equal to the area of the triangle AOB. This result and others of the same type showed that areas of curved regions could have simple values, capable of being exactly calculated and constructed geometrically. Archimedes later showed that the area included between an arc ACB of a parabola and the chord AB is equal to 4/3 of the area of the parallelogram with adjacent sides CD and BD (D is the center of AB, C is the point of contact of the tangent parallel to AB, CD is parallel to the axis of the parabola).

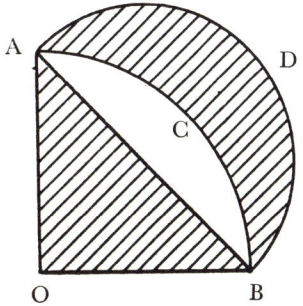
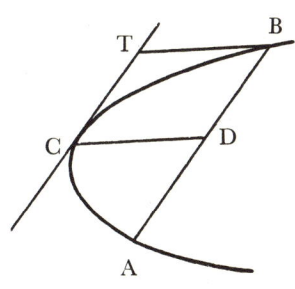

After that, was it not reasonable to seek an analogous result for the circle? The problem was quickly seen to be difficult, but the difficulty of problems has never discouraged those seeking a solution. Was it not compensated for by the allure of the glory?

Doubtless, this is why we find more than one great name on the long list of squarers; first of all, the Cardinal Nicholas of Cusa (1401–1464), considered as one of the greatest minds of the 15th century. As far as the problem which interests us is concerned, he had the merit of being the first to find the expression $(3 \sin x)/(2 + \cos x)$ for the arc x, which was later utilized by Snell as an approximate value. He gave $\pi = 3.1423$ as the exact value and was refuted by Regiomontanus, a distinguished mathematician who was the first to introduce the centesimal division of the circumference.

Orontius Finaeus (Oronce Fine; 1494–1555) seems to have considered $\frac{245}{78}$ to be the exact value of π, but he also stated correct propositions that were later confirmed by Huygens. His book *De rebus mathematicis hactenus desideratis*, published after his death, abounded in wonderful discoveries concerning not only the quadrature problem but also the construction of regular polygons and the trisection of the angle. All this was too good to be true, and it was quickly refuted by Bateon, a student of Orontius, then by Pierre Nonius (Pedro Nunes), who wrote a book entitled *De erratis Orontii* expressly for this purpose.

The same inordinate ambition seems to have animated Joseph Scaliger (1540–1609). A celebrated philologist, author of a great treatise on chronology, *De emendatione temporum*, a Protestant and a notoriously bitter enemy of the Jesuits, he sought to make himself accepted as a mathematician. He made known his proof that the circle can be squared in a book entitled *Nova Cyclometria* (1592), but it was refuted by Viète and Adrien Romain, and Fr. Clavius showed that Scaliger's propositions led to the following consequence: the perimeter of an inscribed dodecagon is greater than the circumference. In spite of this striking rebuttal, Scaliger's assurance and obstinacy enabled him to the end to hold on to some of his followers outside of the narrow circle of true geometers.

A very curious case is that of the philosopher Thomas Hobbes (1588–1679). Of very modest origins, he attracted attention from his early childhood because of his exceptionally brilliant intellectual talents. After studying Scholastic philosophy at Oxford for five years, he was employed as a tutor by some great English families, the Cavendishes and the Cliftons, whose sons he accompanied on long trips to the Continent; later on, he even taught mathematics to the son of the King of England, the future Charles II, when he was in exile at Saint-Germain.

His first work was a translation of Thucydides, but soon philosophy attracted him and he became a disciple of Bacon. The sciences interested him; he learned geometry and frequented medical and naturalist circles, where he followed the first great medical discoveries with enthusiasm. In keeping with the spirit of the times, he was to devote his life to the search for truth; he conceived the grandiose project of a universal science.

During one of his stays in Paris, his daily contacts with Fr. Mersenne enabled him to become acquainted with the ideas and discoveries of Descartes, who was then in Holland. He was violently shocked by them, affirmed himself to be an athiest, declared himself to be opposed to the application of algebraic methods to geometry, and did not accept the existence of a close link between mathematics and the natural sciences. Mersenne invited him to formulate his objections.

His major works followed in rapid succession; in 1642, *De Cive*, which achieved considerable success; then *Leviathan* (1651), *De Corpore* (1655) and *De Homine* (1658). It has been said that, having created the first system of empirical philosophy, Hobbes can be regarded as "the connecting link between the materialism of the ancients and modern positivism" (Frischeisen-Köhler).

Unfortunately, Chapter XX and the following ones of *De Corpore*, which are devoted to geometry and in particular to the problem of squaring the circle, readily lent themselves to criticism and provoked between Hobbes and the mathematician Wallis what philosophers call, with their delicate discretion, an unfortunate dispute. In his *Histoire de la Quadrature*, Montucla expressed himself more crudely: "Hobbes surpassed all his predecessors in ridiculousness in respect to this question; for not only did he believe that he had succeeded in squaring the circle and had found the two mean proportionals, but he defended himself with unexampled bitterness against Wallis, who took the trouble to refute him in several writings. His spite on this account was later turned against all geometers and geometry itself. At first, he had accepted its methods and principles; however, the contradictions in his work pointed out by Wallis led him to deny step by step all the axioms and to undertake total reform . . . ." Today, perhaps, we might agree that Hobbes deserves some respect for this critique of the foundations of geometry; unfortunately, it did not lead him to the conception of a non-Euclidean geometry! In this dispute, which was envenomed by political and religious motives, since Wallis represented the Presbyterians and belonged to those universities which had refused to introduce Hobbes' *Leviathan* into their curriculums. Wallis was so obviously right that Hobbes' credit suffered within the circle of English scientists. However, one cannot withhold a measure of sympathy for his intellectual prowess and his vigorous personality.

Another markedly interesting dispute at about this time was that in which the Flemish Jesuit Gregory of Saint-Vincent opposed Descartes, Roberval and Huygens. Gregory of Saint-Vincent attacked the quadrature problem by employing the classical geometric methods of the time; he manipulated them with incontestable skill; he especially merited respect for his generalization of the problem and for studying a host of areas connected with the conics. He was aware of the fact that the area between the arc of a hyperbola and the corresponding segment of an asymptote is connected with the logarithms of the abscissas of the end points. Thus Leibniz did him justice by giving him an eminent rank among the geometers. Descartes, on the contrary, having observed the inexactness of Gregory's solution to the quadrature problem, and also influenced by his own discovery of analytic geometry, which could not but alienate him from the methods of the Flemish mathematician, considered the work of the latter to be confused and incorrect. Fr. Mersenne also judged it in a very contemptuous manner, but his refutations were wrong. Roberval did likewise, and the dispute took on even greater force because Gregory of Saint-Vincent had found some excellent defenders among his students. Finally, Huygens and Fr. Léotaud, another Jesuit who was a good geometer, struck the decisive blows after a dispute which Huygens thought was purposely prolonged and confused by Gregory and his partisans in the hope that it would remain indecisive.

This much abbreviated list of squarers cannot be terminated without the mention of the name of Sieur Mathulon who, in the 18th century, offered one thousand *écus* to whoever would refute the solution that he had claimed to have found; summoned to court, he saw his claim of a solution refuted and had to pay the thousand *écus*. Squarers have since been more prudent—unfortunately for the mathematicians!

All picturesque details aside, it is incontestable that the work of the squarers, encumbered with details, unorthodox and so frequently inexact, is not entirely without interest. Nicholas of Cusa and Orontius Finaeus prepared the way for Snell and Huygens. Others obliged the true mathematicians to reflect upon and improve their methods and results. Moreover, the repeated failures of the squarers soon gave rise to the idea: is not quadrature impossible?

This suspicion seems to have been formulated first in 1544 in the *Arithmetica integra* of Michael Stifel. A century later, James Gregory (1638–1675), professor at St. Andrews and the University of Edinburgh, after giving a series of remarkable recurrence formulas concerning the areas of inscribed and circumscribed regular polygons, undertook to prove the impossibility of quadrature in his *Vera circuli et hyperbolae quadratura*. But Huygens showed his proof to be invalid. The latter added, moreover, that in his opinion, too, quadrature was impossible, and he pointed out to Wallis that it still was not even known if π was rational. It was Lambert who, in 1761, decided this important question in the negative in his *Mémoire sur quelques propriétés remarquables des quantités transcendantes circulaires et logarithmiques*

(Academy of Berlin). In 1775, the Academy of Sciences of Paris published a resolution which stated that it would no longer examine purported solutions to the squaring of the circle, the duplication of the cube or the trisection of the angle, or plans for perpetual motion machines. In this enlightened epoch, the Academy was deluged by material of this nature. But alongside this practical reason, the conviction that the quadrature problem was impossible was growing more and more firm in scientific circles. Legendre expressed it very clearly in his *Éléments de Géométrie* (1794): "It is probable that the number π is not an algebraic irrational, i.e., not a root of an algebraic equation with a finite number of terms and with rational coefficients. But it seems to be very difficult to prove this rigorously."

How was this perfectly correct opinion able to establish itself in the minds of mathematicians when it was not proved definitively by Lindemann until 1882? Why was it necessary to wait so long for this confirmation? This is no doubt what the reader would like to know.

The calculus, and, more particularly, the theory of infinite series and infinite products furnished remarkable formulas which, while providing new and much more powerful methods of calculation, also detached the number π to some extent from its geometric origins and clarified the essential role that it plays in all of mathematical analysis.

It was the French mathematician Viète who, at the end of the 16th century, showed the way by obtaining the first formula of this type:

$$\frac{\pi}{2} = \frac{1}{\sqrt{\frac{1}{2}} \sqrt{\frac{1}{2} + \frac{1}{2}\sqrt{\frac{1}{2}}} \sqrt{\frac{1}{2} + \frac{1}{2}\sqrt{\frac{1}{2} + \frac{1}{2}\sqrt{\frac{1}{2}}}} \cdots}.$$

This expression for π in terms of a convergent infinite product, making use only of the numbers 1 and 2, arises from the recurrence relation that exists between the areas of regular polygons of 2^n and 2^{n+1} sides.

Wallis should be regarded as one of the founders of analysis for his creation of the calculus of limits. The expansion

$$\frac{\pi}{2} = \frac{2}{1}\frac{2}{3}\frac{4}{3}\frac{4}{5} \cdots, \frac{2n}{2n-1}\frac{2n}{2n+1}\frac{2n+2}{2n+1} \cdots$$

is due to him; it is obtained by computing the area of a semicircle of unit diameter in two ways, (see p. 102), i.e., by writing

$$\frac{\pi}{2} = \int_0^1 \sqrt{x(1-x)}\, dx$$

$$OH = x;\ HM^2 = x(1-x).$$

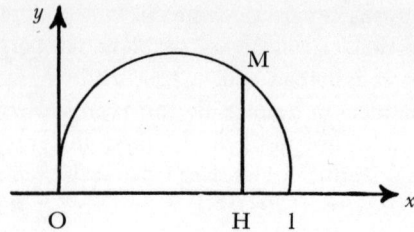

Following Chuquet and the Italian mathematicians, Wallis also contributed to the development of the theory of continued fractions, i.e., fractions of the form

$$a_0 + \cfrac{1}{a_1 + \cfrac{1}{a_2 + \cfrac{1}{a_3 + \cdots}}}$$

We have

$$\pi = 3 + \cfrac{1}{7 + \cfrac{1}{15 + \cfrac{1}{1 + \cfrac{1}{292 + \cdots}}}} = (3, 7, 15, 1, 292, 1, 1, 1, \ldots)$$

The fractions

$$r_0 = 3, \quad r_1 = 3 + \frac{1}{7} = \frac{22}{7}, \quad r_2 = 3 + \frac{1}{7 + \frac{1}{15}} = \frac{333}{106},$$

$$r_3 = 3 + \cfrac{1}{7 + \cfrac{1}{15 + \cfrac{1}{1}}} = \frac{355}{113}, \quad r_4 = \frac{103\ 993}{33\ 102}$$

are called successive convergents, and if a fraction in lowest terms

p/q is closer to π than the convergent $p_n/q_n = r_n$, we have $p > p_n$, $q > q_n$.

Gregory in 1670, then Leibniz independently in 1673, expanded the function arc tangent in a series:

$$\text{arc tan } x = x - \frac{x^3}{3} + \frac{x^5}{5} - \cdots,$$

which for $x = 1$ gives the *Leibniz formula*

$$\frac{\pi}{4} = 1 - \frac{1}{3} + \frac{1}{5} - \cdots.$$

This series is quite remarkable, but its slow convergence makes it inconvenient for the calculation of π; therefore Sharp chose other, more convenient values of x in order to calculate π to 72 decimal places. Newton expanded the arc sine function in a series and applied it to the calculation of π, obtaining 14 decimals. In 1706, Machin combined the Gregory-Leibniz formula with the addition formula for arc tangents and calculated π starting from the equality

$$\frac{\pi}{4} = 4 \text{ arc tan } \frac{1}{5} - \text{arc tan } \frac{1}{239},$$

which allowed him to reach the hundredth decimal place, since only rapidly converging series come into play. Later on, numerous calculators perfected the method in detail; each one confirmed the results of his predecessors and obtained a greater or lesser number of decimal places. The record is held by Shanks who, in 1874, gave the 707 decimal places shown at the beginning of the article. Here is also an expansion of π in the binary system:

$$\pi = 11.00100100\ 00111111\ 01101010\ldots.$$

In addition to the endurance records, let us also list a speed record. In 1775, Euler, a great calculator as well as a profound analyst, calculated π to 20 decimal places in one hour by means of the formula

$$\frac{\pi}{4} = 5 \text{ arc tan } \frac{1}{7} + 8 \text{ arc tan } \frac{3}{79},$$

using the expansion

$$\arctan t = \frac{t}{1+t^2}\left[1 + \frac{2}{3}\frac{t^2}{1+t^2} + \frac{2\cdot 4}{3\cdot 5}\left(\frac{t^2}{1+t^2}\right)^2 + \cdots\right].$$

We shall appreciate the ingenuity of this calculation if we note that essentially only the powers of $\frac{1}{50} = 0.02$ come into play. Another formula established by Euler:

$$\theta = \frac{\sin\theta}{\cos\dfrac{\theta}{2}\cos\dfrac{\theta}{4}\cos\dfrac{\theta}{8}\cdots} \qquad (\theta < \pi)$$

includes that of Viète ($\theta = \pi/2$) as a special case.

The custom of designating the ratio of the circumference to the diameter by π is also due to Euler. The great discovery which enabled us to unlock the mystery of the number π may be traced back to him especially.

This was the discovery of a certain relationship between π and two other numbers e and i, no less dear to mathematicians, as well as that of the relations which exist between the circular functions sine and cosine and the exponential function e^x: the latter is periodic and its imaginary period is $2\pi i$.

These facts are the common terminus of several currents of ideas. The logarithms invented by the Scotchman Napier (1550–1617) are not only important for numerical calculation; the function, zero at $x = 1$ and with derivative $1/x$, gives rise to an especially interesting system of logarithms from the theoretical point of view, the Napierian logarithms. The number which has logarithm one in this system is denoted by e. Its approximate value is

2.71828 18284 59045 23536 02874 71352 66249 77572
                                                                                                47093 66995...

and it was calculated by Boorman to 346 decimal places. Euler showed that it is the sum of the series

$$1 + \frac{1}{1} + \frac{1}{2!} + \cdots + \frac{1}{n!} + \cdots$$

and that the function $y = e^x$, the inverse of the Napierian logarithm ($x = \log y$) admits the expansion

$$e^x = 1 + \frac{x}{1} + \frac{x^2}{2!} + \cdots + \frac{x^n}{n!} + \cdots,$$

which is valid for all x.[1]

On the other hand, the brilliant school of 16th-century Italian algebraists, Tartaglia (c. 1500–1557), Cardano (1501–1576) and Ferrari (1522–1565), had been using the number $i = \sqrt{-1}$ for the solution of algebraic equations. The introduction of this symbol allowed them to solve not only all equations of the second degree, but also those of the third and fourth. In 1629, A. Girard sensed what the fundamental theorem must be: Every algebraic equation must have at least one complex root $a + b\sqrt{-1}$. D'Alembert gave an insufficiently rigorous proof, and it was only in 1799 that it was definitively established by Gauss in his thesis. Note carefully the title of this thesis: *Nouvelle démonstration, que toute fonction rationelle entière d'une variable peut être décomposée en facteurs réels du premier ou du second degré.* (A new demonstration that every rational integral function of one variable can be decomposed into real factors of the first and second degree.) The care that Gauss took to translate the statement of the theorem of d'Alembert into real terms is very significant; it shows that at this time even the greatest mathematicians actually distrusted the so-called "imaginary quantities." This attitude was to disappear at the beginning of the 19th century under the influence of the satisfactory definitions given by Wessel (1799) and Argand (1806) and later by Cauchy and Kronecker, and especially thanks to the truly astounding applications that the complex numbers made possible in both analysis and geometry.

But this distrust makes clear what a stroke of genius it was on the part of Euler to define the functions e^z, cos z, sin z, for complex z, as sums of the series

$$e^z = 1 + \frac{z}{1} + \frac{z^2}{2!} + \cdots + \frac{z^n}{n!} + \cdots$$

$$\cos z = 1 - \frac{z^2}{2!} + \frac{z^4}{4!} - \cdots$$

$$\sin z = \frac{z}{1} - \frac{z^3}{3!} + \frac{z^5}{5!} - \cdots$$

---

[1] Although the definition of factorial occurs in several other articles of this work, let us recall that $n!$ (which is read as n factorial) $= 1 \times 2 \times 3 \times \cdots \times n$. (Note of F. L.L.)

and thus to arrive at the fundamental relation

$$e^{iz} = \cos z + i \sin z,$$

which yields

$$e^{\pi i} = -1.$$

This last formula, so simple and so remarkable, is certainly one of the most beautiful in mathematics, and is the one which more than a century after Euler's death allowed Lindemann to prove the transcendence of π and the impossibility of squaring the circle.

Why again this long delay? According to one's taste, one may believe either that mathematicians do not have very acute minds or that the problem is not easy. Moreover, we can clearly see, a posteriori, that, after Euler, an important step still needed to be taken; a more precise notion still had to be gained of number itself. At the beginning of the 19th century, the existence of non-algebraic or transcendental numbers was only an hypothesis; it was Liouville who definitively established it in 1844. After observing that the algebraic irrational numbers are only very roughly approximated by fractions in lowest terms with a denominator not exceeding a certain limit, he was able to construct numbers which, approximated much more closely, are necessarily transcendental. Such is the case, for example, with numbers of the form

$$\sum_{n=1}^{\infty} \frac{1}{a^{n!}}.$$

Cantor later showed that the algebraic numbers could be ordered in an infinite sequence in which each of them would have a place, whereas the set of all real numbers, algebraic and transcendental, does not have this property; the existence of transcendental numbers is a consequence of this, and these are even much more numerous in some sense than the algebraic numbers.

An important problem then consists in deciding whether some noteworthy number is algebraic or transcendental. The first great discovery of this type was that of Hermite, who established the transcendence of e in 1873; not without some trouble, we may believe, since he wrote a short time later to Borchardt: "I shall not dare to try to demonstrate the transcendence of the number π. If others try it, no one will be more pleased than I at their success; but, believe me, my dear friend, this cannot fail to cost them a great deal of effort." The proof obtained by Lindemann in 1882 is based on methods which do not differ essentially from those of Hermite.

Lindemann established that if $x_1, x_2, \ldots, x_n$ are distinct algebraic numbers (real or complex) and $p_1, p_2, \ldots, p_n$ are algebraic numbers not all of which are zero, then the sum

$$p_1 e^{x_1} + p_2 e^{x_2} + \cdots + p_n e^{x_n}$$

cannot be zero. If we take $n = 2$, $p_1 = 1$, $x_2 = 0$, then we see that e^{x_1} cannot be algebraic for an x_1 that is algebraic and non-zero; in particular, e is transcendental. Similarly, according to the relation $e^{\pi i} + 1 = 0$, the number $i\pi$ cannot be algebraic, and therefore the number π is not either, since the product of two algebraic numbers is algebraic.

This beautiful result of Lindemann was duly admired at the time. But one would be greatly mistaken in imagining that it provoked astonishment or that it attained great fame. With perfect justice, Lindemann himself introduced his work in these terms: "Because of the failure of so many attempts to square the circle with ruler and compass, the solution of this problem has been generally considered as impossible. However, only the irrationality of π and π^2 has been established. The impossibility of quadrature will be proved if it can be shown that π cannot be the root of any algebraic equation ... with rational coefficients; the object of what follows is precisely to provide this proof...."

Following that of Lindemann, simpler proofs were given by Stieltjes, Hilbert, Hurwitz, Gordan, Mertens, Vahlen, Padé and Veblen. Borel and Popken laid the foundations for a classification of transcendental numbers.

And is this story now finished? Not at all; mathematicians know how to give new life to such questions. At the International Congress of Paris in 1900, Hilbert posed a whole series of unsolved problems in a justly celebrated presentation; the seventh concerned the irrationality and the transcendence of certain numbers, in particular, numbers of the form α^β, for example $2^{\sqrt{2}}$. The Russian mathematician Gelfond gave the solution in 1934; if ω and η are algebraic, ω being different from 0 and 1, and if η is irrational, then ω^η is transcendental. Gelfond also established the transcendence of e^π and that of the ratio $\mathrm{Log}\,\alpha/\mathrm{Log}\,\beta$ whenever it is irrational, α and β being algebraic numbers different from 0 and 1. For his part, Siegel showed in 1932 that if the constants g_1 and g_2 are algebraic, then at least one of the periods ω_1, ω_2 of the function p defined by the equation $p^2 = 4p^3 - g_1 p - g_2$ is transcendental. Different extensions of this theorem have been given by Siegel himself and by Schneider.

These very remarkable pronouncements do not clear up all the mysteries. Is *Euler's constant*

$$C = \lim_{n \to \infty} \left[1 + \frac{1}{2} + \cdots + \frac{1}{n} - \text{Log } n\right]$$
$$= 0.57721\ 56649\ 01532\ 86060\ 65120\ 90082\ldots,$$

which has been calculated by Adams to 263 decimal places and which plays an important role in analysis, notably in the study of the functions Γ and $\zeta(s)$, rational or irrational, algebraic or transcendental? The surface of this question seems to have been barely scratched, in spite of the beautiful formulas collected or discovered by Appell, who thought he had proved the irrationality of C in April of 1926 in an article in the *Comptes rendus de l'Académie des Sciences*. But, as Appell himself pointed out a week later, a gross error led to a gap in the proof.

And if one asks when this problem will be solved, I believe that mathematicians will call attention to the principal lesson to be drawn from this long story and will reply: "In its own time."

9

THE PROBLEM OF THE INFINITE: TRANSFINITE NUMBERS AND ALEPHS

by Henri Eyraud

PROFESSOR AT THE FACULTY OF SCIENCES OF LYON

À G. Cantor
Après toi, pélerin, nous irons sur les cimes
D'où, pieux et tremblants, nous verrons l'Éternel
Ordonner et compter dans l'alphabet du Ciel
Les feux du firmament, les germes des abîmes.

THE mathematician G. Cantor (1845–1918) brought a new arithmetic to mankind, the arithmetic of infinite sets. His work is a landmark in human thought and knowledge.

In its continuous effort to understand nature, to grasp and to express its laws, the human mind has been led to construct an edifice of symbols and operations which constitutes mathematics; it is a construction continually enriched by new materials and new structures. The question arises: Just what are the limits of the ability of the human intellect to conceive new notions and to combine them?

This simultaneously poses the question: Supposing that a new notion was able to be born in a human brain by some stroke of genius or supernatural revelation—to what extent is it possible for the finite being who received it to transmit it in finite terms?

Common sense leads us to reply modestly to these two questions: the mind is limited, it can neither conceive of the infinite nor give an account of it in finite terms.

We are certainly tempted to think that when Cantor undertook his research on infinite sets, he glimpsed the possibility of a totally different response. Perhaps in his hours of enthusiasm and in the intoxication of some notable success, he was even able to say to himself: "There is no transfinite which can serve as a limit to the operational possibilities of the mind, nor which is not capable of being given a finite mode of representation."

It was a student of Cantor who stated the famous theorem which bears the name of Zermelo's theorem: "Any infinite set can be well ordered."

The notion of transfinite number is due to Cantor. The transfinite cardinal numbers serve to number infinite sets.

Cantor taught us that there are as many even integers as there are integers and that there are as many integers as fractions. The number of these sets is the cardinal number aleph null (the letter aleph is the first letter of the Hebrew alphabet).

In contrast to this, he showed the existence of a higher cardinal number; this is the number of points of a line, which is equal to the number of points of a square, of a cube or even of all of three-dimensional space.[1]

This number is itself less than the number of discontinuous functions of one or several variables.

In more scientific language, Cantor extricated the notion of "cardinality" of an infinite set of objects or mathematical entities.

Two sets have the same cardinality if we can establish a one-to-one correspondence between the elements which belong to one set and those of the other. Every cardinality is characterized by a cardinal number, which Cantor denoted by an aleph. The alephs thus extend the finite numbers with which we are familiar.

In addition to the notion of cardinal number, Cantor discovered a much more subtle and delicate notion, that of ordinal number. To exhaust an infinite set by removing its elements one by one while assigning to each a rank to denote the order in which it is selected is an operation to which it is not easy for the mind to become accustomed. It poses a preliminary problem which is itself not without its hazards: to construct the transfinite sequence of ordinal numbers which are to serve for this enumeration. Cantor gave us, in addition to the ordinals of the first class:

$$1, 2, 3, 4, \ldots$$

some of the ordinals of the second class which succeed them:

$$\omega, \omega + 1, \ldots, 2\omega, 2\omega + 1, \ldots$$

---

[1] This equality extends to the case of spaces of more than three dimensions.

The ordinals of the second class are those which allow us to enumerate a set of cardinality aleph null, no matter in what order we may decide to select the elements of this set.

Here a remarkable fact, completely unknown in the enumeration of infinite sets, presents itself. It is possible to choose an order of enumeration for any set of cardinality aleph null so that it will end at any one of the cardinals of the second class, and this property extends to sets of any cardinality whatsoever. Thus there exist classes of transfinite ordinals in conformity with the following scheme:

First class: $1, 2, 3, \ldots$

Second class: $\omega, \omega + 1, \ldots, 2\omega, 2\omega + 1, \ldots$

Third class: $\lambda, \lambda + 1, \ldots, \lambda + \omega, \ldots, \lambda^2, \lambda^2 + 1, \ldots$

Each ordinal of the second class represents a process of enumerating a set of cardinality aleph null. Let us give an example: we take the set of natural numbers and begin by removing the odd numbers in increasing order, then the doubles of the odd numbers, then the quadruples of the odd numbers, and so on; this mode of enumeration leads to the ordinal number ω^2. No one has yet succeeded in giving a mode of enumerating the set of points of a line or the set of discontinuous functions. This problem, for which Cantor was not able to find a solution and with which his students struggled in vain, bears the name of the continuum hypothesis.

It is a perfectly unambiguous problem which will one day be solved; it seems reasonable that the enumeration of the points of the line (or of space) could be accomplished by an arbitrary (transfinite) ordinal of the third class; however, no correct proof has yet been given. Thus we still speak of the cardinality of the continuum, and not of the cardinality aleph one, when we wish to designate the cardinality of the set of points of a line.

Cantor showed that it was possible to construct sets of higher and higher cardinality, with no transfinite cardinal as an upper bound. The transfinite ordinals also constitute a series whose end we cannot imagine. We do not wish to lay stress upon the difficulties of a logical nature to which this question can give rise; we would rather linger a while on the technique itself used in constructing the transfinite ordinals. It involves two fundamental operations:

(1) The finite recurrence which consists in adding one to an ordinal in order to obtain the ordinal which immediately follows it;

(2) The transfinite recurrence or diagonal method.

This operation allows us to cross the boundary that cannot be reached by finite recurrence. We shall give only a simple example in order to explain the mechanism of this method.

Let us seek to transcend the series of increasing functions of an arithmetic variable ω (ω takes the integral values $1, 2, 3, \ldots$). We shall represent each function by the sequence of numerical values (integers) that it takes on for the values of the variable. We shall have the following succession of functions obtained by finite recurrence:

$$1, 1, 1, 1, --$$
$$2, 2, 2, 2, --$$
$$3, 3, 3, 3, --$$
$$4, 4, 4, 4, --$$
$$-- -- -- -- --.$$

The application of the diagonal method consists in having this infinite sequence of constant functions succeeded by the function

$$1, 2, 3, 4, \ldots,$$

obtained by following the descending diagonal of the above table. It is this diagonal which supplies the transfiniteness.

The systematic application of recurrence procedures to arithmetic functions allows us to successively construct all the (transfinite) ordinals of the second class; applied to functionals, it allows us to construct the (transfinite) ordinals of the third class and so on.

The diagonal (or transversal) method certainly has a bearing outside of the domain of mathematics. Are not the pedagogical methods which allow young minds to skip over the successive periods of the intellectual development of our ancestors essentially a transversal technique?

What we know of the development of the embryo makes us think that even the creation of living beings proceeds by some sort of transversal method.

Another conclusion is suggested by the very technique of attaining a transfinite ordinal by a bounded number of operations on a finite number of symbols.

The stockpile of operations available to the mind is transfinitely unbounded, but, thanks to it, any accessible notion can be presented in finite terms.

Cantor's metamathematics has opened to the mind the dazzling perspective of its own infiniteness.

B. SPACE

Geometry, the science of space, is more vast, more diverse and more difficult to comprehend within a unitary view than arithmetic, the science of number; this is due essentially to the properties of multidimensionality and continuity of space, properties that number has tried hard to imitate, but always tardily and never in a manner conducive to the complex, lightning attacks of the intuition. Many mathematicians prefer to say "geometries" rather than "geometry," thus avowing the difficulty of grasping its majestic unity and resigning themselves to the description of the separate provinces. We have very modestly limited ourselves in this first series to two of these regions, chosen because they are indispensable to the understanding of modern relativistic physics: the question of a space of four dimensions and its close but distinct relation, curved space, to which André Saint-Lagüe and René Thiry have each devoted an article.[1]

The organizer of the mathematics section of the Palais de la Découverte, André Saint-Lagüe, has already frequently discussed there what might be called the classic aspects of the controversy over space of four dimensions. You will find in his article, which reads like a scientific novel, the examples and arguments which should be a part of the common equipment of the educated man of the 20th century.

A pupil of Paul Appell, whose legendary clarity he has inherited, and of Henri Villat, René Thiry, director of the Institut de Mécanique des Fluides of Marseilles, is one of those French scientists who have remained in close contact with the details of the development of mathematics. With an elegant simplicity all his own, he reveals to us the ways which enable us—without leaving a relatively small region of space—to force an admission of its curvature. The last

---

[1] In the second collection [never published], our intention is to discuss more fully the domain of geometry, first by recalling some of the always charming aspects of classical geometry (synthetic, as created by the Greeks; analytic, according to the Cartesian revolution; the so-called modern, according to the discoveries of Poncelet, Chasles, Grassmann, von Staudt); then we shall submit for judgment some of the most surprising constructions of recent years: algebraic, projective and finite geometry and, above all, topology, to which we are so impatient to offer the place that it merits.

part of his study demonstrates convincingly that the most extraordinary imagination is by no means the exclusive possession of artists.

Just as for the notion of number, the individual soundings to be taken gain by being preceded by a more general introduction. No one was more qualified to undertake a tour of this vast domain than Maurice Fréchet, the first pioneer to explore the stratosphere of the geometric skies: the abstract spaces. He explains the possibility and the utility of their study from both the metric as well as the topological point of view. The air there is very rarefied, and one would not expect to remain there very long, but what enormous vistas we can discover from the height of the icy regions where the pure streams which descend to fertilize the habitable regions are born!

F. LL.

10

FROM THREE-DIMENSIONAL SPACE TO THE ABSTRACT SPACES

by Maurice Fréchet

PROFESSOR AT THE SORBONNE

FROM SETS OF NUMBERS TO SETS OF POINTS

THE task of treating the notion of function is reserved to others in this work. However, this concept will help us to show how we can pass by a very natural transition from the notion of number (which we treated in our first article in this work) to that of space, or rather from sets of numbers to sets of points.

A variable number u is a function of a variable number x when there corresponds to each fixed value of x a fixed value of u. For example, we can take for u: $x + 1$, $2x$, x^3, 10^x, etc.; these are so many distinct functions of x. But u may also be a function of several variables; we might take for u: $x + y$, xy^2, $x^2y^4z^3$, etc., where x, y and z are independent variables. If we consider x, y and z as the coordinates of a point M, then we see that we can consider u equally well as a function of the three numerical variables x, y and z or of the variable point M. The introduction of the point achieves an economy of notation, therefore an economy of thought; it leads to clearer and more effective thinking, i.e., to more exact and bolder reasoning.

This transition being so readily carried out, we are now going to treat directly the notion of a set of points and, as a consequence, that of space.

EXTENSION OF THE NOTION OF SPACE

Euclidean Space. Euclid's geometry contains a detailed study of the figures contained within a line, a plane or the whole of Euclidean space, that is, the figures contained with a space of one, two or three dimensions, respectively.

One of the reasons for the latter designation is that in order to define the position of a point of one of these three spaces, it suffices to know, respectively, one, two or three lengths which we call the coordinates of the point. For example, the surface of a sphere is two-dimensional, because a point on this surface is determined by its two geographic coordinates: latitude and longitude.

Any property of a figure in n-dimensional space, for $n = 1, 2$ or 3, is translated into a property of systems of numbers, and conversely; this is the principle at the base of the "analytic geometry" founded by Descartes.

n-*Dimensional Space.* At this stage the idea arises naturally of extending this relationship to the case where $n > 3$. But, in this case, it would become the definition of an n-dimensional space for $n > 3$.

The generalization is not just a matter of words. The use of this language would have the advantage of suggesting geometric analogies which would be translated into properties of systems of n numbers. Poincaré said (we quote from memory): "Mathematics is nothing more than a well-constructed language."

For example, in the theory of probability it was necessary to calculate averages expressed in the form called "multiple integrals of n variables." Their calculation either eluded the first mathematicians who became interested in them or were made only after long and tedious computations. When someone had the idea of considering the n variables as the coordinates of a point in n-dimensional space and these integrals as volumes or masses in this space, then the computations were rendered simple and intuitive by following geometric analogies.

In the study of liquid mixtures or in the more general so-called ergodic problem, we start out with the principle that the movement of a material body subjected to a given field of force is known when the initial positions and initial velocities of the points of the body are known. In the case in which the totality of these positions is determined at a given instant by the knowledge of a finite number ν of arbitrary quantities (called parameters), the movement is determined by knowing the initial values of these ν parameters and also those of their rates of change (called the derivatives of these parameters). The set of these $n = 2\nu$ quantities is called the *initial phase*. From here on, the movement of the different parts of this system in our three-dimensional space, often very complex, will be exactly represented by the fictitious movement of a "point" of the n-dimensional space which represents the phase of the system. In the study of the ergodic problem, the language, notation and computations are enormously simplified by operating directly in this space.

Moreover, in cases other than the preceding ones, in which the notion of a space of more than three dimensions seems to be a convenience, it may become a necessity. It is well known, of course, that in the physical space (as conceived in the theory of relativity) the knowledge of three lengths is not sufficient; time must also be added, so that this physical space has four and not three dimensions.

Functional Analysis

We may arrive at other extensions of the notion of space by another means.

The first equations studied by mathematicians were obtained by setting two functions of the same real variable equal to one another (or, what amounts to the same thing, setting another function—their difference—equal to zero). For example, such an equation is obtained by setting the trinomial $ax^2 + bx + c$ equal to zero; here a, b, c are known quantities and x is an unknown number. Systems of equations in several unknown variables $x_1, \ldots, x_n$ had also to be considered. The unknown could then be said to be a point with coordinates $x_1, \ldots, x_n$ in n-dimensional space.

But less simple unknowns had also to be considered. For example, if we wish to determine the trajectory of a planet in the sun's gravitational field, then the unknown is no longer a number but a curve. In order to determine it, we have to write down what is called a differential equation in which the unknown is a real variable, itself a function of another real variable; for example, the distance of the planet to the sun is a function, to be determined, of the angle between the line which joins the two bodies and a fixed direction. Thus we have to deal with an equation in which the unknown is no longer a number but a function.

Similarly, in the theory called the calculus of variations, we consider problems such as the following, whose statement is of very ancient origin: Using a cord of given length placed on the ground, form the boundary of a region with the largest possible area. The unknown is therefore again not a number but the position and the form of the curve whose shape the cord assumes (it is known that the solution is a circle).

Thus, the unknown in numerous problems is no longer a number or a system of a finite number of numbers, but a curve or a function.

We could also cite examples from physics or mechanics in which the unknown is a surface or a function of several real variables.

Historically, therefore, it was first necessary to determine from the equations considered unknowns which were numbers or systems of numbers, or, what amounts to the same thing, points of a space of $1, 2, 3, \ldots, n$ dimensions; following that, curves, surfaces and functions. The unknown in each case is to be selected from a certain class of elements. Just as this class is called a space of $1, 2, 3, \ldots, n$ dimensions in the simplest cases, it can be called a function space in the other cases.

In order to evoke the geometric intuition whose advantages we have already discussed, we shall consider a set of elements which are functions of the same type to be a more general type of space in which each of the functions plays the role of a point.[1]

But for this set to have some analogy with a space, it is necessary that we establish some order in this space, that we have some way of recognizing neighboring points in it. We shall succeed in doing this if, in particular we can define a distance in it, i.e., if we can associate with every pair of functions y_1, y_2 of x a number which plays the role of a distance between these two points.

For example, this is what we do when we take the distance to be the maximum absolute value $|y_1 - y_2|$ of the difference of the values of y_1, y_2 for the same value of x.[2]

Abstract Analysis

Abstract Spaces. But if it is possible to study the properties of a space as complex and as general as a function space, one may ask whether it is also possible to make further generalizations. Instead of constructing several parallel theories of 1-dimensional, of 2-dimensional space, of

---

[1] In order to give some conception of the range of the ideas summarized here briefly, let us note that Vito Volterra, the great founder of functional analysis, has undertaken, in collaboration with Pérès, an exposition of this new branch of mathematics in a three-volume treatise (publisher: Gauthier-Villars).

[2] We remind the reader that the notation $|a|$ represents the absolute value of a, that is, its numerical value after its sign is suppressed.

certain function spaces, would it not be possible to include all these in one theory of spaces whose "points" are elements of any kind?

This is actually the goal of a rather recent theory (it goes back to 1904): the theory of abstract spaces.

At first glance, it might seem an absurd or foolish enterprise to want to construct a geometry in a space whose "points" are not well defined, when we do not know whether they are numbers, curves, surfaces, functions, series, sets, etc. We would be going beyond the scope of this article if we attempted to prove, on the one hand, that such an enterprise is possible[3] and, on the other hand, that its interest is not merely mathematical or philosophical, but that there have already been numerous applications in a variety of areas.[4] However, we shall try to throw some light on these points. Note first of all that, contrary to what our first thought might be, the idea of reasoning mathematically with abstract elements is far from being new. The equality or the inequality of two numbers are two properties relative to two sets of elements whose nature does not play any part; even a savage takes it for granted.

Descriptive Definitions. One of the best ways of reasoning with abstract elements is to use descriptive definitions in place of constructive definitions.

Generalization of the Notion of Distance. For example, we may designate as the distance between any two elements a, b of an abstract set a number $(a, b) = (b, a) \geq 0$ which satisfies the following conditions:

(I) $$(a, b) = 0,$$

if and only if b is the same point as a;

(II) $$(a, b) \leq (a, c) + (c, b).$$

This is a descriptive definition. On the other hand, when we are dealing only with elements of a known type, it is better to use a constructive definition. For example, for the set C of continuous functions, we generally adopt as the distance between two of these, $y_1(x)$, $y_2(x)$, either the quantity mentioned above or the minimum (y_1, y_2) of the absolute value of the difference

$$|y_1(x) - y_2(x)|.$$

[3] See, for example, our book *Les espaces abstraits*, Gauthier-Villars, 1928.
[4] Cf. *Mélanges mathématiques*, Vol. I of the *Comptes rendus du Congrès International des Mathématiciens*, Oslo, 1936.

It is easily seen that this distance satisfies conditions (I) and (II). For the set L_2 of functions $f(x)$ defined on the segment (α, β) whose squares are integrable over (α, β), we take as the distance between two elements $f(x), g(x)$ of L_2 the quantity

$$(f, g) = \sqrt{\frac{1}{\beta - \alpha} \int_\alpha^\beta [f(x) - g(x)]^2\, dx}.$$

It can be proved that it satisfies condition (II); as for (I), it is clear that (f, g) is zero if $f(x) = g(x)$. In order to make certain of the converse property, we are led to consider two functions as the same if they are equal "almost everywhere," i.e., if the set of points at which they differ, if any exist, can be contained in a denumerable set of intervals the sum of whose lengths can be taken to be as small as we wish.

In order to give an idea of the kind of problems about sets of abstract elements which can be posed and solved, we offer the following example:

What condition must the distance defined on an abstract set E satisfy in order that we can find a mapping which preserves the distance between the elements of E and the points of a given set F of a simple type, where F is, for example, a line, a plane or Euclidean space or the set of numbers < 2, etc.?

An entire geometry has been constructed on the sole basis of conditions (I) and (II) by K. Menger and his students, who have sought to preserve what can be preserved of Euclidean geometry. In many cases, the same definition from Euclidean geometry can be generalized in several ways, and it becomes necessary to investigate the relations connecting the various generalizations obtained. Such is the case for the notions of convexity, curvature, etc. A very complete exposition of this abstract metric geometry will be found in a clearly written work of L. Blumenthal,[5] which is not as well known as it deserves to be. It is one of those rather rare works which introduces the reader to the most modern discoveries without requiring a very extensive knowledge of mathematics (except for a few non-essential points).

Topological Spaces. However, classical geometry is not restricted to metric considerations. It also studies the properties of figures which are unchanged by continuous deformation; this is the subject matter of *topology*. The essential notion here is not that of distance, but that of

[5] *Distance Geometries*, The University of Missouri Studies, 1938, 145 pages.

limit or that of neighborhood. We can even say that an abstract space is not actually defined until there has been associated with a certain abstract set E an unambiguous definition of limit, or of neighborhood, or of some notions which allow a continuous mapping to be defined.

If a distance has previously been defined on this set, it is natural to define the limit of a sequence $a_1, a_2, \ldots, a_n, \ldots$ of elements of E as follows: An element a (if it exists) of E is called the limit of the above sequence if the distance (a, a_n) tends to zero as $n \to \infty$; it is also natural to define a neighborhood of a point a of E as the set of elements b of E such that $(a, b) < \epsilon$; the neighborhood becomes smaller as ϵ decreases.

But problems of analysis sometimes lead us to consider spaces in which limit or neighborhood is defined in a way which cannot be expressed by means of a distance. For example, this is the case for the function space of the so-called Baire functions.

Therefore, if a "topological" space is defined (that is, if it is possible to define continuous mappings in it—this leads to spaces far more general than the Hausdorff spaces to which some authors limit the application of the adjective topological), it may or may not have the property of being *metrizable*, i.e., such that the notion of limit or of neighborhood (or of the notion which plays an analogous role there) can be defined through a distance.

For example, let us consider the space E_ω in which each element X is determined by an infinite sequence of numbers $x_1, x_2, \ldots, x_n, \ldots$ called its coordinates, and in which we consider a sequence of elements Y to converge to X if, for each index n, the coordinate y_n of Y converges to x_n.

If we set

$$(X, Y) = \sum_{n=1}^{+\infty} \frac{1}{n!} \frac{|x_n - y_n|}{1 + |x_n - y_n|},$$

then we see that, on the one hand, (X, Y) is a distance (it satisfies (I) and (II)) and, on the other hand, the prior—very natural—definition of convergence adopted for E_ω can be expressed by means of this distance—the expression for which is much less natural and is certainly not the only one which could be chosen for this purpose.

On the more general basis of the notion of neighborhood, a topological theory of abstract spaces independent of the notion of distance can be developed.[6] A slightly more general notion is that which starts

[6] See pp. 172–185, 222–224, 277–278 of our *Espaces abstraits* (cf. footnote 3).

out from the idea of "closure" of a set, from which we can deduce that of the "accumulation point" of a set.

To every set G of elements of a topological space there corresponds a set $\bar{G}$, called the closure of G, and subject only to the requirement of containing G. An accumulation point of G is by definition an element a of the space which belongs to the closure of $G - a$.

A continuous transformation of G is a mapping of each element b of G into an element b_1 such that if b belongs to the closure $\bar{g}$ of a subset of G, then b_1 belongs to the closure $\bar{g}_1$ of the set g_1 of images of g.

Given all this, it is possible to generalize the notion of number of dimensions in the following way.[7] We shall call a one-to-one bicontinuous transformation a *homeomorphism*. We shall say that the number of dimensions of G is less than or equal to that of F

$$(dG \leq dF)$$

if there exists a homeomorphism between G and F or a subset of F These two dimensions will be said to be equal if we have both

$$dF \geq dG \quad \text{and} \quad dG \geq dF.$$

One of the advantages of this definition is the following. Heretofore, all the spaces of infinite dimension used to be classed in the same category. The preceding generalization permits us to create a hierarchy of infinite dimensions after we have found that several exist. At the same time, it throws a light on the topological affinity which exists between the principal function spaces considered in analysis by showing that they have the same dimension.[8]

We shall describe yet another very useful generalization. Grassmann and other authors had founded a theory of abstract vector spaces in which the notion of continuity is lacking, whereas the preceding topological theory made this a fundamental notion. Contrary to this being a failing, it is one of the great successes of these two theories to have been able to be used simultaneously by Wiener and Banach in order to give rise to the very useful theory of metric vector spaces, later generalized in the form of topological vector spaces.

---

[7] There are, moreover, other generalizations of this same notion, in particular the definitions of Brouwer, Urysohn and Menger, all three deriving from that of Poincaré. See Bouligand, *Les définitions modernes de la dimension*, Hermann.

[8] See pp. 30–113 of *Espaces abstraits* (cf. footnote 3).

By making use of metric vector spaces we are able to define[9] the differential of an abstract continuous transformation in a simple way (and in conformity with its intuitive meaning).

In contrast with this, it is interesting to note that the first definition[10] of the integral of a real function of an abstract element was given without using any topological notion.

Applications. If it is now understood that it is perfectly possible to reason with abstract elements, there might still remain some doubt about the utility of such general theories.

Let us immediately do away with a prime objection by noting that an abstract element is here considered as an element whose specific nature is not being taken into account but which may be perfectly well understood and have the most concrete significance. Thus, the number of elements of an abstract set maintains its value when apples, francs, etc., are substituted for its elements.

For this reason, the theory of abstract spaces and functions can be used equally well in any of the branches of mathematics.

For example, the above generalization of the notion of distance was utilized by Kürschak in number theory, and the abstract algebra introduced by Emmy Noether is a generalization of classical algebra which, while not using the same notions as abstract analysis, is derived in a similar way. The very modern theory of topological groups makes great use of the notion of "compactness" borrowed from the theory of abstract spaces. The same is true of the theory of "normal" families of analytic functions developed by Paul Montel.[11] In the calculus of variations, the existence theorems of Tonnelli, Menger, Bouligand, *et al.*, make use of the notions of compactness and distance.

The integral over an abstract set and the abstract integral of footnote 10 are indispensable in the theory of probability, the former for the representation of the probability of random events depending

---

[9] "La notion de différentielle dans l'Analyse générale", *Annales de l'École Normale Supérieure*, XLII (1925), pp. 293–323. The results are further generalized in "Sur la Notion de différentielle," *Journal de Mathématiques*, XVI (1937), pp. 233–250.

[10] "Sur l'intégrale d'une fonctionnelle étendue à un ensemble abstrait," *Bulletin de la Société Mathématique Française*, XLII (1915), pp. 248–265. Generalized still further in "L'intégrale abstraite d'une fonction abstraite d'une variable abstraite...," *Revue Scientifique*, 82 (1945), pp. 483–512.

[11] See the article of P. Montel in this book on normal families of functions and their role in mathematical analysis.

on the form of a curve, of a surface, etc., the latter for the representation of the averages of such random events.

In hydrodynamics, Leray and Shauder use the generalization of the notion of differential applied to a transformation, etc. We shall limit ourselves to these examples, which suffice, however, to give the reader an idea of the variety of the applications of abstract analysis.

11

A JOURNEY INTO THE FOURTH DIMENSION

by André Sainte-Lagüe

PROFESSOR AT THE CONSERVATOIRE NATIONAL
DES ARTS ET MÉTIERS

For some years now the idea of a fourth dimension of the universe, or even of several dimensions beyond those known to us, has been implanted in the minds of numerous researchers. We find on analysis that this idea has two very different aspects, which, if they sometimes interpenetrate one another, are nevertheless essentially distinct.

The Mathematician

Let us first say a few words about the point of view of the mathematician. Mathematicians are people who spend their lives puzzling out abstractions, never seeking, at least in so far as they are mathematicians, any relation between their theoretical ideas and the real world, even though it constantly surrounds them. Here I cannot help thinking of the prologue to Eddington's book *Space, Time and Gravitation*. There we see a discussion between an "experimental physicist," a "pure mathematician" and a "relativist."

The physicist says to the mathematician, who has just admitted that he is unable to form a conception of magnitude: "Yours is a strange subject. You told us at the beginning that you are not concerned as to whether your propositions are true, and now you tell us that you do not even care to know what you are talking about." The mathematician is in perfect agreement: "That is an excellent description of pure mathematics, which has already been given by an eminent mathematician."[1]

---
[1] See the beginning of Borel's article. (F. LL.'s note.)

Since the reading of Eddington's prologue provides us with an opportunity for it, let us see what the mathematician thinks of the fourth dimension. Regarding time, he has just said: "It is only necessary to consider time as a fourth dimension. Your complete natural geometry will be a geometry of four dimensions." The physicist then asks him: "Have we then found the long-sought fourth dimension?" And the mathematician replies: "It depends what kind of fourth dimension you were seeking. Probably not in the sense you intend. For me it only means adding a fourth variable, t, to my three space-variables x, y, z. It is no concern of mine what these variables really represent .... The four variables may for all I know be the pressure, density, temperature and entropy of a gas; that is of no importance to me. But you would not say that a gas had four dimensions because four mathematical variables were used to describe it."

The Man in the Street

The second point of view, which could be described somewhat simplistically as the point of view of the man in the street, is entirely different. The man in the street scoffs at the theoretical constructs of the scientist and wishes only to know whether a four-dimensional space really exists. When he is told that no one knows and that everything happens for us as if it did not exist, he is disappointed but, lacking anything better, clings to side issues. He asks: "But if it really existed, don't you think that some day, aided by further scientific progress, we would discover it? And even if we could never know of it, what would it be like for beings that inhabited the fourth dimension? What geometry would these beings have? What strange aspects would it have for us? Have not Einstein and his commentators spoken of the fourth dimension and even of many others?"

In spite of the relatively small value or interest of certain of these questions to scientists, we shall try to answer them in order not to overly disappoint those of our readers who may have this interest. That is why, moreover, that we have chosen "A Journey into the Fourth Dimension" as the title for these pages; but first we are going to say something about the geometry of four-dimensional space.

This is evidently not the place to discuss analytic geometry and the extremely simple way that it affords of introducing this fourth dimension. However, we remind the reader that in the old geometry of Descartes with its two coordinate axes for the plane, and three for space, it is said that $ax + by + c = 0$ is a line and $ax + by + cz + d = 0$ is a plane; it is also said that $x^2 + y^2 = R^2$ is a circle and that $x^2 + y^2 + z^2 = R^2$ is a sphere; further examples could be given. Therefore, why not say that $ax + by + cz + dt + e = 0$ is a hyperplane and $x^2 + y^2 + z^2 + t^2 = R^2$ is a hypersphere? It is by these methods based on the addition of another variable that the geometry of the fourth dimension is introduced.

Thus we are led to add a fourth axis perpendicular to each of the three axes Ox, Oy, Oz, to study parallelism, orthogonality, rotation and symmetries of this generalized space. Here we distinguish between "absolutely perpendicular" planes which have only one point in common and perpendicular planes in the ordinary sense of the word. We also have the possibility of rotating a figure about a plane.

The generalization of the regular polyhedra here gives rise to 5 regular polyhedroids, easily studied by means of the appropriate descriptive geometry. Thus the octahedroid has 16 vertices, 24 square faces and 8 bounding "cells" which are cubes. We also find in this way that the hexacosihedroid has 1200 faces which are equilateral triangles, etc.

Our Spatial Intuition

But let us abandon these studies that are of interest only to specialists and return to everyday language.

It is a very simple and very obvious fact, certainly known to La Palice and even before him, that we cannot sense or even imagine more than three dimensions in space. Just as it is possible to cover an entire surface, no matter how large, by placing side by side equal rectangles, which have only two dimensions, length and width, it is also possible to fill all of space with bricks first piled side by side and then above one another; and these bricks have only three dimensions, length, width and height.

The three-dimensionality of space has been studied with great expenditure of effort by numerous scientists, in particular by Poincaré; in his works we find alongside very perceptive and profound remarks on the twofold knowledge, muscular and visual, that we have of the universe, some faintly ironic observations such as those concerning the three semicircular canals of the ear, whose orientation evokes the idea of three mutually perpendicular coordinate planes and which seem to have been expressly constructed for the use of mathematicians. "The three pairs of canals seem to have as their sole function to inform us that space has three dimensions," says Cyon. "Japanese mice have only two pairs of canals; it appears that they think that space has only two dimensions, and they manifest this opinion in the strangest way; they form a circle, each putting its nose under the tail of the one in front, and, having done this, begin to go rapidly round and round." Moreover, it seems that if they are put into a flat region with a rim, they do not know how to get out of it. Poincaré adds: "The lampreys, which have only one pair of canals, think that space has only one dimension, but their behavior is less tumultuous!"

We fear that we should not have complete confidence in some of the interpretations made on the basis of such facts, although they are exact, but it is necessary to note that the three semicircular canals of man and the higher animals, again placed in three mutually perpendicular planes, seem to be the basis for our sense of orientation, at least for the determination of the positions that we must take; moreover, certain diseases affecting these canals cause vertigo and a sensation of disequilibrium in the body.

How it Feels in the Fourth Dimension

As we have said above, mathematicians, all the while knowing that space as we see it, as we feel it, has, or at least shows us, only three dimensions, have found it convenient to conceive of a space of four or even a much greater number of dimensions as a place in which to lodge the cumbersome products of their imagination. Whether a space with four dimensions exists or not, it is possible with a little effort to see what it is like or—and this is sufficient—to suggest to ourselves that we have seen what it is like. Let us try to make this point a little clearer.

We first draw a square on a sheet of paper, then beside it another square oriented in the same way and partially superimposed on the first, if we wish; then we contemplate this figure without attempting to conceive of both squares as being altogether in the same plane. We shall be able without too much effort to imagine that the second square is above the first and that the two together define a cube seen in perspective; this last image will be clearer, moreover, if we draw lines joining the corresponding vertices of the two squares. Everyone is in accord with what we have done up to now, but, unfortunately, what follows is more complicated.

Now let us consider a cube in space, for example a die, better still a cube formed by using twelve pieces of iron wire for its edges. Beside this first cube and very close to it, we take a second identical cube oriented in the same way. We can try to imagine that the second cube is in a different space from the first, just as a while ago the second square, as we saw it, was detached from the paper. If we draw lines joining the eight pairs of corresponding vertices of the two cubes, we shall then have the twelve plus twelve plus eight, or if we wish, the thirty-two edges of a hypercube or octahedroid constructed in a four-dimensional space.

Perhaps with a little practice we could come to develop a sort of intuition in ourselves about what the fourth dimension can or could be. Poincaré claims that, if this intuition is so rarely found, it is first of all due to the rapidly growing complication that the use of an additional dimension causes: "Do we not encounter students in our schools who are very strong in plane geometry but have no spatial intuition?" But it is also and especially because we do not have the habit of using such a new dimension and that it requires an effort.

"And moreover," Poincaré says further, "in order to imagine a figure in space, don't we all imagine in succession the various possible perspectives presented by the figure?" A solid in space that we have seen turn slowly before our eyes, and of which we have thus seen a variety of different aspects, then forms itself in our imagination as a sort of image that is unreal, but imagined, which uses all the materials brought to it by our view of the exterior.

The biologist who sees thin slices made by a microtome pass under the lens of his microscope, sees the organ he studies form before his eyes and build itself up into three-dimensional space. After seeing only a series of pictures representing cuts made millimeter by millimeter into an orange, seeds included, an eye a little accustomed to imagining a solid when seeing only a picture would give a complete spatial representation in detail of an orange, of the thickness of its skin, of the number and placing of the sections, of the positions and the location of the pips, etc.

Those who, when seeing a series of three-dimensional solids varying in a continuous and imperceptible manner, would like to see them as neighboring three-dimensional slices of a four-dimensional solid, will have to make analogous mental efforts.

Our optimism is shared by certain authors, notably by Howard

Hinton, but not by all, for Eric Temple Bell boldly asserts: "No one outside of an insane asylum can succeed in imagining a four-dimensional space." Therefore let us be prudent in making assertions.

Planar Animals

The most convenient method, though a limited one, however, of trying to imagine what a four-dimensional space is like, is, as we have always done, to proceed by analogy and to ask ourselves what totally flat animals, living on a surface that we shall take to be an infinite plane, would be like with respect to us.

We can suppose them to be formed of a single layer of molecules which includes all their cells; we shall come back again to this question of thickness. Thus they are living protoplasmic strips with fixed exterior contours if it is a question of higher animals, or with retractable tentacles if it is a question of lower animals. We may, moreover, suppose, if we please, that they are endowed with an intelligence equal to our own and in possession of an intellectual or social life as complex as we wish. They have senses analogous to ours, which allow them, for example, since they are able to judge distances, to readily discern the contours of the other flat beings which surround them and with which they live in a society.

Hypotheses analogous to those that we envisage at present have very often been utilized in order to clarify or to illustrate complicated questions such as those arising in non-Euclidean geometries.

Plane Non-Euclidean Geometries

We know how this question arises in plane geometry. Euclid, after showing that a parallel to a given line can be drawn through a given point, then asserts that only one can be drawn.

Lobachevsky assumed that several could be drawn. He thought that by making this assumption, which he believed to be false, a point of departure, he could establish this falseness by obtaining a contradiction, which, in fact, never materialized.

The geometry of Lobachevsky evidently contains theorems of a totally different form from the classical or Euclidean form. The most important is the one according to which the sum of the angles of a triangle is less than two right angles; the amount by which the sum differs from two right angles is proportional to the area of the triangle.

As a result of this, we cannot construct a figure similar to a given figure but of a different size.

In addition to Lobachevskian geometry, we must take note of another non-Euclidean geometry named Riemannian geometry after its creator, the mathematician Riemann. In a way it is the counterpart of the preceding one. Through a point no parallel can be drawn to a given line. The sum of the angles of a triangle is greater than two right angles; the difference between the sum and two right angles is proportional to the area of the triangle, etc.

We see that here, contrary to what Euclid had shown, there does not exist even one parallel to a line. This seeming contradiction arises from the fact that, at least in certain cases, Riemann abandons another axiom of classical geometry according to which only one line can pass through two given points.

In order to give full meaning to planar Riemannian geometry, it is convenient to take up our hypothesis of planar animals again and to suppose that the world in which they live is the surface of a sphere. By hypothesis, they can conceive of only two dimensions and—this point is fundamental—they call the two-dimensional surface of a sphere of large radius on which they live a plane, since they can only conceive of motion to the right or to the left, forward or backward, never up or down, and so they have no possibility of imagining the curvature of their "plane" in the third dimension of space, which is inconceivable to them.

We must immediately take note of an essential point. For these animals, thinking beings and geometers, their universe, although finite, is without limits, for they never encounter any borders which prevent them from going further; but the area of their "plane" is finite and consists of a certain number of square kilometers. These animals will naturally call the shortest path joining a pair of points or, as mathematicians say, a geodesic of their plane, a straight line. It will be, therefore, what we, three-dimensional beings, call a great circle of the sphere.

In this geometry we see that, although in general only one line passes through a given pair of points, there is the exceptional case in which the points are diametrically opposite and an infinity of lines pass through them, i.e., the halves of the great circles joining them.

Let no one exclaim that these hypotheses are absurd! Suppose that this sphere is the terrestial sphere, but an ideal terrestrial sphere, completely smoothed out and without unevenness, with its

20,000-kilometer-long meridians, and that, in addition, our little animals are a hundredth of a millimeter long. The theoretical observation that two lines always intersect at two points separated by 20,000 kilometers—a distance two million million times greater than their own bodies—would have no practical interest for them, and all their geometrical constructions, their engineers' working drawings, would be those which would result from the application of Euclidean geometry.

As we see, we are in the process of describing and of giving a concrete image of the Riemannian geometry of our plane. Riemann's two-dimensional geometry is, in fact, nothing other than Euclid's spherical geometry. Riemann's rectilinear trigonometry is what we call spherical trigonometry.

To allow the possibility of a contradiction in this Riemannian geometry, in which the words "straight lines" and "circles" are exactly what Euclid would replace by "great circles" and "small circles," would amount to allowing the possibility of a contradiction in Euclidean geometry itself, something which no one has as yet maintained.

We become aware of the impossibility of proving Euclid's parallel postulate by noting that Riemannian plane geometry, which does not satisfy this postulate, cannot contain an internal contradiction.

We can now return to the geometry of Lobachevsky and justify it in the same way. It would suffice, keeping to our example of the small planar animals, to have them live no longer on a sphere, but on a pseudo-sphere, a surface of constant negative curvature. This is the name given to the surface of revolution generated by rotating a tractrix, or curve with equal tangents, about its asymptote. This curve, whose form is roughly that of a cissoid, received this name because the length of the segments cut from a tangent by the curve and the asymptote is constant.[2]

Four-Dimensional Beings

Let us now return to our planar animals, this time, however, supposing that the plane in which they move is really our Euclidean plane and without troubling ourselves any further about what their geometry would be.

We have already said that the words "above" and "below" would

[2] See note 25 of Le Lionnais's article "Beauty in Mathematics" in this collection, Vol. II, pp. 121–158.

have no meaning for them. If, for example, we were to put the tip of our finger, or a thread or a hair, etc., into their universe, they would see an unexpected miracle in this apparition. The principle of conservation of matter will not exist for them unless we consent to it and do not introduce any new object into their world or remove something from it.

If one of these beings were to lock up some treasure in a strong box with double bolts, it would suffice for us to pluck it with our fingers, which are in three-dimensional space, in order to steal it from him, and none of their detectives could understand it.

In a completely analogous way, if there existed a fourth dimension, the beings who lived there would be absolutely invisible and non-existent as far as we were concerned. They would appear mysterious and incomprehensible to us. They could come and pull our ears without our seeing the tips of their fingers. If we stretched out our arm in the direction of the fourth dimension, it would disappear from our sight. In his *Journey to the Country of the Fourth Dimension*, the humorist Pawlowski tells how the hero of his novel found that he had the ability to move in an unknown space. After placing some love letters in a box with the intention of never looking at them again, he locks it, carefully winds ribbon around it and seals it with beautiful red seals. A few seconds later, not aware of what he is doing, and absorbed in the thought that he perhaps forgot one letter, he plunges his hand into the box, takes out this letter, verifies that there is no mistake, puts it back and only then; becoming cognizant of what he has just done, he looks at the box with astonishment and observes that it is still closed and that the seals are intact!

You or I would say that he dreamed it, but the author deduces from this that the hero then became aware that he had the ability to move in the fourth dimension.

Some Surprising Facts

We can inform the reader of many extraordinary, although logical, results without entering the domain of fantasy. For example, we may call a knot in two-dimensional space the point in which a curve intersects itself, for instance the loop formed by a thread which passes over itself and is held between two glass plates. If the ends are glued to the frame, such a knot can be moved between the plates, but can never be undone, since in order to do that we would have to lift the thread into the third dimension, something which the glass plates will not allow.

Similarly, for us, if the ends of an ordinary knot are affixed to the walls of the room, it cannot be undone. We can only move the position of the knot from one end to the other. But it has been established by mathematical reasoning that the knot can be easily undone by anyone who uses the fourth dimension to move the cord about. It has been claimed that certain mediums were able to untie such knots. If such a fact could be verified, then it would constitute a serious argument in favor of the existence of the fourth dimension.

All the cells of which the small, planar animals of whom we spoke are composed would be open to our inspection, and we would be able to see all of the interior of their bodies. In the same way, a four-dimensional being would see all of our organs simultaneously, and perhaps would even know our thoughts, if indeed they correspond to visible movements of certain nerve particles. He could remove a bone from our skeleton, femur or scapula, without any traumatism or slit being formed in our skin, that "leather sac" which envelops us.

Better yet, suppose that we had taken one of the flat creatures and replaced it gently in its own planar universe after having turned it over; it would occupy a position exactly symmetric to its original one but would continue to live as before, since the single layer of molecules which composes it is the same on both faces and the animal has neither back nor underside. Thus, if a four-dimensional giant were to pick up one of us gently between his thumb and forefinger and put him down after having turned him over, the unfortunate (?) fellow after that would have his heart on the right and no longer on the left.

The Thickness of Our Space

Of course, we say in mathematics that a point has no dimensions, but the physicist is not able to take this literally. For him a point does not exist unless it is material and has dimensions, small certainly, but not zero. Similarly, a curve or a plane has a measurable thickness, no matter how thin we may suppose it to be.

For example, our flat creatures would be, as we have already said, formed of juxtaposed molecules or cells, and when we say that they are two-dimensional, it is the same as saying that their thickness in the third dimension is extremely small, a thickness, moreover, of which they are not aware.

In the same way, if there is a four-dimensional region in which our three-dimensional universe floats, it is because every particle of our universe, each part of ourselves has a very tiny thickness in this fourth dimension, even though we have no consciousness of it. Later we shall indicate various ingenious corollaries which have been deduced from these premises.

But first, in order to better understand the meaning of this hypothesis, by way of introduction we shall recall for those of our readers who are not acquainted with them the amazing adventures of Mr. Barnstaple among the "Men Like Gods," as told by the novelist H. G. Wells.

At first we see the hero of the novel, Mr. Barnstaple, as he is leaving for his vacation, alone, traveling in his small car toward the Great North Road. Passed by a big limousine, he dashes off in pursuit and, coming to the bend which had hidden the horizon, he no longer sees the limousine; this appears incomprehensible to him. Suddenly his own car skids so violently that he loses his head for a moment or two. Later he remembers having heard a cracking sound "like the snapping of a lute string, which one hears at the end—or beginning—of insensibility under anaesthetic." The most extraordinary thing was that he no longer recognized the road or the countryside, but as a compensation he did see the big limousine again.

Mr. Barnstaple had been transported into another world, that of the "men like gods!" In spite of its interest, let us pass over all that part of the book which describes to us the thoughts of the "men like gods," their life in "Utopia," their great civilization and the happiness to which they were able to attain, and say simply that at the end of the novel Mr. Barnstaple is sent back to earth, a fact which saddens him, by the way.

He is given back his car, he is put back on the road and he crosses the prophetic line drawn to mark the frontier between Utopia and the earth. He closes his eyes with emotion during the passage, and with a bump finds himself with his auto in a field. As he had promised the Utopians, he puts a red flower brought from Utopia at the precise spot where he made contact with the earth and then: "Something happened very quickly. It was as if a hand appeared for a moment and took the flower. In a moment it had gone. A little eddy of dust swirled and drifted and sank. . . . It was the end."

Piled-Up Sheets of Paper

Let us try to understand what has happened, and in order to do this let us again take up our classical interpretation, which consists in assuming that somewhere there are two-dimensional animals, small, flat creatures who think, however, and who have no idea of what is meant by thickness or height. At present, they are moving on a page of a partially open book; they have gone over all the letters, all the brightly colored pictures; thus they have traversed great stretches of black or white, blue or violet, or of a beautiful golden yellow. They zigzag from the top to the bottom of the page before our amused eyes, seeking to discover the philosophy and the deeper meaning of the printed world which is theirs.

But the preceding page, which is bent slightly by the breeze and which shakes from time to time, just like the following page and many others, is likewise peopled by flat creatures more or less identical and more or less advanced in their evolution and in their degree of comprehension of the miracles of the volume which is the world. There, side by side, exist a number of worlds in miniature which we, beings of a superior nature, or at least in possession of one more dimension, watch evolving, not without some pity—worlds which will always remain ignorant of one another, will never be able to interact and to know one another unless . . .

Unless we bend one of the pages so that its edge touches the center of the following page and, perhaps with a little shake, cause one or two of our little animals to cross over to the following page; and there we have Mr. Barnstaple changing universes.

Why not imagine that two-dimensional worlds live side by side, perhaps like different pages of a book or like sheets of paper piled up one above the other, and that as a result of some unknown and exceptional circumstances the edge of one sheet can touch another sheet or lie beside it in such a way as to create a means of passage from one universe to the other?

Perhaps also we are in a three-dimensional world which is extremely flat in a fourth dimension, just as the pages of a book have a thickness which is barely perceptible in the third dimension.

Thus the universe is perhaps like an enormous file in which each sheet is a world analogous to the one that we know, with its stars, its moons and earths, without any normal means of communication between the neighboring more or less evolved worlds, which are perhaps peopled with dull brutes or repulsive monsters, or perhaps also with supermen, those supermen that we would like to be but are not.

Four-Dimensional Space and Physical Phenomena

While still limiting ourselves to a purely spatial conception of the fourth dimension, we should like to ask ourselves whether an answer is possible to certain questions: Has space four dimensions? If so, is the higher dimension inhabited? Who inhabits it?

If there is no higher dimension, or if there is one and its inhabitants, supermen or not, are peaceful, quiet people, who do not try to bother their neighbors, none of the mild pranks that its inhabitants could play on us will ever take place, and it seems that we could go on asking ourselves this question without its ever receiving an answer. In particular, no affront to the principle of conservation of matter, or if we prefer, no sudden appearance or disappearance, will ever by observed. If one day we saw a prisoner suddenly disappear from a well-guarded prison, we could muse about the fourth dimension. But if no such event does happen, we shall be able neither to affirm nor deny the existence of the fourth dimension.

However, even in such a case, science would perhaps not be completely unable to treat such a problem. In fact, people have asked themselves whether certain laws of physics could not be explained by the introduction of a new spatial dimension, and in saying that, we do not mean even to consider the theories of Einstein.

Here is an example of the speculations that have been made on this subject, which W. W. Rouse Ball speaks of in his *Mathematical Recreations and Problems.*

Let us make use once again of our hypothetical, minuscule animals, who will now live on the rim of a discoid earth, comparable, if you like, to a coin without perceptible thickness. Thus, miniscule people without thickness are to be walking about upright on the edge of this coin. Each of them is to continue his walk until he finds himself face to face with another of his fellow creatures, in which case a tiny flying machine, or perhaps the strength of his legs, will allow him to jump to the other side and to continue his walk.

Our earth-disk can be placed on the table, where, if we have good eyesight, we shall see these little people in silhouette as they live, act and move about, their feet on the rim of the disk and their heads in the direction of the exterior, exactly as we ourselves stand on the surface of the earth.

On this same table, a little further away, we shall place another disk, no less flat, but golden and shining with light; this is the sun-disk. This planar world of two dimensions, earth-disk and sun-disk, thus is placed on a solid, three-dimensional table; this table and its third dimension are not known to our little people. All the molecules of this universe, earth and sun, thus touch the wood of the table.

Now if some kind of vibration, for example radiation, took place in this sun, it would shake the supporting, three-dimensional table, and in doing this, would affect the molecules of the earth-disk. In this way the attraction of the sun for the earth, universal gravitation, could be exerted without any material basis apparent to the physicists of our earth-disk. Clearly, they would continue to be perplexed by such phenomena. Perhaps our three-dimensional earth is in the same way nothing but a disk in the fourth dimension, i.e., with an imperceptible thickness in this new and mysterious direction of space. Let us further suppose that the same is true of the sun and that these two spheres, which, in a sense, are also disks, are attached to an immense three-dimensional plaque, which itself has no perceptible thickness in the fourth dimension.

If we suppose that this plaque has uniform elasticity, then we would find that the amount of energy transmitted by the sun would be exactly inversely proportional to the square of the distance, precisely what takes place in reality, and it would no longer be necessary to conceive of an ether in order to explain the passage of something from the sun to the earth.

Jouffret makes other hypotheses about the molecular layer which our universe would represent; in his conception, actions having their source in the fourth dimension can affect this molecular layer. If we examine the component normal to our universe arising from such an action, then we are led to effects which he shows to be comparable to those caused by an increase or a reduction of the temperature.

Better still, he distinguishes between the chemical mixture of two substances, for example oxygen and hydrogen, in which the atoms are placed side by side in some way, and a chemical combination in which they are stacked one above the other. This means that the pile, although of very small height and contained within the thickness of our universe, nonetheless has its greatest dimension oriented in the direction of the fourth dimension of space. Upsetting this pile corresponds here to a chemical decomposition.

Almost certainly, these are only singular hypotheses. If I have nevertheless insisted on discussing them, it is in order to show that, even if the higher dimension is uninhabited, it could happen that one day science will establish its existence or, speaking more cautiously, as scientists always do, that people will say that everything happens as if it existed. The simplest explanation of the laws of the universe would then be the one which assumed the existence of such a dimension, inhabited by supermen or not.

A Journey Through Time

Many things have already been said about four-dimensional space, considered as a space of which we are acquainted with only a small part: the three-dimensional space in which we live.

But it is more interesting and at least novel to introduce time into our discussion. Likening time to a fourth dimension of space, moreover, gives rise to some complicated, not to say frightening problems.

Here again we are going to allow Wells to introduce our subject. In his *Time Machine*, he tells us of his "Time Traveller" and has this bold explorer say: "There is no difference between Time and any of the three dimensions of Space except that our consciousness moves along it." Some people, he continues, have concluded from this that there had to be a four-dimensional geometry, three-dimensional images representing a perspective of four-dimensional objects. "Here," he goes on, "is a portrait of a man at eight years old, another at fifteen, another at seventeen, another at twenty-three, and so on. All these are evidently sections, as it were, Three-Dimensional representations of his Four-Dimensioned being, which is a fixed and unalterable thing.... Here is a popular scientific diagram, a weather record. This line I trace with my finger shows the movement of the barometer.... Surely the mercury did not trace this line in any of the dimensions of Space generally recognised? But certainly it traced such a line, and that line, therefore, we must conclude was along the Time-Dimension."

Here is an explanation of what has gone before. We are three-dimensional, but we are constantly rising, as would an elevator or a balloon, along the fourth dimension with a motion that we can suppose to be rectilinear (I dare not say uniform, because I am not too sure what this word means now). The present is then the place where we are now, the past is in some sense below us and the future above.

The Fourth Dimension and the Future

In order to study all this better, let us again make use of the frequently employed hypothesis of planar animals with no consciousness of a third dimension, not possessing, moreover, any thickness, and moving about on what we may suppose to be an infinite, horizontal plane. Instead of imagining, as we did a while ago, that the plane in which these animals live is constantly rising in going from the past to the future, let us suppose—and this clearly amounts to the same thing—that their plane is fixed and that the great river of time flows peacefully around them with an overall movement which causes their future in some way to fall upon their heads, while the past moves further and further away beneath their planar world.

Let us examine some historical event in the life of these creatures, for example, in order to be clearer, an orange crossing through their world, which we shall call the "orange phenomenon," supposing that this crossing is possible without molecular collision or destruction of matter. It is easy to foresee how this phenomenon would appear to the flat creatures whom we are studying. They would first see a little yellow patch which continues to enlarge, then a white patch at the center and so on in this way as the center of the orange approaches. Each seed makes in turn a more or less yellow patch which grows in diameter, diminishes, then disappears. In turn the large patch bordered by a bright yellow circle diminishes, becomes entirely yellow, then fades away.

Suppose now that the passage of the orange is a catastrophe for these creatures, since it causes floods which drown them. If one of our flat animals has invented a time machine, what is he going to do? He launches his machine, which acts as the balloon, bravely into space and goes searching for an "orange phenomenon" in the future. As soon as he finds one, he makes it disappear with a stick of dynamite. Such a traveler will therefore change future history, but he would not be able to affect past history, for if he goes to blast an orange in the past, that will not revive those who may have been drowned by its passage through the world.

In this conception, once again so bizarre that it would scarcely ever be imagined, the time explorer meets other living creatures like himself in the future, and in some way he blends his own future, as time explorer, with their present. The two coexist simultaneously and live side by side until he takes the machine back. The life of the world is like a film which unrolls before our eyes without our having the possibility of changing anything that the operator does unless we have a time machine in our possession. Seated in our chairs, we watch the living world, and us with it, on the screen; the past is constituted by the reels that we have already seen, the future by those that are going to be shown to us. If we know how to travel in time, we can make a mark, a hole, in some future frame, and we shall find it again when it becomes the present, but tearing a past frame will change nothing that has already happened.

The main trouble with these conceptions is that they are much more complex than the preceding discussion seems to indicate. In fact, let us take another example of a phenomenon passing through our planar world, observed curiously by our little flat creatures. We are going to light a wooden match!

We are not interested in the past of this match, but only in its present and future. In our two-dimensional world, our match, laid flat, is a small, very elongated wooden rectangle, its end painted with inflammable material. But this match extends indefinitely into the future, in which it thus forms a strip of indefinite height, as would a very high wall. In fact, if no one touches this match, if no cause for its destruction or displacement arises, it will necessarily be there indefinitely, and at every minute, at every hour, on every day, we shall see this "match" wall along which we gradually move as we grow old.

In fact, if a future phenomenon is one day going to destroy this match, the strip will be deformed or will suddenly stop after a certain height. But at present, as in the case of the "orange phenomenon," it is only a question of a future prepared to crash down upon our universe and in some way immutably predetermined in a world subject to the laws of the most inexorable fate.

Let us complicate things a little by supposing that one of the flat creatures who live in our universe displaces the match-rectangle, the only one which he sees and is able to imagine; immediately the entire strip of which it is a section is displaced at the same time, since in the future it will always be at the new place. If someone lights the match, its carbonization and destruction occur simultaneously throughout its entire altitude, as we would see by stopping the combustion and noting that the remaining length of the match must be the same in the present and in the future.

Moreover, to speak of the instantaneous carbonization in the future of the match which we see burning in the present perhaps simply means that we conceive solely of an extremely rapid propagation of this incandescence throughout its height, since the future of the match at the moment at which it passes through the plane of the present will never again show us the match intact. If this propagation were very slow, our time explorer, by hurrying a little, would arrive in the future in time to see the match still intact, although he had seen it with his own eyes completely burnt in the present. Then would there exist in our own world dead people still living in the future, but toward whom the death set in motion here is moving with lightning speed?

Perhaps the author should apologize for raising such problems in discussing the fourth dimension, but he wanted very much to show how great a role the fourth dimension plays or could play in the research into the eternal verities which the men of science breathlessly pursue generation after generation.

12

ON THE CURVATURE OF SPACE AND THE POSSIBILITY OF FORMING A CONCEPTION OF IT BY ELEMENTARY MEANS

by René Thiry

DIRECTOR OF THE INSTITUTE OF FLUID MECHANICS, PROFESSOR IN
THE FACULTY OF SCIENCE AT MARSEILLES

SPECULATIONS on spaces of more than three dimensions have long been considered as simply mental constructs, fruits of the natural tendency of mathematicians to generalize without the slightest interest in applicability to the real world.

The possibility of such extensions was opened up the moment that Descarte's genius conceived of substituting for direct geometric reasoning algebraic calculations based on the coordinates of the points constituting the figures to be studied. Geometry, under the name of analytic geometry, was then reduced to the study of sets of three numbers—in mathematical language, of three-dimensional manifolds. Clearly, it then became possible to apply the same computational procedures to the study of sets of four numbers, five numbers, etc., i.e., to undertake the study of n-dimensional manifolds, where n is any positive integer. Naturally, all these generalizations were in the beginning based on the classical geometry of Euclid.

In the same way that plane geometry can be regarded as included in three-dimensional geometry, the latter can in its turn be studied within a larger manifold, for example, a four-dimensional manifold. However, in three-dimensional space there exist other two-dimensional manifolds besides the planar manifolds; these are constituted by surfaces distinguished from the planes by a *curvature*; similarly, we can find three-dimensional manifolds within a four-dimensional manifold distinct from the Euclidean manifolds.

Thus we are led to conceive of three-dimensional spaces which possess a *curvature*; we shall leave the meaning of this word a little vague until it is more convenient to make it precise. Naturally, one would define in a similar fashion curved spaces with more than three dimensions. The sphere is clearly the simplest of the curved surfaces; since it is identical to itself at all its points, its curvature is therefore constant. Moreover, since it closes on itself, its area is finite, although no boundary line encloses this area.

Therefore one ought to be able to consider a *spherical* three-dimensional space with analogous properties, closed upon itself and thus having a finite volume, although there is no boundary enclosing it; this would be the simplest example of a curved three-dimensional space.

At first all these considerations remained strictly in the domain of mathematicians; those scientists interested in mathematics, not so much for itself as for the aid that it affords them in their own research, were far from thinking that these generalized geometries were capable of future application.

However, one fine day a bold conception arrived on the scene which associated time with the three spatial dimensions and transformed physics into the geometry of a four-dimensional manifold, the *space-time* of relativity theory. And far from appearing chimerical, these new ideas advantageously reconciled what had seemed up to then to be disparate facts and threw a brilliant light on entire areas of physics. We can say that none of the minds most reluctant to accept these revolutionary ideas was not deeply affected by the unity and harmony that they were capable of bringing to our knowledge of the world. Later, the physicists working on the theory of relativity even came to consider space-time as non-Euclidean, possessing a *curvature*, and thus we find the concept of a *curved, finite and unbounded* universe in modern works on mechanics and astronomy.

However, a rather unsatisfying point persists in the summary exposition that we have just furnished: we have not succeeded in giving a physical reality to the curvature of the real space in which we live, other than through the algebraic fiction of a higher-dimensional manifold in which this space is supposed to be imbedded. But mathematicians have also succeeded in studying non-Euclidean spaces in themselves, and in particular, in defining their curvature in terms of their inherent geometric properties. Their speculations are naturally rather refined and require a highly developed mathematical apparatus. It is nevertheless still possible to give someone with only a very elementary mathematical training an insight into the basic ideas which will allow him to form a conception of such a finite, unbounded space, however paradoxical these terms may appear at first view. This is what I would like to attempt in this modest article.

Geometry is looked upon in totally different ways by the physicist and the mathematician. The latter seeks to construct a logical edifice based on conceptions which do not exist in actuality, although they might be the product of his contact with the external world. In nature we find neither the point, the line nor the surface of geometry. Starting from these abstract concepts, the mathematician constructs geometry, taking as his basis a certain number of properties which he assumes and which he calls postulates. It was thus that Euclid constructed his geometry, taking in particular among his premises that celebrated statement, Euclid's postulate, which asserts that in a plane it is possible to construct exactly one line through a given point which does not intersect another given line. Geometers have always made an effort, moreover, to reduce the number of postulates to a minimum, trying to prove that certain of them were consequences of the others. Innumerable attempts were made in vain over a long period to demonstrate Euclid's postulate, the necessity of which did not appear to be self-evident. Then some geometers of genius showed that this postulate was in fact not at all necessary and that by rejecting it logical structures could be built which were just as consistent as the Euclidean structure. Two geometries were constructed which bordered on Euclidean geometry. In one it was assumed that an infinity of lines could be drawn through a point in a plane which did not intersect a given line; this is the geometry of Lobachevsky. In the other, the one in which we shall be more interested, it is assumed that two lines in the same plane always *intersect* no matter what their positions are; this is the geometry of Riemann.

The physicist starts out from a completely different viewpoint. For him, geometry is an experimental science; it has to do with the real world in which we live, on which he measures lengths, surfaces and angles. We shall call natural geometry the structure built up of these elements. It does not present, of course, the perfection of that of the mathematician; the measurements of the physicist are always estimates with their inherent errors, but he considers himself satisfied if he knows the limits of error. He then asks the mathematician to point out to him the geometry, among all the geometries that the latter has imagined, that is most easily reconciled with natural geometry. For a long time no one even asked this question; it was assumed as a matter of course that the classical geometry of Euclid satisfied this wish and that it would always be in complete accord with natural geometry, no matter how precise the latter's measurements were. It is in this sense that we shall say that the physicist then regarded the universe as Euclidean.

It is precisely this conception that the theoreticians of relativity claim to have invalidated. For them, the accord between natural and Euclidean geometry is only an approximation, and they assert that the physicist would observe this fact if he improved the accuracy of his measurements sufficiently.

It is neither my intention nor within my capabilities to give an exposition and a critique here of the foundations of the theory of relativity, nor even to find fault with the likening of time to a fourth dimension; I would only wish to show how a conception of a spherical three-dimensional space can be formed without a complex mathematical apparatus.

In order to arrive at this goal we shall begin, using an illustration suggested by Einstein, by making some very simple observations on the geometry of the surface of a sphere. We shall imagine this surface to be inhabited by completely flat creatures, who crawl upon its surface and have no idea of what goes on outside of it; thus this constitutes their only perceived universe. Let us try to put ourselves in the place of these creatures, who we assume are endowed with an intelligence analogous to our own and are trying to construct a geometry. One of the fundamental properties of the sphere is that it allows the translation of a figure drawn upon it without any modification of its shape or size, and with the same freedom as if it were a plane figure sliding on the plane. A geometry, often called the science of translations, can be built upon it as easily as our customary plane geometry is constructed, which is certainly not true for any surface. The line, which is the shape assumed by a string stretched between two points in space, will be replaced here by an arc of a great circle, which would be the shape assumed by a string stretched upon a sphere. These curves, which are in actuality the shortest paths in going from one point to another while remaining on the surface of the sphere, are what mathematicians call *geodesics*. We shall still call them "lines," putting them in quotation marks, however, in order to stress the conventional character of this name. After having chosen a fixed arc of a great circle as the unit of length, the inhabitants of the sphere will define, as in Euclidean geometry, the length of an arc of an arbitrary curve. They will observe that their "lines" are closed curves, that they all have the same finite length, that two "lines" always have two points in common which divide each of them into equal parts, and, finally, that all the "lines" issuing from a point will pass through another point which the inhabitants will call the antipode of the first and which is, in fact, the farthest point on the surface from it. They will also be able to draw a circle around a central point, and by dividing the circumference into equal parts they will construct a sort of protractor, which will allow them to measure angles on the surface of the sphere. If they limit their studies to a very small portion of their space, their geometry will naturally be very close to Euclidean, since a small spherical cap differs little from a small disk; thus they will be able to construct figures of small size approximating squares, and the smaller the figure the better the approximation. Using the methods of integral calculus and beginning with these elementary squares, they will succeed in forming a notion of the area of a domain of arbitrary extent and they will observe in particular that their universe has a finite area. The more refined their measuring procedures become, the more they will become aware of the divergence of their geometry from that of the geometry of the plane. Thus, for them, the "ideal" square will be formed by four consecutive, equal arcs of great circles, joined together in pairs at the same angle. Basically, it will be the curve of intersection of the sphere with a regular tetrahedral angle having its vertex at the center of the sphere. They will observe—as evident to us who see the dihedral angles of this tetrahedral angle—that an angle of this "square" is greater than a right angle. Therefore they will not be able to pave their sphere with the aid of these "squares," as we pave a plane by placing squares alongside of each other. By noting the impossibility of doing this, they will become aware of the curvature[1] of their space.

[1] As I have already said, I do not intend to give any mathematical proofs here. Let me indicate, however, that if we denote the angle at a vertex of one of these "squares" by A (expressed in radians), its area by S and the radius of the sphere on which it is drawn by R, it would be very easy by elementary geometrical methods to demonstrate the formula
$$S = (4A - 2\pi)R^2.$$
Since the creatures inhabiting the sphere are capable of measuring A and S,

ON THE CURVATURE OF SPACE

Now suppose that we introduce a tangent plane touching the sphere at one of its points A, and that we place a point source of light at the antipode S which will cause anything that lies on the sphere to cast a shadow on the plane. The transition thus realized from the sphere to the plane is nothing other than what mathematicians call a transformation; it is the classical stereographic projection of elementary geometry.

If we now place on the sphere a circular disk of very small radius— or, more exactly, a small spherical cap—its shadow on the plane will be bounded by a circumference whose radius will be practically equal to that of the disk if this shadow lies in the neighborhood of A, but will be greater if it is distant from this point. It is very easy to show that, if the shadow is found at a distance r from the point A, the ratio of its radius to that of the disk will equal $1 + r^2/4r^2$ (R always denotes the radius of the sphere). Now let us imagine that this plane is inhabited by inanimate objects and living beings whose size increases in the preceding ratio as the distance of their position from the point A increases, somewhat in the manner of objects with the same coefficient of expansion in a region in which the temperature varied from point to point. It is clear that these beings would construct a natural geometry identical at every point with that created by the inhabitants of the sphere. As for their changes in size, these would, of course, be apparent only to us; the beings would find it impossible to perceive them, since all of their measuring instruments would be formed of material obeying the same law as their own bodies. Thus, they would arrive at the conclusion that they inhabited a non-Euclidean universe which was finite and unbounded, and whose *radius*, moreover, they would be able to calculate in the way we have indicated above. In particular, their "lines" would be those circumferences of the plane with respect to which the point A would have a power equal to $-4R^2$. If these creatures are endowed with visual perception, it would be natural for them to assume in addition that light follows the shortest path in going from one point to another in their universe, i.e., that the light rays would coincide with the "lines" just mentioned. For us who continue to hold to the Euclidean notion of distance in the plane to be able to form a conception of this fact, it is sufficient that we imagine these planar creatures to be living in a

they thus would have the means of computing the *radius* of their universe by using these quantities. This would only require strictly local measurements; they would have no need of traveling over the entire sphere.

transparent world where the index of refraction varied in the appropriate way with the distance from the point A.[2]

Moreover, it is essential to note that, if the point A seems to us to play a very special role, this is not at all the case for the inhabitants of the plane; in their eyes their universe appears to be completely homogeneous and identical to itself in all its parts, as was the case for the sphere.

This planar representation of Riemannian geometry may seem unnecessarily complicated in comparison with the much more easily conceived spherical representation, but its great advantage is that it now allows us to extend the definition of the Riemannian universe to a three-dimensional space. It will suffice to imagine that the variations in size of the contents of the universe mentioned above now apply in accordance with the same law in all directions in space around the point A, and not only just in a plane passing through this point. The beings who inhabit this universe will then have the impression that they live in a three-dimensional, finite and unbounded universe; the natural geometry which their physicists will use will have to take the form of Riemann's geometry. We can even imagine the universe to be inhabited by two distinct races; one (us, for example) with the Euclidean conception, the other (composed of the creatures of whom we have just spoken) with the Riemannian conception. Each of the beings of the two races will consider himself, like the dog Riquet in Anatole France, "always the same size wherever he is," and will not be able to make the others understand that it is not the same for them. And if the Euclideans try to make the Riemannians grasp the notion of the infinite extent of the universe by paving it with an infinity of cubes, all equal in volume in their eyes, the Riemannians will only see in it a paradox analogous to that of Zeno, who claimed that Achilles could never catch the tortoise in a finite time by covering an infinite series of intervals of distance which became smaller and smaller in size. In the same way the proof claimed by the Euclideans would be reduced in the eyes of the Riemannians to the obviously false assertion that a finite volume is infinite because we can always place within it a larger and larger number of elementary solids whose volumes tend toward zero.

A test, however, would serve to bring the dispute to an end; if they both wanted to preserve Fermat's principle, rays of light in a vacuum

---

[2] It would be easy to show that this index would have to be inversely proportional at each point to the coefficient of *expansion* met with above.

would have to follow the geodesics of the space, and the latter are not the same in the two geometries. If these beings are able to move freely over vast regions of their universe and if the curvature of the Riemannian space is large, then the crucial test will be easy to make; but if they have access to only a limited part of their universe and if the curvature is slight, then the observation of the divergence between the two families of geodesics will require extremely precise measurements. This is what happens in our own universe; its curvature, as it is predicted by the theory of relativity, is so small that even by extending our perceived world to the extreme limits possible in astronomical observation, the region thus determined is still too small for us to be readily able to sort out the effects of its supposed Riemannian character.[3]

It is possible in addition to conceive of finite, unbounded universes in a totally different way without abandoning the Euclidean notions of distance. But these universes no longer have anything to do with physical theory, and I only speak of them as of an amusing curiosity.

Let us again take up our example of completely flat creatures who inhabit a plane, but who this time have adopted Euclidean geometry, and let us suppose that the disposition of the objects in their universe is periodic, i.e., can be divided into parallel strips in such a way that one strip is identical with any other after translation. Suppose now that we are traveling through this region; it is very clear that we shall have the same impressions as if we were on a cylinder. If we had some doubts about this, we would make some identifying marks on the strip on which we were situated in order that we could recognize it later. However, suppose that a very clever being external to our universe, a sort of Maxwell's demon, was watching our efforts and repeated our marks exactly in the neighboring strips. Our doubts would certainly be dispelled and we would be *certain* that we lived on a cylinder; if someone came along and told us that we could just as well believe that we lived on a plane if we entertained the simultaneous hypotheses of periodicity and the external being who amuses himself by deceiving us, we would clearly reject this theory as an unnecessary complication.

[3] In fact, the situation is really much more complex; in the theory of relativity the curvature of space at a point is intimately connected with the presence of matter, so that this curvature is not constant and we would have to replace the sphere we have considered by a more or less dented surface. This is why astronomers have sought to carry out the crucial test on rays of light passing in the neighborhood of a significant mass, in fact, on a ray emitted by a star whose light passes close to the sun; according to relativistic ideas, this ray ought to deviate from a straight line.

Instead of imagining a distribution of objects in our plane which is periodic in dimension, let us now suppose that we have a doubly periodic distribution, i.e., that each of the preceding strips is in turn divided into identical rectangles; the cylinder will now appear to close up on itself, and we shall have the impression that we live on a sort of torus. Now, this conception can clearly be extended to a three-dimensional domain; we may suppose that the space is cut into identical rectangular parallelepipeds which are glued together and filled with objects in the same way, an identity carefully maintained by Maxwell's demon. Let us imagine ourselves not just transported into such a space, but having always lived there, as did our ancestors. This universe will appear closed to us; we could compute its volume, but would find no boundaries; it would be connected in the same way as what we might call a hypertorus and locally it would be rigorously Euclidean.

Such a domain would have strange properties, in which the integers would play an important role. To mention only one of them, suppose that the inhabitants constructed a wall parallel to the plane of one of the faces of the parallelepipeds and tried to extend it as far as possible in all directions; since their movements would be faithfully reproduced in the neighboring boxes, after a while they would find stones which they believed that they had placed there themselves, and finally they would believe that they had constructed a closed surface. However, this partition would not prevent them from having access to any part of their universe, and if they are on one side of the wall, they will be able to see what is going on on the other side without having to cross it; they will only need to go and examine the other side in one of the neighboring boxes.

A reader interested by these speculations will readily find other properties of these spaces. He could also make use of his powers of imagination to develop analogous ideas applied in different directions. He could, for example, try to imagine the impressions that two-dimensional creatures living on a sphere would have, if the sphere were covered with objects symmetrically placed with respect to the center and a Maxwell's demon were to amuse himself by repeating at the antipode of the creature's position the marks made by any of these creatures or their positions. This is a conception which it would even be possible to extend to a three-dimensional, spherical Riemannian universe.

C. FUNCTION[1]

We now come to function, that proud daughter of number and space (this allusion to its natural origins is in no way meant as a judgment on its individual mathematical form). Foreshadowed in antiquity (in the study of geometric loci) and during the Renaissance (in the study of algebraic equations), this notion became aware of its strength in the 17th century, first under the impetus given it by Descartes, who related algebra and geometry, next thanks to the independent but convergent labors of Newton and Leibniz, in their discovery of a scalpel ideally suited to its shifting and subtle nature: the calculus. Having won its autonomy, the theory of functions was greatly enriched in the 18th century thanks to the audacious strokes of an army of mathematicians led by Euler, later by Lagrange. Rising with Gauss, Fourier, Cauchy, Abel, Jacobi, Riemann and Weierstrass from the algebraic to the transcendental, from the real to the complex, from the continuous to the discontinuous, it reached that Himalaya which dominated all the mathematical thought of the 19th century: the notion of analytic function. Georges Valiron's article is devoted to the origins and the evolution of this sovereign theory. His study, like an arrow, possesses a point: the automorphic functions, whose value resides in their ability—discovered by Henri Poincaré—to uniformize the algebraic functions. Valiron, the creator of authoritative works on meromorphic (analytic) functions (which arise within the framework and as an extension of his own article), is especially well suited to give this presentation. Its remarkable documentation will be appreciated; it is in the same vein as the documentation which furnishes such great additional interest to the technical value of his recently published Cours d'Analyse mathématique.[2]

This bird's-eye view of the historical development of analytic functions sees them from the classical point of view, i.e., in a sense as individuals. It was also

---

[1] Those of our readers with little mathematical preparation will find it valuable to refer to the notes of Brunet's article. Certain elementary definitions are given there (roots of equations, functions, derivatives, integrals, series, rectification and quadrature of curves, etc.) which will help them greatly to understand the articles in this section.

[2] In two volumes: I. *Théorie des Fonctions*; II. *Équations fonctionelles; Applications*, Gauthier-Villars.

worthwhile to devote a study to classes of functions and, in particular, to those "normal families of functions" which shed so much light and are so serviceable in the most diverse branches of analysis. No one is better qualified to speak of these than Paul Montel. The noteworthy article which he submitted to us contains, in fact, only one gap, but it is an important one: for Montel has neglected to inform us that he is not only the author of important and varied works on differential equations, the zeros of polynomials, polynomial approximation, descriptive geometry and finite geometries, but also the father of those "normal families" whose serviceability he allows us to admire.[3]

The theory of analytic functions, however, does not exhaust the enormous range of the domains of functions. There are still mysteries to be unraveled within the theory of real functions, closely linked to the theory of sets introduced in a preceding article. Jean T. Desanti, a young philosopher possessed of true mathematical learning, is an assurance that the French school of the philosophy of mathematics will live again in spite of its decimation by the war. He centers his study on the rise, after Euler, of the modern theory of functions of a real variable from the period of Cauchy to that of Riemann, thus clearly marking off the boundary separating his study from that of Valiron, and stressing the operation of integration whose development necessitated the theory of sets.

Arnaud Denjoy is one of those most knowledgeable at present in the area of trigonometric series, the basis of the renaissance in analysis, and has contributed notably to the creation (and also in a sense to the obliteration) of the gulf separating functions of a real variable from functions of a complex variable. He is one of the creators of the "geometry of sets" and the father of that "total abstraction" which in 1912 consummated the successive extensions of the notion of the integral. (We shall mention this discovery again when discussing Perrin's article on Lebesgue.) Denjoy has just published an exciting work, L'Énumération du Transfini, in which we find that typical combination of the highest competence with a lyrical inspiration which makes his article so absorbing. He takes up in what seems to be a definitive manner the analysis of the ancient problem of Achilles and the tortoise and shows us what good service Cantor's famous number ω (omega) renders here, and how, far from putting new obstacles in the path of thought, it offers a convenient and valuable way-station within the sequence of numbers for the reconciliation of the intellectualism of mathematics with the reality of motion. Encapsulating the contents of Valiron's article (which begins with the Middle Ages and ends with the 19th century), the short cut taken by Denjoy allows him to start out with the initial Greek attempts at the analysis of the infinite, which precedes Valiron's subject matter, while

---

[3] See Godeaux's article, p. 292.

terminating at the heart of Cantor's revolutionary ideas, which go beyond it. Such transverse pathways clarify and give life to a question infinitely better than the approaches of what is usually a worn-out academicism.

F. LL.

13

THE ORIGIN AND THE EVOLUTION OF THE NOTION OF AN ANALYTIC FUNCTION OF ONE VARIABLE

by Georges Valiron

PROFESSOR AT THE SORBONNE

IN the beginning the notion of a magnitude or quantity which is a function of another magnitude, i.e., depends on the value of this magnitude, was geometrical and practical; a century before our era Hipparchus gave a table of chords for a certain number of arcs of circles; in his work which has come down to us in Arabic translation as the *Almagest*, Ptolemy explained the construction of a table of chords. The Hindu Āryabhaṭa calculated the first table of sines in the 5th century of our era; at the end of the 10th century the Arab Abû'l Wefâ set up a new table of sines and introduced the tangent. The goal that these ancient analysts had in mind was a convenient way of locating the stars by means of spherical trigonometry. Their works, which did not penetrate western Europe until about the middle of the 15th century, were popularized by the astronomer Purbach; his student Müller, called Regiomontanus (1436–1476), was the first to use the decimal system in trigonometric tables. A century later Joachim Rhaeticus (1514–1576), a student of Copernicus, constructed a table of trigonometric sines for arcs at ten-second intervals with the values of the sines computed to fifteen decimal places. It is beyond doubt, even if they did not express it, that these clever calculators had a very clear idea of what we call the continuity of the trigonometric functions, and that they were perfectly aware of the degree of approximation of the values given in their tables.

The definition of logarithm was later to add to this conception of numerical functions defined geometrically the much more modern idea of a function defined directly in terms of a fixed correspondence between variable and function. In his *Triparty* (Lyons, 1484), Chuquet simultaneously considered the arithmetic progression $1, 2, \ldots, n, \ldots$ and the geometric progression $a, a^2, \ldots, a^n, \ldots$, and observed that by making the terms in the same place in these two progressions correspond, the sum of two numbers of the arithmetic progression would correspond to the product of the two corresponding numbers of the geometric progression. Chuquet's observation, developed by Stifel in 1544, led to the definition of logarithms. Napier constructed the first table of logarithms (Edinburgh, 1614); in order to do this he introduced intermediate progressions to permit a sufficient degree of approximation; these progressions were generated by fluxions, i.e., by continuous motion. Here again we find the notions of continuity and approximation.

Another discovery which was to play a fundamental role in the evolution of the theory of functions was made by the algebraists of the Bologna school. Toward the end of the 15th century, Scipione del Ferro (1465–1526) solved the incomplete equation of third degree by algebraic methods, but communicated his method only to a select few. It was not until 1535 that Cardano and Ferrari simultaneously published del Ferro's result along with the necessary completions that they had added: the solution of the general equations of the third and fourth degree and the introduction of new numbers, which Descartes called imaginary, but which we prefer to call complex; these allow the (real) solutions of such equations to be found in all cases. Bombelli gave the rules of operation for these new numbers in his *Algebra* (1572); he also introduced the use of parentheses in algebraic calculations and the use of letters to designate the coefficients of the powers of the variable in equations. He had a very clear idea of the notion of number.

The magnificent upsurge of mathematics in the 17th century was greatly aided by the works of the mathematicians of the 16th, whose importance has been brought out in the preceding remarks. It was aided by the symbolic notation in algebra systematically used by Viète in his treatise (1591), then by Descartes, in the unification and simplification of the signs for the operations, which was carried out step by step until it reached the form adopted today. An entire area of the new mathematics was still openly pragmatic: Descartes was the creator of mechanistic physics; Cavalieri, a student of Galileo, composed a geometry of indivisibles (1635) which contained the first hints of the integral calculus; Newton created rational mechanics. The successes obtained in the mathematical explanation of the phenomena of nature increased the zeal and the faith of the researchers tenfold.

The functions considered by Descartes in his *Geometry* (1637) were primarily algebraic functions defined by simple geometric curves, and the same was true of those treated by Fermat in the same period. After Descartes the idea spread that the methods of analytic geometry would apply not only to all the elementary functions—polynomials, rational functions, trigonometric functions and their inverses, the logarithmic function and its inverse the exponential—but also to general algebraic functions and to finite combinations of these functions. An algebraic function of x is the function $y(x)$ obtained by solving the equation $P(x,y) = 0$ for y; here $P(x,y)$ is a polynomial in x and y. With the exception of certain special values of the variable, called singularities, every function considered by Descartes has a geometric representation in intervals between these singularities by arcs of curves which admit a tangent. If we construct two rectangular axes Ox and Oy in the plane and if at each point P of Ox with non-singular abscissa x we erect a vector PM whose length y along Oy is the value of the function corresponding to this x, the locus of the points M is the representative curve. The practical determination of the ordinate y when the abscissa x is given introduced the notion of the indefinitely close approximation of numbers defined by certain algebraic formulas; in the case of square roots, as early as 1626 Cataldi considered the expansion of these numbers in infinite series and continued fractions. The arithmetic notion of the convergence of an infinite sequence of numbers to a known number is due to Wallis (1655); Newton, Leibniz and Jean Bernoulli introduced and utilized a few series, but in the 18th century many analysts used infinite sequences and series with great skill, while supposing the existence of a limit if the series converged and in most cases neglecting to prove this convergence. We had to wait until the beginning of the 19th century for the reappearance of rigor in these matters. The theorem on the existence of real or complex roots for an algebraic equation in number equal to the degree of the equation, which we call the theorem of d'Alembert, was known to A. Girard (1629), but a rigorous proof was first given by Gauss in 1799.

At its origin, the infinitesimal calculus also appealed to geometric or mechanical considerations. The mechanistic conception was clearly evident in the invention of the differential calculus by Newton (1669);

the function in his terminology is a fluent, i.e., a quantity which flows along with time; the derivative is a velocity or fluxion, useful in studying the variations in the fluent.[1] Leibniz, who in 1675 gave the definitions and the rules of differential calculus, and a little later the differential notation, took at first the geometric point of view; he considered the geometric elements to be attached to curves. The work of the Marquis de L'Hospital, whose efforts made the infinitesimal calculus known in France, was entitled: *Analyse des infiniments petits pour l'intelligence des lignes courbes*. However, the notion of function was extricated from secondary considerations at the end of the 17th century by Leibniz and Jean Bernoulli and given an analytic form which is still vague in the works of Jean Bernoulli, but which is made somewhat more precise in those of Euler. A function of a variable quantity for Jean Bernoulli is a quantity somehow composed of this variable quantity and constants (1718). For Euler, a function is an analytic expression composed in some way of the variable and constants.[2] In his *Introduction à l'analyse infinitésimale* (1748) Euler distinguishes among elementary functions of a variable those which are algebraic from those which are transcendental. The former include in particular the polynomials and the rational functions and may be either implicit or explicit: an algebraic function of x is said to be explicit when its value y is obtained as the result of a finite number of sums, differences, products, quotients, powers with rational exponents and extractions of roots involving x and constants; otherwise the function is said to be implicit. The transcendental functions are those functions which are not algebraic; such functions can be obtained by taking logarithms or exponentials, by raising the variable to irrational powers and by carrying out certain integrations. A function is single-valued if to each value of x there corresponds exactly one value of y, otherwise it is multiple-valued. In addition, Euler does not limit himself to the case in which the variable is real: coordinating and extending the work of his predecessors and contemporaries, he introduced a complex variable into the elementary functions and presented the algebraic theory of these generalized functions in a systematic form; the relation between the generalized exponential and the trigonometric functions is made completely clear in the formula

$$e^{ix} = \cos x + i \sin x, \qquad i^2 = -1.$$

[1] See the article by Pierre Brunet in this work. (Note of F. LL.)
[2] See the article by Jean T. Desanti in this work. (Note of F. LL.)

Euler showed that the exponential thus defined possesses the fundamental multiplicative property, and his formula contains that given by A. De Moivre in 1706. Euler developed the ordinary functions into power series, infinite products, series of rational fractions and continued fractions by methods whose rigor at times leaves something to be desired, but his virtuosity as a calculator is without peer.

According to Euler's definition, a function is an analytic expression. A function can be defined in terms of a power series in x, for example:

$$a_0 + a_1 x + a_2 x^2 + \cdots;$$

a series can be used to find the solutions of a differential equation; Euler thus integrated differential equations of the second order which are a special case of the Bessel equation. He introduced a cosine series

$$a_0 + a_1 \cos x + a_2 \cos 2x + \cdots;$$

in a paper on the inequalities of Jupiter and Saturn (1748), but this series is the expansion of a function which still belongs to the class of those that since Lagrange (1797) we call analytic functions. A function of one variable is said to be analytic if it can be expanded in a power series in $x - a$ in a neighborhood of each value a at which it is defined.

It was again a practical problem, the study of the vibrating string in musical instruments, which led Euler to consider functions more general than analytic functions. D'Alembert had shown in 1747 that the solution of the problem of the vibrating string depended on two arbitrary functions subject to certain restrictions. A year later, Euler discussed d'Alembert's result and found a solution which satisfied the known initial conditions given in terms of curves drawn in an arbitrary way and no longer the analytic curves of Cartesian geometry. Euler applied to the functions thus defined the methods previously restricted to functions given in terms of an analytic expression, without troubling himself to find out whether they allowed such a representation. When Daniel Bernoulli in 1753, using an earlier result of Taylor, gave the solution of the vibrating string problem by means of an expansion in a trigonometric series, the comparison of his method with Euler's gave rise to the question of whether a function defined by an arbitrary curve could be represented by a trigonometric series. In spite of the interest aroused by this question, the violent reactions of d'Alembert against the extensions of his method made by Euler, and the efforts of Lagrange (1762–1765), no significant progress was made until the beginning of the 19th century. Accustomed to operating successfully on functions defined analytically, the most illustrious mathematicians of the 18th century had no difficulty in conceiving of a continuous function which coincided in successive intervals with an arbitrary function, but they did not go beyond that and they would not accept the possibility of representing such a function by a single analytic expression—in the present case, a trigonometric series, In 1807 Fourier, generalizing formulas obtained by Euler and Clairaut for cosine series, determined the coefficients of the trigonometric series

$$a_0 + a_1 \cos x + b_1 \sin x + \cdots + a_n \cos nx + b_n \sin nx + \cdots$$

capable of representing a function $f(x)$ defined between 0 and 2π. These coefficients are given by integrals involving $f(x)$. Fourier was able to assert, without meeting any contradiction, that any function of a variable x defined in an interval was representable by an analogous series, which is not quite correct. Lejeune-Dirichlet rigorously proved in 1829 the statement made by Fourier, assuming conditions which were later to be made weaker, but Dirichlet's basic result was already enough to imply that the same analytic expression could represent distinct analytic functions in two different intervals, and it allowed mathematicians to conceive of functions which did not coincide in any interval with an analytic function, But to establish rigorously the existence of continuous functions which were not analytic in any interval, no matter how small, such functions had to be constructed. In 1854 Riemann, while discussing another question, gave an example of a continuous function which did not admit a derivative at any value of the variable represented by any fraction in lowest terms with an even denominator; such a function cannot be analytic in any interval. Going further, Weierstrass in 1861 considered the function defined by the cosine series

$$\sum b^n \cos (a^n x),$$

in which a is an odd integer and b a positive number less than 1, and showed that its sum is a continuous function for all x, but has no derivative at any value of x, provided that $2ab > 3\pi + 2$; he established in 1885 that any function continuous over a closed interval is the sum of a series of polynomials which converges on this interval. Euler's definition of functions was thus justified as far as continuous functions are concerned, the only ones considered in Euler's epoch; the antimony between this definition and the definition by means of an arbitrary curve was thus eliminated, but continuous functions were capable of much greater diversity than Euler and his contemporaries had imagined. These results of Riemann and Weierstrass gave rise at the end of the 19th century to the theory of functions of a real variable, in which we study and classify the most general continuous functions as well as the discontinuous functions. The definition of Dirichlet is used here: a function y of x is defined for a set of values of x, if to each number of the set values of x there corresponds a fixed value of y. The analytic functions constitute only a special class of functions; their definition can be extended to complex values of the variable, allowing their properties to be more completely elucidated, and they are frequently designated under the title of functions of a complex variable.

The complex variable was introduced by Euler in his research into the points of intersection of two algebraic curves (1748); the general theorem on the number of common points, which equals the product of the degrees of the two curves, was given by Bezout in 1766 and was later to play an important role in the study of algebraic functions. Euler contributed in another way to the construction of this theory of algebraic functions by his works on the elliptic integral:

$$u = \int_0^x \frac{dt}{\sqrt{(1-t^2)(1-k^2t^2)}},$$

in which k is a constant. The transcendental function defined by this integral arises in the study of the length of an arc of an ellipse as well as in numerous problems of mechanics; it had already been considered by Wallis (1665) and Jean Bernoulli (1691). Beginning in 1786, Legendre, who foresaw the importance of this kind of research, undertook an intensive study of this integral u and other elliptic integrals, integrals involving a rational function of the variable and of the square root of a polynomial of third or fourth degree. Even if his results were fragmentary and usually complicated, he had the satisfaction of keeping the question alive because of his efforts and of inspiring the research of Jacobi. H. Abel in 1826 had the idea of replacing the study of the integral u by that of its inverse function, considering x as a function of u and allowing these quantities to be complex, all of which, thanks to a formula of Euler, allowed him to to discover the double periodicity of this inverse function. This brilliant discovery gave rise to the theory of elliptic functions; these are doubly periodic functions of a complex variable which are analytic except at isolated points, around which they behave as rational functions; more exactly, if a is the value or point at which the function is not analytic, in a neighborhood of a it is equal to the sum of a rational function whose denominator is a power of $x - a$ and of a function analytic at the point a; a is said to be a pole of the function. Almost simultaneously, Jacobi, inspired by the works of Legendre, systematically constructed a theory of these functions, expressing them in terms of fundamental functions which play a role similar to that of the sine and tangent in the theory of simply periodic functions; the fundamental functions of Jacobi are defined in terms of rapidly converging series expansions. Later on, beginning in 1861, Weierstrass introduced simpler fundamental functions than those of Jacobi and extended the theory. The elliptic functions allow the integration of numerous functions and of new types of differential equations; their introduction has more than doubled the power of analysis. Their study played a considerable role during the second half of the 19th century in the development of the general theory of functions of a complex variable, whose present form was basically set up by Cauchy and Riemann.

The origins of the general theory of functions of a complex variable are to be found in the more practically oriented works of the great mathematicians during the second half of the 18th century. Clairaut, in 1743, in his theory of the shape of the earth, gave a condition which insures that the line integral

$$\int P\,dx + Q\,dy$$

computed on a curve joining two arbitrary points, does not depend on the curve, but only on the two points. In 1752, d'Alembert, in his studies on the resistance of fluids, considered a complex function $u(x,y) + iv(x,y)$, $i^2 = -1$, of two real variables x and y, and indicated a condition insuring that this function will have a derivative as the point with coordinates x and y approaches a given point, this derivative to be independent of the path taken. One need only consider the function $u + iv$ as a function $f(z)$ of the complex variable $z = x + iy$ in order to derive the Cauchy-Riemann condition: the function $f(z) = u + iv$ has a derivative if

(1) $$\frac{\partial u}{\partial x} = \frac{\partial v}{\partial y}, \qquad \frac{\partial u}{\partial y} = -\frac{\partial v}{\partial x}.$$

The celestial mechanics which had been constructed on the basis of Newton's law of universal gravitation also led Legendre, starting in

1782, and Laplace, starting in 1785, to study the attraction of spheroids and the equilibrium shapes of the planetary bodies. For this purpose they introduced the harmonic functions of three variables X, Y, Z, solutions of Laplace's equation:

$$\Delta V \equiv \frac{\partial^2 V}{\partial X^2} + \frac{\partial^2 V}{\partial Y^2} + \frac{\partial^2 V}{\partial Z^2} = 0.$$

The intensive study of functions of a single variable linked with harmonic functions, which were discovered by Legendre, Bessel (1816), Gauss, and Lamé (1837), and the research into representations of the most general harmonic functions by means of series formed in terms of the special functions [of a single variable], contributed to the broadening of the conceptions of the mathematicians of the first half of the 19th century and furnished a model for a great number of problems of mathematical physics. If they have second derivatives, the functions u and v considered by d'Alembert satisfy the two-dimensional equation of Laplace:

$$\Delta u \equiv \frac{\partial^2 u}{\partial x^2} + \frac{\partial^2 u}{\partial y^2} = 0, \qquad \Delta v = 0.$$

These are harmonic functions of two variables.

The cartographer's problem, which consists, in the simplest case, of representing the surface of a sphere on a plane in such a way that the representation is conformal, i.e., so that the angle of the image curves of any two intersecting curves drawn on the sphere equals the angle of these curves, can be solved in several ways. The simplest representation is stereographic projection, which is a geometric inversion. We can obtain all the other maps by starting with this one and performing all the conformal mappings of the plane onto itself. Such a mapping transforms a point with coordinates x, y into a point with coordinates $u(x, y)$, $v(x, y)$, and the conformity of the mapping implies that the relations of d'Alembert hold; u and v are then harmonic functions connected by these equations. Euler, then Lambert in 1772 and Lagrange in 1779, had treated important problems relating to maps; the general problem of the conformal representation of a plane on a plane was solved by Gauss in 1825. On the other hand, Wessel in 1797, then Argand in 1806, had given geometric significance to the complex numbers by letting a point with coordinates x, y correspond to the complex number $x + iy$. From this it follows that if the function $Z = f(z)$ is a function of the complex number $z = x + iy$, where Z has the form $u + iv$ and u and v are functions of x and y, then the function $Z = f(z)$ sets up a correspondence between the point with coordinates x, y and the point with coordinates u, v. If the function $f(z)$ has a derivative, then the conditions of d'Alembert are satisfied and the point transformation defined by the function is conformal.

Thus, when Cauchy, who had already given the first rigorous treatment of the theory of elementary functions of a complex variable in his *Analyse algébrique* (1821), in which the notions of limit, continuity and convergence are thoroughly clarified, undertook the general study of the theory of differentiable functions of a complex variable in 1825, two approaches were open to him: to consider the complex function $u + iv$ as a set of two harmonic functions of x and y related by conditions (1), or to consider it directly as a differentiable function of $z = x + iy$. His studies of the notion of fluids on the one hand, and of the series expansions of Lagrange on the other, caused him to avoid separating the function into its real constituents. The relations

$$\int u\, dx - v\, dy = 0, \quad \int v\, dx + u\, dy = 0,$$

consequences of Clairaut's condition, and valid along every closed curve C contained within a simply connected part of the representative plane of $z = x + iy$ in which $f(z) = u + iv$ has a continuous derivative, are transformed into the fundamental formula

$$(2) \qquad \int f_C(z)\, dz = 0.$$

From this formula Cauchy deduced another which allowed him to set up his remarkable theory of residues. This method of residues is applicable to the computation of a large quantity of definite integrals and to the research into the expansion of functions in series of rational functions; the theorem of d'Alembert on the number of roots of an algebraic equation is obtained in turn as a special case of the general formulas. Moreover, the above formula shows the identity of the definitions of functions of a complex variable either as differentiable functions or as analytic functions in the sense of Lagrange. From 1831 to 1850, Cauchy, followed by Liouville, developed and applied the new theory and method. In 1841 he founded his powerful theory of the calculus of limits, which we prefer to call today the method of majorizing functions; it shows that the solutions of analytic implicit equations and analytic differential equations are analytic; one cannot leave the domain of analytic functions by operations of this nature and we see why non-analytic functions were only able to appear in problems relating to partial differential equations. Refined by Briot and Bouquet in 1856, then by Picard, Weierstrass and Poincaré (1892), this method also demonstrates the continuity of the solutions of the equations considered under slight variations in the coefficients; it justifies the use of differential equations in mechanics and physics. Although Cauchy was mainly interested in single-valued functions, principally those that are called meromorphic, whose only singularities are poles, Briot and Bouquet considered multiple-valued functions, integrals of these functions and of simple irrational functions; and Puiseux (1850) studied algebraic functions in the neighborhood of their singularities, which are poles and points called critical algebraic points,[3] around which the function interchanges its various values. But the rigorous definition of multiple-valued functions requires an extension of the methods of Cauchy, which was supplied by the notion of analytic continuation.

The first idea of the notion of analytic continuation is found in Gauss, but it was first actually put to use by Weierstrass (1842) and Riemann (1851). According to Cauchy, a function $f(z)$ defined in terms of a power series in z is analytic at any point a within the circle of convergence C of this series; $f(z)$ is then also equal in C to the sum of a power series in $z - a$ which converges in a circle C' with center a which is not contained within C. If circle C' intersects circle C, the series in $z - a$ defines the analytic continuation of the function $f(z)$ in the portion of C' external to C. By continuing from one circle to the next, indefinitely if possible, we define the analytic continuation of $f(z)$; the function thus extended, which we continue to call $f(z)$, may be single- or multiple-valued; it is now defined in its natural domain of definition. Moreover, if the extended function is multiple-valued, we may suppose, in conformity with the ideas of Riemann, that the analytic continuation is carried out not in the single plane of Argand and Wessel, but in a system of superimposed planes defined in conformity with the type of extension. The function $f(z)$ is single-valued on this system of infinitely close planes which constitutes the Riemann surface of $f(z)$. For example, the function $\sqrt{z}$, $z = x + iy$ is single-valued on a system of two superimposed planes which are joined along

---

[3] As a simplification, we denote by a also the representative point of the number a in the Argand plane; point and number are synonymous.

the length of the half-line $y = 0, x > 0$; one of the pair of values of $\sqrt{z}$ is mapped into points of one of the planes which one can follow continuously to points of the other planes where the other values are mapped. Here we have a Riemann surface of two sheets, for which the point $z = 0$ and the point at infinity constitute algebraic critical points. Finally, utilizing a result of Goursat and following the formulation of Poincaré and his successors, we may say that an analytic function of one variable is a single-valued, differentiable function on the Riemann surface that it defines; the function has derivatives of all orders and can be expanded in a power series in $z - a$ around each point a at which it is defined; the points z are taken on the same sheet as a. The singular points of the function are the points b of the sheets at which the function is not defined, even though it is defined at points arbitrarily close on the same sheet. In general, those singular points which are poles or algebraic critical points defined as above are incorporated into the surface.

In his thesis (1851) and in his theory of Abelian functions (1857), Riemann, who had in mind the study of algebraic functions, only considered surfaces (Riemann surfaces) with a finite number of sheets and algebraic critical points (the only other allowable singularities being poles). By incorporating the singular points into the surface a closed surface is obtained, and Riemann studied the connectivity of this surface. To do this he removed a small portion of one sheet; the surface was then bounded by a curve C, and he determined the number of successive cuts that could be made without separating the surface into fragments by cutting along the length of a curve beginning with and ending on C; this operation was then carried out on the new surface thus obtained, and so on. The connectivity of the surface was then taken as the number of possible cuts plus one; the genus is this number of cuts divided by 2; it is a positive integer. The genus of an algebraic function $z(u)$ defined by $P(z, u) = 0$ is the genus of the corresponding Riemann surface; we also say that it is the genus of the [real] algebraic curve whose equation is $P(z, u) = 0$. The genus is invariant under birational transformation of the curve $P(z, u) = 0$ into another curve.

A passage from one sheet of the surface to another may be carried out by moving along lines issuing from critical points. Lüroth and Clebsch (1871 and 1873) showed that the construction of the surface can be simplified by drawing these lines in the appropriate manner. Their method allows the establishment of a topological equivalence between the surface and a surface of two sheets, then between the latter surface and a planar disk in which the two faces are distinguished and which is pierced by p holes, if p is the genus of the surface; following Clifford (1877), the disk may then be dilated in space. A surface of genus one, for instance, is topologically equivalent to a torus.

The consideration of the canonical curves into which the curves of genus zero and one are carried under birational transformation makes it possible to secure readily a result obtained by Clebsch as early as 1864. Every curve of genus zero is unicursal; the coordinates of any point on a curve of genus one may be expressed rationally in terms of elliptic functions of the same period. These propositions possess converses. The general problem of the parametric representation of curves of genus greater than one by means of single-valued functions could not be solved, assuming that it could be solved at all, by use of the simple functions known at the time; one of Poincaré's most praiseworthy achievements was his solution of this problem by means of the remarkable creation of the Fuchsian functions.

Riemann had constructed the theory of functions of a complex variable in his thesis not only by taking Cauchy's point of view, but also by decomposing the function into its harmonic elements. He stated and proved, although with insufficient rigor,[4] that harmonic functions satisfy a theorem which he called Dirichlet's principle: there exists a harmonic function defined in the interior of any given closed curve which takes on any set of values on C provided they insure that the function will be continuous on C. He deduced from this the theorem of conformal representation for functions of the complex variable z which can be stated in the following manner: Given a domain D of the simple plane bounded by a curve C, a point A of this domain and a direction T at this point, there exists a function $Z = f(z)$ which is analytic in D and which maps the domain D conformally and bi-uniquely into the interior of the circle of radius 1 centered at the origin of the plane of the complex variable Z; under this mapping A is sent into the origin and T is sent into the direction of the positive real axis. This celebrated theorem founded the geometric theory of functions of a complex variable; all the conformal representations obtained by translating A and the direction T may be obtained from one another by homographic transformation on Z. To every simple closed curve C bounding a domain D there corresponds an analytic

---

[4] Concerning this insufficient rigor, see the beginning of P. Montel's article. (Note of F. LL.)

function which effects the conformal transformation of D onto the above-mentioned circle and which is defined to within a homographic transformation.

Dirichlet's principle serves as the basis for a part of Riemann's paper on Abelian functions. Abelian integrals are integrals of the form

$$\int R(z, u)\, dz,$$

in which $R(z, u)$ is a rational function of z and u, and z and u are related by an algebraic equation $P(z, u) = 0$; these integrals are attached to algebraic curves. Abel had given for these integrals a proposition which extended a theorem of Euler on elliptic integrals; Abelian integrals had formed the object of study in some of Jacobi's works. Riemann's work on these integrals allowed him to solve, at the same time as Weierstrass, an inversion problem which introduces functions of several variables, called Abelian functions. They represent generalizations of elliptic functions. However, Riemann also solved another inversion problem; he showed that, given a priori a Riemann surface with a connectivity analogous to that of the surfaces which correspond to algebraic functions, i.e., a system of planar sheets ramified around a certain finite number of points, there exists a family of algebraic functions for which the given surface is the Riemann surface.

In another basic paper on the hypergeometric functions, solutions of Gauss' equation, Riemann, on the basis of his studies of the behavior of the solutions around their singularities, expressed these solutions in terms of definite integrals (called hypergeometric integrals) of these functions.

One of Riemann's immediate successors, Schwarz, defined the functions which give the conformal representation of the interior of a polygon on a half-plane or a circle (1866). A little later he showed that the function $Z = f(z)$ which gives the conformal representation of a half-plane on a domain bounded by a finite number of arcs of circles is given by the ratio of the solutions of a second-order linear differential equation with polynomial coefficients. In the case of a domain bounded by three arcs of circles, the differential equation is a Gaussian equation, and if the angles of the circular triangle satisfy certain conditions, the inverse function to $Z = f(z)$ is a function $z = \phi(Z)$ whose natural domain of existence is a half-plane or a circle. This function remains invariant under transformation of Z by elements of a group of homographic substitutions members of which are the products of two basic transformations. The given circular triangle is the fundamental domain of the group; it is transformed into analogous triangles whose union covers the half-plane or circle considered. This result may be applied, in particular, if we take for $f(z)$ the ratio of the periods of the elliptic integral

$$\int \frac{du}{\sqrt{u(u-1)(u-2)}}.$$

The function $\phi(Z)$ is then one of the modular functions which were earlier introduced in the study of the equivalent periods of elliptic functions; they were studied in some noteworthy articles of Hermite and were to be the object of direct treatment by Hurwitz (1881). These first examples of functions which, contrary to Euler's opinion, existed only in a portion of the plane, astounded mathematicians.

Around 1870 Schwarz and Neumann rigorously proved certain of the theorems that Riemann had based on Dirichlet's principle. Schottky studied the conformal representation of multiply connected domains (1875); Fuchs extended some of Riemann's results on Gauss' equation to general linear differential equations. These works, as well as those devoted by Klein starting in 1874 to modular functions and to the study of transformation groups, prepared the way for Poincaré.

In his courses and in his works on the theory of functions published in 1876, Weierstrass, as Méray also did, defined analytic functions a priori starting from power series expansions and analytic continuation; there is nothing more here than a difference in approach. Besides the results already given previously and a theory of algebraic functions independent of that of Riemann, rigorous studies on the convergence of sequences of functions, we are especially indebted to Weierstrass for a theorem on the decomposition of entire functions into products of factors and a pioneering study of analytic functions in the neighborhood of essential singularities. An entire function $f(z)$ is a single-valued function other than a polynomial which has no singularities in the finite plane; it is defined by a power series which converges for every finite value of the variable. Weierstrass extended to these functions the theorem on the decomposition of polynomials into linear factors involving their roots. If a is a zero of the function, he introduced a factor of the form

$$\left(1 - \frac{z}{a}\right) e^{g(z)},$$

the function $g(z)$ being chosen in such a way that the product formed of these various factors is convergent. The function $f(z)$ equals the product thus formed multiplied by an exponential factor. An entire function can be constructed which vanishes at predetermined points; a meromorphic function can be expressed as the ratio of two entire functions which may reduce to polynomials. This theorem of Weierstrass, extended in various ways by his successors, served as a model to Mittag-Leffler in his construction of a meromorphic function with predetermined behavior around its poles (1884).

The extension to entire functions of certain algebraic theorems relating to algebraic equations led Laguerre (1882) to single out the classes of entire functions whose decomposition into factors includes only exponential factors with polynomial exponents; Poincaré (1883) exhibited relations between the degree of these polynomials and the rate of growth of the entire function and between the rate of growth and the coefficients of the series expansion. He thus opened up a whole branch of the modern theory of functions.

Weierstrass' theorem on essential singularities was almost immediately surpassed by the celebrated theorem deduced by Picard in 1879 from the properties of the elliptic modular function. In its simplest form, this theorem asserts that any entire function assumes every finite value, with the possible exception of one value, an infinite number of times. It may be extended to general essential singularities; by an essential singularity we mean either an isolated singularity other than a pole, around which essential singularity the function is given by its Laurent series, or a singular point in the neighborhood of which the function is meromorphic. Picard in 1883 gave his uniformization theorem: An algebraic curve of genus greater than one cannot be represented parametrically in terms of single-valued functions of a parameter at least one of which possesses an essential singularity.

Poincaré's attention was drawn by the works of Fuchs to the problem of integration of linear differential equations with rational or algebraic coefficients. Guided by the theory of elliptic functions, he invented the functions that he called Fuchsian functions at one stroke (1881). These functions are more often called automorphic functions today. They are meromorphic functions within a certain circle, called the fundamental circle, which are unchanged by certain of the homographic transformations which leave the circle invariant; these transformations form a group called the Fuchsian group; Schwarz's functions constitute the simplest Fuchsian functions. Poincaré utilized some conceptions of non-Euclidean geometry of the Lobachevskian type—the geometry of Poincaré—in his research on Fuchsian groups and their fundamental domains, which are bounded by arcs of circles orthogonal to the fundamental circle. The construction of Fuchsian functions is carried out by taking functions analytic in the fundamental circle and multiplying them by simple factors, the transformations of the group; a Fuchsian function is the quotient of two of these theta-Fuchsian functions. The Fuchsian functions which correspond to the same Fuchsian group are connected by an algebraic relation whose genus, which may be arbitrary, depends only on the group. Conversely, Poincaré showed that the coordinates of any point on an arbitrary algebraic curve may be expressed in terms of Fuchsian functions of one parameter which correspond to the same Fuchsian group; the uniformization problem for algebraic functions was solved. The Abelian integrals attached to a curve are single-valued functions of the parameter. Moreover, the inverse function of a Fuchsian function is the ratio of two integrals of a second-order linear differential equation with algebraic coefficients. Such an equation, which Poincaré called a Fuchsian equation, may be integrated by means of Fuchsian functions. Poincaré and Klein proved that, given any linear differential equation in z with algebraic coefficients, say (E), a Fuchsian equation (E') can be associated with it such that z is a Fuchsian function of the ratio x of the solutions of (E'), and the solutions of (E) are single-valued functions of x which undergo linear transformations under the substitutions of the corresponding Fuchsian group; these are the zeta-Fuchsian functions. This result resolves to a certain extent the integration problem for linear equations (E).

Continuing his triumphal march, Poincaré announced in 1883 his general uniformization theorem for multiple-valued analytic functions. Given an analytic function $Z = f(z)$, there exist two single-valued analytic functions $z(t)$, $Z(t)$, such that for all t for which these functions are defined we have $Z(t) = f(z(t))$ and such that every part of the Riemann surface of $f(z)$ is described by the values of $z(t)$. It was only in 1907 that Poincaré, at the same time as Koebe, gave a complete proof of his theorem. His method was based on a general proposition which contained Riemann's theorem on conformal representation. Thanks to Poincaré, all the fruits of Riemann's wisdom were gathered and the geometric theory of analytic functions was definitively founded. After some later effort on the part of Poincaré's successors, the general notion of multiple-valued function, i.e., of analytic continuation or Riemann surface in the most general cases, now rests on a firm foundation.

Should we deduce from this that the notion of analytic function will not evolve further? The study of singularities other than poles and algebraic critical points which are incorporated into the Riemann surface has not been completely carried out. Picard's theorem only constituted a first step in the study of essential singularities, the simplest case; Borel, who established the relation between the researches of Poincaré and those of Picard, instituted a method of investigation which has led his successors to results of an unhoped-for precision, but the question of the nature of the corresponding singularities of inverse functions has not been carried that far. Painlevé has developed the notion of singularity in the general case in his studies on the solution of differential equations; their originality is comparable to that of Poincaré's most elegant work; however, some unexpected complications have arisen. It is possible that we shall be constrained to single out some restricted classes of analytic functions in which singularities of a somewhat prodigious nature do not arise. These functions separated out would be no less analytic. From the opposite point of view, it appears necessary to retain the generality of the notion of differentiable function of a complex variable given by Borel beginning in 1895 and the creation by Denjoy and Carleman of the theory of quasi-analytic functions. An evolution is possible in the direction of an expansion of concepts while preserving the property of cohesiveness relating the values of the function. But in a theory thus expanded, the functions introduced naturally in the course of centuries when mathematicians were studying analytic functions without knowing it would clearly retain their preeminent role.

14

THE ROLE OF FAMILIES OF FUNCTIONS IN MATHEMATICAL ANALYSIS

by Paul Montel

THE ACADÉMIE DES SCIENCES

THE study of classes of functions has been carried on for about forty years. We are now in a position to take a backward glance and take account of the results obtained, the role that this concept has played and the influence it has exerted on the development of modern analysis.

Perhaps the origin of the theory may be found in Riemann's famous error concerning Dirichlet's principle.[1] The harmonic function, which takes values on a given closed curve, minimizes the integral of the square of the length of its gradient extended over the planar domain bounded by the projection of the curve in the plane of the variables. Riemann took for granted the existence of a function which gives the integral its minimum value. Let us consider a sequence of functions defining surfaces which pass through the curve and giving to the integral values closer and closer to the minimum. If this infinite sequence of functions has a limit, then this will be the harmonic function sought. But this is not always the case, and it is important to find conditions which permit us to assert the existence of a limit.

This example shows once again the value that an error can have for scientific research. The importance of the role of error in the progress of human knowledge has not been contested since the discovery of America. La Fontaine made it the subject of his fable *Le laboureur et ses enfants*; Jules Romains devoted a perceptive and enjoyable comedy, *Donogoo*, to it. Restricting ourselves to contemporary mathematics, we need only recall the work of Henri Poincaré on

[1] Concerning this, see a passage in Valiron's article, p. 161 (F. LL.'s note).

divergent series, which was utilized profitably by astronomers who thought them convergent; the discovery of the Lebesgue integral which had its origin in the falsity of the assertion that every surface that can be mapped upon the plane is ruled; the research stimulated by the incorrect intuitive views of Painlevé on analytic functions which admit a perfect, disconnected set of singularities and on the number of asymptotic values of an entire function.

The fecundity of error in mathematics is natural enough. A correct proposition often exhausts the question of the relations among the elements it brings together. An incorrect theorem leaves the way open for a more penetrating analysis of the phenomenon and to the further development of the notions involved therein.

An infinite set of points always admits an accumulation point or limit point, i.e., we can always extract from the set an infinite sequence of points which converges to a unique limit. This property is the basis for proofs of a large number of properties of functions of points. However, there does not always exist an accumulation function for an infinite set of functions. Thus, from the very beginning, a profound difference separates the analysis of functions of points from the functional analysis which treats functions of lines or surfaces. But the proofs become easy in this area when the arguments, lines or surfaces possess the property of admitting a limit argument.

The research of Ascoli and Arzelà on equicontinuity, the specialized studies of Arzelà, Hilbert and Lebesgue on the Dirichlet problem, and those of the latter on the Plateau problem long possessed the character of isolated results and mathematical curiosities. But these researches involved sequences of functions which admitted a limit function. It therefore appeared natural and useful to learn to recognize whether a family of functions possessed this property; such a family is called a *normal family*.

The theory of normal families was therefore created in order to determine criteria allowing the assertion that a family is normal. But it has showed itself to be a particularly fertile source of ideas outside the functional domain in which it found its origin, and the study of conditions for normality is only a part, although an important one, of its task. This is due to the fact that the functions of a normal family are united by a common nature which imposes on the elements of the collection common properties made evident by the theory. This uniformity has shown itself to be especially remarkable in the case of functions of a complex variable.

Thus, the theory of normal families consists of the totality of properties which result from the fact that every infinite sequence has an accumulation element and from this fact alone. As a result of this, every criterion of normality leads to the same consequences; theorems of analysis, which seem to be consequences of special hypotheses, actually derive solely from the fact that the family is normal. This is the case in particular for the theorems which constitute what is called the Picard cycle. There are as many analogous cycles as there are criteria of normality.

Normality is a local property. If a family of functions is normal in a domain, it is normal at each point of the domain, i.e., in a small circle with its center at this point, and conversely. If a family is not normal in a domain, there are points in this domain at which it is not normal. These are *irregular points*. These irregular points possess many properties relative to the set of functions of the family which belong to the singular points of an individual function. These are the *collective singular points*; the natures of their singularity are distributed over the functions of the family.

The notion of normality for continuous, bounded functions of a real variable is equivalent to that of *equicontinuity*, which expresses the fact that a small variation of the argument leads to a small variation in each of the functions of the family.

The normal families of continuous functions of real variables have been used in the problems of the calculus of variations, in particular in those involving geodesic lines, Plateau's problem and Dirichlet's problem. They allow an intuitive form to be given to theorems on the existence of solutions of ordinary and partial differential equations. They play a useful role in the study of the almost periodic functions of Harald Bohr. Let $f(x)$ be an almost periodic function of the real variable x, defined and continuous for all values of x. Bohr's definition is equivalent to the following: the family $f(x + h)$, where h is any real number, is normal in the interval $(-\infty, +\infty)$ and this definition extends to almost periodic functions of a complex variable.

Other normal families of real functions involve harmonic or subharmonic functions or the positive set functions introduced by de la Vallée Poussin in his works on the "sweeping out" method in the solution of the Dirichlet problem.

It is in the study of complex functions that the theory has given the most surprising results.

The criteria for normality of a family of meromorphic functions in a domain are related to the existence of lacunary regions, i.e., regions never entered by the representative points of the values of the functions. The oldest criterion concerns the uniformly bounded functions; we can deduce from it that the existence of a lacunary region, no matter how small, leads to normality. But we can go even further: the existence of three lacunary points, i.e., three exceptional values, is enough to assure normality.

We are thus led to a classification of analytic functions based on the existence of exceptional values. The general theory concerns functions which have no exceptional value in common. Then comes the study of functions having a single exceptional value in common; since we can always suppose that this exceptional value is infinity, we see that it is a question of functions holomorphic in a domain. Next come the families of functions with two exceptional values in common. Finally come sets of functions having three exceptional values in common which always form normal families.

We can go even further. If we consider the value of the variable and the corresponding value of the function as the Cartesian coordinates of a point in the projective plane, meromorphic function represents an analytic curve, and the set of three exceptional values can be considered as a cubic [curve] consisting of three parallel lines. To say that the function never assumes these values is to say that the curve does not intersect the cubic. We can replace the cubic by any third-degree curve or even by an algebraic curve having at least three branches. We are thus led to the following criterion: The meromorphic functions whose representative curves do not intersect a given algebraic curve having at least three branches form a normal family.

Can we distribute the exceptional values between the function and its derivatives? For instance, is a family of holomorphic functions which do not assume the value zero and whose first derivatives do not assume the value *one* a normal family? For a long time it seemed that the answer should be yes, but no one succeeded in giving a proof. Florent Bureau and Miranda were able to establish it on the basis of their research and this question is the subject of recent articles of Valiron and Milloux.

An important application of the theory of normal families relates to the study of iterates of a given function. This theory has allowed us to pass from the study of local properties, the only type previously undertaken, to a global study and—thanks to the fundamental work of Fatou and Julia—to solve the problem completely in the case that the function iterated is rational. Fatou systematically introduced the set of irregular points of the sequence of iterates. It is a perfect set which coincides with the derived set of the fixed points of repulsion. It forms the boundaries of the domains, finite or countably infinite in number, in each of which the sequence converges to a constant; these are the domains of attraction. These boundaries may have the most diverse forms; they may be formed of analytic lines or of point sets no part of which is analytic, or even possesses a tangent.

Fatou's work on iteration led him to introduce the notion of normal family into the study of automorphic groups. He obtained the following result: In order for a group of homomorphic substitutions to be actually discontinuous it is necessary and sufficient that the family of these homomorphic functions be normal and that all its limits be constants.

When we consider the iteration problem, we naturally consider a family of functions, that of the iterates. The situation is not the same if we wish to apply the theory to the study of the indeterminacy of a uniform function around a singular point. It is then appropriate to substitute for the function in question an infinite sequence of related functions. This is arrived at by a method of partitioning the plane of the variable which allows us to distribute the set of values that we seek to classify over the functions of a sequence.

Landau's theorem-types can be roughly summarized by saying that a function cannot resemble a constant in a sufficiently wide region without being itself a constant. It was natural to suppose that a function cannot resemble a polynomial in a sufficiently large region without being itself a polynomial. It was found useful in the clarification of this question to introduce quasi-normal families of meromorphic or holomorphic functions. A *quasi-normal* family is a set of functions such that every infinite sequence of functions of this set contains a subsequence which converges uniformly except at a finite number of points.

The theory of entire functions has been enriched for a long time by research relating to the moduli of the zeros of these functions. The distribution of the arguments of these zeros was not the subject of a general study prior to the introduction of normal families. On the basis of this theory, Julia was able to undertake this study and to show that any entire function takes on all values with at most one exception in any angle, however small, which has a given half-line as interior bisector.

Such a half-line does not always exist for a meromorphic function. Ostrowski has shown that certain exceptional meromorphic functions of genus zero do not admit a half-line of condensation.

These are the features which characterize the individual physiognomy of each function and in some sense determine its personality. These functions have similar properties which it is useful to know in regions where they are analytic. The theory of normal families has also permitted us to explore these regions of analyticity from the point of view of both single-valued and multiple-valued functions.

We shall consider only two of the many extensions of the notion of normal family.

The first relates to the *complex families*, whose elements consist of systems of p-functions. To say that a meromorphic function admits an exceptional value is to say that the system of two entire functions whose quotient it is admits an exceptional linear combination which has no zeros. We would then go on to study the complex families which admit exceptional linear combinations. This research has allowed the examination of algebroid functions from the point of view of their exceptional values or, more generally, systems of numbers not in involution with the different determinations of the algebroid. Following the work of Painlevé, this theory was elaborated by Rémoundos, Varopoulos and Ghermanescu, and seems to have found its definitive form in the recent work of Dufresnoy.

The second extension concerns the functions of several variables. The majority of propositions relating to families of functions of a single variable are applicable to these, and Cacciopoli has established a fundamental theorem which facilitates this extension: A family of functions of several variables which is normal with respect to each of these is normal with respect to the set of variables.

We have reviewed here some of the progress that has been achieved in various branches of modern analysis on the basis of the theory of families of functions.

The proofs generally utilize indirect reasoning: we establish the impossibility of the negation of the proposition. This may doubtless be explained by the fact that we do not know all the criteria for normality and it is often necessary to carry on our reasoning without their aid.

We also should remark that the greater part of the results are of a qualitative nature; the reason for this is easily discerned. An algorithm or method of reasoning which is applicable in every case cannot attain the level of specificity desirable in each of them. Thus, expansion in a Taylor series, which is applicable to any analytic function, only rarely permits us, and even then with difficulty, to recognize the individual properties of each of these. Similarly, each of the qualitative propositions obtained in the theory of normal families corresponds to a complementary theorem of a quantitative nature, still unknown in a great number of cases, which challenges the insight of researchers. Much remains to be done along this line.

15

FROM CAUCHY TO RIEMANN, OR THE BIRTH OF THE THEORY OF REAL FUNCTIONS

by Jean T. Desanti

EULER wrote the following lines in his *Institutiones Calculi differentialis* (1755):

> Differential calculus gives rules which allow us to find the first differential of a given quantity. And since the second differential can be found starting with the first, the third starting from the second by the same operation, and so on, differential calculus possesses a method for finding all the differentials of any order.
>
> In the same way that we look for the differential of a given quantity in differential calculus, we develop a new calculus in order to find the quantity whose differential we know. This calculus is called integral calculus.
>
> The origin and nature of integrals can be explained as clearly as that of differentials in terms of the theory of finite differences which was developed in the first chapter. Indeed, after we showed how the difference of any quantity could be found, we posed the inverse question and showed how, being given the difference, we could find the quantity of which it was the difference. And taking note of its difference, we called this quantity the sum. Now when the differences become infinitely small, they end by becoming differentials. Under the same conditions the quantities which we called sums take the name of integrals [§§ 138 and 139].

Euler thought that he had thus resolved the difficulties raised by the infinitesimal calculus from its inception. His confidence in the all-powerful nature of pure operational technique allowed him, he thought, to escape the impasse in which the Leibnizian metaphysics of continuity had become trapped. However, the development of this very technique was soon to pose urgent problems which would force a reconstruction whose end product was modern analysis. We propose to give a brief exposition of one of the aspects of this development.

At the end of the 18th century it was apparent that neither Leibniz's metaphysics nor Newton's empirical dogmatism allowed a complete justification of the methods of the new calculus. If space is an absolute (*sensorium Dei*, as Newton and, after him, Clarke were to say), how are we to understand this infinite divisibility which constitutes its essence? We shall always stumble over the obstacle of the geometric reality of extension and the arithmetic reality of whole number. This is what Berkeley saw clearly; it is impossible to "imagine" this "flowing quantity"—it cannot "exist"; therefore it would be possible to justify the infinitesimal calculus only by its technical effectiveness, by a system of "compensating errors." This was also the point of view of Lazare Carnot, who wrote a pamphlet entitled *Métaphysique du calcul infinitésimal* at the end of the century. If, on the other hand, we take Leibniz's view that the infinite divisibility of space is linked to its ideal nature, this union can only be justified if space appears as a coexisting abstract order in which an absolute ideal law may be realized completely, the law of order and of continuity. But nothing is gained in seeking the absolute nature of space, if we must find it in the law by which the space is constructed; this law would then have to enable us to understand, beyond the abstract and relative continuity of "mathematical points," the real and absolute continuity of real objects, of the "metaphysical points." From this moment on we run into the classical difficulties of monadology and we again see arising that absolute nature of space that we had wished to reject as unintelligible: If there exists a multiplicity of monads, if these monads are real objects and if every multiplicity implies the coexistence of the elements of that multiplicity, then in the view of God, who embraces *uno intuitu* the existing infinite totality of the universe, there will exist an absolute order of coexistence of monads, i.e., by definition, an absolute space. It appears from this that the attempt to reify the law of continuity leads to an internal contradiction; the law of continuity is that of a process which is carried on indefinitely always according to the same ordering principle; as such it implies the ideality of the domain that it defines. This domain cannot be said to exist other than as a consequence of the totally ideal law which serves to define it. As soon as we take it as an absolute, we negate it, because if we suppose that the domain to which it applies has been realized as an existing totality, then we negate the ideality of this domain.

Therefore, if infinitesimal calculus is to find its justification, this cannot be found in a metaphysics of the infinite, in which it would reject the principles of its origin, the very principles which are the source of its power. Only a turning back toward the mathematical meaning of the operations that it defines, an effort to grasp them with complete rigor, could possibly allow a justification to be found.

Cauchy was to be the initiator of this movement. We cannot say even now that its goal has been completely attained.

The first notion that was to offer itself for remolding was that of continuity and, through that, the notion of functional correspondence. Nothing was more ambiguous in Euler's time than the expression *functio continua*. This was because in practice geometers since the time of Newton and Leibniz had worked with analytic expressions (algebraic and transcendental), to which they applied the operations of differentiation and integration and which they knew how to represent by means of continuous curves (except at a few points). They could not conceive of functions other than the ones they thus defined and manipulated, and they called these continuous functions, doubtless wishing to indicate that they were always well defined, infinitely differentiable, capable of expansion in a Taylor series, integrable and capable of being represented graphically by an algebraic or transcendental curve. To them the domain of analysis was clearly defined and its operations justified by the way in which they were carried out. This was, for example, the point of view of Lagrange. But it was no longer possible to maintain this empiricist view of continuity at the end of the 18th century, which, as we saw, turned its back on Leibniz's metaphysics. It began to be clear that the notion of functional correspondence contained much more than was apparent in the analytic relation in which it usually found expression.

This extension became necessary from the moment that the study of the equation of the vibrating string led mathematicians to show that certain non-continuous functions could be represented by means of a convergent trigonometric series of the form

$$\sum_{n=0}^{\infty} (a_n \cos nx + b_n \sin nx),$$

in which it was possible to determine the coefficients a_n and b_n. Such a result, due to Fourier, did not find its source simply in the methods of calculus, whose essentials were already known to Lagrange and Euler. It introduced a new general idea into analysis: Expansion in a trigonometric series allows one to represent a more general class of functions. It thus posed a new problem: Under what conditions is the trigonometric series associated with a given function convergent?

Now, an answer to such a question implies, as Dirichlet put it, "a substitution of ideas for computations," and, first of all, a definition of functional correspondence independent of any type of analytic expression. This is the definition that we find in the work of Cauchy and, later, Riemann. One says that y is a function of x if to every definite value of x there corresponds a definite value of y, no matter what form the relation connecting x and y takes. After this, in order to establish properties of functions in a completely general and rigorous manner, the particular form of relation, algebraic or transcendental, by which the function may be expressed can never be allowed to enter into the argument concerning these properties. However, by this step the notion of curve has likewise been generalized; the "mechanical curves" will have to be included in the domain of analysis. At the same time the notion of function is liberated from its exclusive recourse to geometrical intuition. There followed the rigorous definition of continuity, due to Cauchy, which is found in all the textbooks of analysis: A function $f(x)$ is continuous at the value x_0 of the variable if for any $\epsilon > 0$ we can find a number η, depending on ϵ, such that if $|h| < \eta$, then

$$|f(x_0 + h) - f(x_0)| \leq \epsilon.$$

We shall say that the function is continuous in the interval (a, b) if it is continuous for any value of x in (a, b), i.e., if the correspondence between ϵ and η is independent of x_0 (uniform continuity). This separation of the definition of continuity from geometric intuition was to achieve full fruition from the moment that Weierstrass gave an example of a continuous function without a derivative at any point. The expression

$$F(x) = \sum_{n=0}^{n=\infty} b^n \cos(a^n \pi x),$$

in which b is a positive number less than one and a is an odd number, does not admit a derivative for any value of the variable. Such a discovery not only revealed the autonomy of the methods of analysis with respect to geometry, it also showed at what point the spirit of rigor introduced by Cauchy gave rise to the risk of a new crisis. From then on analysis lacked the support of geometric intuition; it would have to be based on its own principles and use its own means to delimit its subject matter. If geometric intuition was no longer sufficient to guarantee freedom from absurdity, where was there to be found a sufficient criterion for clarity and reasonableness? From this crisis the theory of sets was to arise.

In fact, from 1807 (the date of Fourier's first work on trigonometric series) to 1873 (the date of Cantor's first work on the set of real numbers) two convergent movements led to the theory of sets. One, developing through the work of Cauchy, Dirichlet, Fourier and Riemann, concerned itself with the technical problem of the integration of functions. The other, developing through the work of Bolzano, Dedekind, Weierstrass and Cantor, concerned itself with the principles of analysis and the delimitation of its proper subject matter.

As far as the first point is concerned, limiting ourselves to only one example, we know how Riemann was led to generalize the operation of integration. Following Dirichlet, he asked himself under what conditions a function $f(x)$ can be represented by a convergent trigonometric series. To the three conditions of Dirichlet (1) the function must be infinite throughout the interval; (2) it must have a finite number of maxima and minima in it; (3) it must have only a finite number of points of discontinuity—Riemann added a fourth condition: Since the function must still agree with the series at the points of discontinuity, we must require that the function equal the average of these two limits at the points of discontinuity. However, in order to understand the relation between this new condition and the preceding ones, it was necessary to make a revision of the very principles of analysis. This was because none of the three conditions of Dirichlet was necessary and sufficient; it was therefore necessary to transform them into more general statements and, in order to do this, to refashion the theory of integration. This was the goal of the second part of Riemann's paper. Cauchy had made this possible by redefining integration without the use of geometric intuition: after having partitioned the interval (a, b) into subintervals δ_i by means of increasing numbers $a_i (i = 1, 2, \ldots, n)$, the sum

$$\sum_{i=1}^{i=n} (\delta_i f x_i)$$

is to approach a limit L as the maximum of the subintervals δ_i approaches zero. This limit is to be the definition of the definite integral of the function $f(x)$ over the interval. This definition applies to continuous functions, but it can also be extended without difficulty to discontinuous functions. If c is a singular point of a function, we can write:

$$\int_a^b f(x)\, dx = \lim_{\epsilon = 0} \left[\int_a^{c-\epsilon} f(x)\, dx + \int_{c+\epsilon}^b f(x)\, dx \right]$$

on condition that each of the integrals on the right-hand side approaches a definite limit as ϵ approaches zero. Riemann makes use of Cauchy's partition method and forms the two sums:

$$S = \sum M_i \delta_i$$
$$s = \sum m_i \delta_i$$

subject to the conditions:

$$M(b-a) \leq s \leq S \leq M(b-a),$$

in which m_i and M_i represent respectively the greatest lower and least upper bound of the function in each of the intervals δ_i. Each of these sums approaches a definite limit when the maximum of the intervals δ_i approaches zero. Let L and l be these limits. We set

$$L = \overline{\int_a^b f(x)\, \delta x}$$

(upper integral) and

$$l = \underline{\int_a^b f(x)\, \delta x}$$

(lower integral). If these two limits are equal, then one says, by definition, that the function $f(x)$ defined in (a, b) is integrable and that its integral is

$$\int_a^b f(x)\, \delta x.$$

Such a definition does not involve any special properties of $f(x)$; it is totally independent, in particular, of the mode of analytic expression of the function. However, in order to know the extent of its value we must ask ourselves under what conditions an arbitrary function can be integrable in this sense. Now, if we can no longer have recourse either to geometric intuition or to a fixed analytic expression whose properties can be examined, there remains only one way to carry out such an investigation; this is to examine the way in which the values of the function are distributed throughout the interval (a, b).

It would be sufficient for this to study the condition that $L - l$ approaches zero as δ_i approaches zero. It is clear that this condition is always satisfied if the function is continuous in (a, b). In fact,

$$S - s = \sum (M_i - m_i)\delta_i \leq (b - a)(M_i - m_i).$$

If we call the quantity $M_i - m_i$ the oscillation of the function in each interval δ_i and denote it by ω_i, then $S - s$ must always be less than or equal to $(b - a)\omega_i$. But to say that the function $f(x)$ is continuous in the interval (a, b) is to say that there exists a number η such that for $\delta_i < \eta$, ω_i is less than any given positive number. In particular, we can always choose η so that ω_i will be less than $\epsilon/(b - a)$. $S - s$ will then be less than ϵ. Such an argument certainly implies a deepening of the definitions of Cauchy; it gives a real meaning to the notion of continuity in the way that he introduced it. From now on continuous functions will no longer enjoy privileges of a theological nature in analysis; if they are integrable, it is no longer, as Leibniz believed, because the same law of continuity rules over the organization of the totality into which the infinitely small differences are integrated. If this were the case, then differentiation and integration would be completely symmetric and every continuous function would be infinitely differentiable, which is not at all the case. From now on, on the contrary, integrability will appear as an expression of the set of properties of the domain of values of the variable—continuity is one of its properties among others. Under these circumstances, it is understandable why we form the two sums $\sum M_i \delta_i$ and $\sum m_i \delta_i$ a priori. It is because the two quantities M_i and m_i allow us to establish valuable properties; if we suppose that under successive divisions the interval δ_i tends to reduce to a point, then each of the numbers m_i and M_i approaches a unique limit λ and Λ, respectively, that we call the minimum and the maximum of the function at the point x_i. Their difference $\Lambda - \lambda = \omega(x_i)$ defines the oscillation of the function at this point. From this point on, Cauchy's definition of continuity appears as a specialization of a more general definition, that of the magnitude of the discontinuity at a point. The quantity $\omega(x)$ can be only positive or zero; if it is positive and equals ϵ, then we say that ϵ is the magnitude of the discontinuity at this point. A continuous function then appears as a member of a special class of discontinuous functions, those in which the magnitude of the discontinuity is zero. Now the condition that $\lim (\zeta - s) = 0$ as the maximum of the δ_i approaches zero expressly depends on the mode of distribution of points in (a, b) at which $\omega(x)$ is greater than ϵ. In particular (and this is almost the very language of the condition for integrability stated by Riemann), if we denote by σ the total length of the intervals (taken to be non-intersecting) in which the total oscillation is greater than or equal to ϵ, it is necessary and sufficient that σ approach zero, ϵ being fixed, as the intervals δ_i approach zero. Thus discontinuous functions will then be integrable if the set of their points of discontinuity can be contained in a finite number of segments, the sum of whose lengths is as small as we please. In conformity with the language introduced by Paul du Bois Reymond, we say that such a set constitutes an integrable group. Here we see the meaning of the operation of integration as defined by Riemann; it is the synthetic translation of a system of properties which local analysis of the set of points of the interval over which the functions varies allows us to state. But the tool of this local analysis is no longer, as in Leibniz's time, differentiation (which always leads to a definite analytic expression with its corresponding geometric representation); this tool is to be a rigorous study of the properties of the function in the neighborhood of the points of discontinuity, correlated with a study of the way in which the set of these points is contained in the interval over which the function varies. This is to say that such an extension could not be complete until it was possible to find a new method which allowed us to study, a priori, in a given interval the relations of inclusion or exclusion found by analysis within the generally infinite set of points which constitutes this interval. Thus the definition of the notion of integral led to the theory of sets.

But such an extension implied a new kind of research; the domain of analysis had to be defined without any appeal to geometric intuition about continuity. As early as 1799, Bolzano, examining the proofs of the theorem, "Every rational algebraic function of a variable can be decomposed into real factors of the first or the second degree," had introduced the lemma: "Let M be a property of the numbers less than the number u within the interval (a, b). Then it is possible to find a number U in (a, b) which is greater than all numbers such as u." If $u + v$ is a number which does not have the property M, and if we form the sequence

$$u, u + \frac{v}{2^m}, \quad u + \frac{v}{2^m} + \frac{v}{2^{m+n}}, \ldots,$$

either the number U is one of the terms of the sequence or it can be defined as the limit of the sequence. In the second case it is necessary to show that the sequence is really convergent and that its limit is U. Bolzano thought he had proved the result by showing that it is possible to calculate the limit of the sequence to within an arbitrary approximation. But the sequence of approximations must itself form a convergent sequence in this case. How are we to show this convergence? It seems that an appeal to intuition is necessary to avoid a vicious circle.[1]

Hence the conceptions of Dedekind, which were to follow a rigorous meaning to be given to the idea of the limit of a convergent sequence.

Let E be the set of rational numbers. This set forms a "field," i.e., if we combine two elements of this set by addition, subtraction, multiplication or division, the result of the operation is still a member of the set (except if the divisor is zero).

Beyond this, the set is characterized by the three following axioms: (1) Any element b of the set is contained between two elements a and c of the set, one preceding and one following b, and we have the relation $a < b < c$; (2) given an element a_i of the set, it is always possible to choose an element a_{i-1} in the neighborhood of a_i such that $a_i - a_{i-1} < \epsilon$, where ϵ is any given positive number, no matter how small; (3) each element separates the set into two classes such that any element of the first class is less than any element of the second.

Let us consider the operation defined by the last of these axioms. If we also take axiom (2) into account, it allows us to construct a set E' each element λ of which possesses the property of dividing the set E into two classes A and B defined by axiom (3) in such a way that A has no element greater than all the other elements of A, and B has no element less than all the others in B. λ, the number which separates these two classes, constitutes a cut in the set of rational numbers; we say that it defines an irrational number. It is a consequence of axiom (2) that the set E' of cuts constitutes a field. But it is impossible to add a new element to E' by this procedure without having it coincide with one of the elements previously defined. One can define the same operations on E' as on E. Thus, if we add to the field E the field E', we obtain a new unextendible field defined by the above three axioms. This field is the set of real numbers.

Now we can always show the existence of the limit of a convergent

[1] See Jean Cavaillès, *Remarques sur la formation de la théorie abstraite des ensembles,* Hermann, pp. 12 and 13.

sequence without having recourse to geometric intuition. Let the sequence $u_0, u_1, u_2, \ldots, u_n, \ldots$ be convergent.

If $n > N$, and N is allowed to be arbitrarily large, u_n approaches a number λ which possesses the following property: It constitutes a cut separating two classes of real numbers, the first consisting of the set of numbers less than u_n for $n > N$, the second consisting of all the remaining numbers. The number λ is therefore well defined.

Thus, after a long detour in the theory of functions, analysis returns to its starting point: the discovery of the irrationals. What scandalized the Pythagoreans now appears as the essential justification of the fundamental operation of analysis: the passage to the limit. But if in order to attain this rigor it was necessary to escape from both the deceptive evidence of spatial intuition and the fetters of the analytic expression, it was still essential to forge a new instrument which would permit us to thoroughly investigate and unambiguously determine the properties of the set of real numbers. The creation of the theory of sets as a means of analyzing the infinite became necessary beginning with the last third of the 19th century. In it the ideas of Cauchy, Riemann and Dirichlet were extended by those of Cantor. But far from settling all the problems, Cantor's discovery was to engage mathematics in a new adventure.

16

THE INNATENESS OF THE TRANSFINITE

by Arnaud Denjoy
PROFESSOR AT THE SORBONNE,
MEMBER OF THE ACADÉMIE DES SCIENCES

THE infinite presents itself to us under diverse aspects, more or less capable of being given a concrete representation.

First of all there is the infinity of boundless magnitude.

Because the immense mutual separation of the stars is known to each of us, the most uneducated man, the man most lacking in numerical intuition, is able to discover in the spectacle of the clear sky on a moonless night, between the luminous rays which connect his eyes to the glittering stars, the presence of bottomless abysses in which his gaze loses itself without ever encountering a visible object. The infinity of space is a concrete infinity, perceptible and immediate.

Another infinity of magnitude, this one of a more abstract nature, and yet equally commonly the object of meditation of the most uneducated, is the infinity of time, the eternity "which had no beginning and will have no end."

These two infinities, that of space and that of time, inspire the soul with emotion, doubt and fearfulness; a human being and a human life are felt to be so crushed under their dimension, their duration, that both become imperceptible points in the scale of the universe and its epochs! A feeling of insignificance and of the vanity of our existence weighs us down. We would be consigned to inconsolable distress if—Pascal tells us—thought, which inflicts these desperate certainties upon us, did not offer us at the same time a refuge which allows us to live with the pathos of our fate.

The mathematician gives to these infinities of magnitude a more rational significance, more precise and, contrary to the above, completely stripped of any pathetic or emotional overtones. For him, the infinity of the universe is stated in the following way: Whatever unit of distance is adopted, meter, kilometer, circumference of the earth or diameter of its orbit, light-year, no matter how great the unit, however great the number of lengths placed end to end that we can conceive, each equal to the unit chosen, the depth of celestial space will always exceed the greatest distance that our imagination evokes.

And similarly, the infinity of time is to be understood thus: If we choose an arbitrarily long unit, hour, year, century, the millions of centuries of the geological periods, if we add up a series of intervals of duration equal to the unit agreed upon and as large in number as we are able to conceive, the double infinity of time means that in the course of epochs this immense period of time extended into either the future or the past will finally be exhausted.

But there is another infinity without end.

It is that of smallness, not one which is suggested to us by our senses, but one to which Pascal's genius leads even minds to which the conceptions of numerical analysis are most foreign, and that by means of images whose dazzling intuitions are confirmed to some extent by modern physics. After having sketched a picture of stellar immensities, the magnitude of planetary revolutions carried out around stars serving as so many suns, Pascal invites his reader to gather up his thoughts thus deployed far and wide in order to concentrate on the imperceptible drop of fluid contained within the leg of a mite. He will discover in this tiny particle a repetition of the stellar universe with its suns and their cortèges of planets.

But—and here the mathematician reappears—in these planets of inconceivably small dimensions Pascal asks us to see once more the habitat of a living fauna, reproducing on this vertiginously reduced scale our terrestrial animals. And we are to conceive of the same vision within a mite of this new domain. This is the progression decreasing without limit familiar to geometric reasoning.

The infinity of magnitude, also conceived as a limitless succession of repeated extensions, eventually exceeds everything and is exceeded by nothing. The infinity of smallness abandons all things progressively; nothing remains within it. Outside of the infinity of greatness, within the infinity of smallness, there is only emptiness.

But finally we have a third kind of infinity, the one which will serve as the preface to our study. We characterize it by means of an adjective whose use here may be surprising. This is the *bounded* infinity, the least intuitive of the three, a logical infinity which has been the basis of many paradoxes; the most celebrated of these will become for us the fundamental argument for the theory of the transfinite. Ought we to be astonished that the limit of an infinity merits the name and plays the role of a transfinite, i.e., of an infinity beyond infinity? Let us recall the specious reasoning of Zeno of Elea.

Achilles, the fleetest of the Greeks, is pursuing a tortoise that is crawling away from him. Zeno asserts that the hero will never catch up with the animal; since (he says) at the instant of starting the man finds himself a certain distance from the animal, a second instant must elapse before he can halve this distance. Then he will have to wait for a third instant to elapse for the distance, again halved, to be one fourth of the initial distance. And then a fourth instant when the distance from Achilles to the tortoise becomes one eighth of the initial distance. Here we say: *et cetera*.

This *et cetera* expresses several ideas: first, the assertion that we possess a perfect conception of the entire sequence of the natural numbers; next, an admission of our weariness even though we have scarcely begun the task of examining the entire sequence; finally, the profession that we consider the entire project to be absurd.

Zeno's argument is analogous to the following:[1] In order to catch the tortoise Achilles must pass through a succession of instants, each of which has an index indicating its rank in the succession. Furthermore, these indices are the increasing sequence of natural numbers considered as ordinals. The successive positions of Achilles would therefore also correspond to all the natural numbers and the runner would have to pass through all of these in order to catch the tortoise.

Since the distance from Achilles to the tortoise is cut in half at each successive instant at which this length is measured, and if we agree that the speed of both the runner and the tortoise does not vary, the lengths of these instants must decrease by half at each stage. If we were able to photograph instantaneously the tableau presented by the man and the animal at each of these instants, we would note that the movements of the two appeared to slow down more and more, to become more and more congealed at each stage. And if we enlarged each of these photographs twice as much as the preceding, the distance from Achilles to the tortoise on the print would remain constant, but the dimensions of the two partners in the action would be doubled each time.

[1] To be precise, Zeno grants infinite divisibility to space but denies it to time

Clearly, we would very quickly have to break off these operations because of their physical impossibility. The interval of time separating two successive snapshots, cut in two each time, would soon become less than the time required by the most rapid film available to record an impression. This last observation already attracts our attention; its full significance will appear later on: the infinite succession of natural numbers, which the mind cannot conceive, is doubtless also unrealizable in nature.

Moreover, Zeno's reasoning shows the logical necessity of Achilles' complete immobility, for, no matter how close we placed the tortoise to Achilles at the starting time, this millimetric distance separating the warrior from the animal would always be divided into an infinity of successive stages, and the difficulty of covering this endless series of paces would always, according to the paradox, keep the pursuer from attaining his goal. There is no point, however close to Achilles, which he could reach. The hero is therefore incapable of motion.

Does modern physics lead us away from this conception of the incompatibility of nature and infinity? On the contrary, it seems to lead us back to it. Certain of the most illustrious theoreticians now ask themselves whether movement is not carried out via a discontinuous succession of distinct positions. According to them, the impression of continuity is an illusion which the observations of our eyes are not sufficiently perceptive to dispel. Moreover, relativity leads astronomers and physicists to conceive of the universe as finite. The continuum and also infinity would then be fictions created by our minds and refuted by reality.

The Marquis de Laplace, having just completed the great treatise in which he developed his sytem of the formation of the planets, presented the work to Napoleon. The latter, having skimmed through this cosmology, said ill-humoredly to the astronomer: "What use have you made here of the hand of God?" "Sire," Laplace replied, "I had no need of that hypothesis."

People who believed in the truth of Euclid's axiom hoped to arrive at the absurdity of its negation by adopting the latter as an axiom and then pursuing the logical consequences of this counter-axiom. Not only were their hopes of finding a contradiction disappointed, but later on mathematicians were able to construct geometries in which Euclid's postulate is false. Moreover, these geometries are perfectly consistent, for they amount to transposing the properties proved in

Euclidean geometry for certain families of curves by changing the name of the latter to "lines," in the same way as we would change key in music. These linear properties (i.e., those belonging to lines) are then found in non-Euclidean geometry.

An atheist would find it interesting to collect and then to submit to logical analysis the properties, those humanly conceivable traits, which God must possess as soon as his existence is postulated.

Since nature presents to us the spectacle of astronomical immensities, populated, though not uniformly, by molecules of unimaginably tiny dimensions, our philosopher would observe that the problems of enumeration presented to us by the universe would involve integers of a magnitude sufficient to confound our imaginations. Yet this nature seems to be finite both in its gigantic as well as in its minuscule aspects. Thus he would conclude:

The mind of God dallies in the innumerable, but it can never escape the finite.

Human imagination tries in vain to assault the innumerable. It quickly drops exhausted. In order to evade this challenge it has created infinity and the continuum.

The discontinuous and the finite are the modes by which God accomplished his task. The continuous and the infinite are the modes resorted to by our intellects, which are incapable of investigating the gaps in nature and of imagining the excessively numerous accumulations of its building blocks.

Let us return to our discussion, to Achilles and his tortoise. Let us suppose that we have a clear idea of the unending sequence of natural numbers—numbers of an ordinal character, since each determines the rank of an instant in a succession ordered in time. Achilles passes through the entire infinity of stages, since it is evident that he will catch the tortoise. Suppose that within the sequence of instants which we have considered up to now, we want to denote by a symbol the rank of the instant at which Achilles catches the tortoise, but has not yet passed him. This rank must follow all the ranks indexed by finite integers: the first, the second, the third, . . ., the tenth, . . ., the nth (as mathematicians say), etc. The rank of this instant of success is the first to follow all the ranks of the instants marked off in the course of the pursuit. Following the geometer Cantor, we shall denote this rank with the aid of a fictitious number, also of an ordinal character, which will be by definition *the first transfinite number*. It is transfinite because it follows all the finite ordinal numbers. And it is the first because no other instant marked off before it follows all the instants marking the stages of the pursuit. It is customary to denote this ordinal number by the Greek letter ω (omega). Achilles catches the tortoise at the ωth instant considered.

But once analysts have learned to carry out an operation they never cease to apply it; for them, too, only the first step is difficult. We now place a second tortoise in front of the one that Achilles has just caught; as in the pursuit of the first one, Achilles will have to pass through an infinity of instants to catch him. The ordinal numbers which mark these intermediate instants are $\omega + 1$, $\omega + 2$, $\omega + 3$, etc. These instants will be the $(\omega + 1)$st, the $(\omega + 2)$nd, etc. And the instant at which Achilles catches the second tortoise, which comes after all these instants just specified, will have the rank "ω plus ω," written $\omega + \omega$ or $\omega \times 2$ (two times ω).

But while Achilles was pursuing his first tortoise, we could have strewn his path with smaller and smaller tortoises at each of the stages where we had observed him: At the original halfway mark, a tortoise half as large as the object of the pursuit; at a distance one fourth of the initial distance from the main tortoise, a tortoise one fourth as large, etc. We reduce by half the dimensions of the new tortoise placed at a given point each time that we reduce by half the distance that Achilles has to cross.

While he strives to catch the principal tortoise, Achilles will have the task of pursuing successively all the intermediate tortoises. As the Eleatic philosopher has taught us, we shall interpolate an infinity of instants between the instant of Achilles' start and the one at which he passes the position of the first small tortoise. These instants will accumulate on approaching the first stage, which is now the instant of rank ω. Starting from this point, we interpolate yet another sequence of instants as we approach the second small tortoise. These will accumulate as we approach the second stage. This instant will now have the rank "two times ω." Then, as we move past the following tortoises, we shall find instants with ranks denoted by "three times ω," "four times ω," etc. And these instants tend toward the instant at which Achilles catches the principal tortoise, the one which was the original goal of his pursuit, and behind which, just as Hop O' My Thumb scattered his pebbles, we placed an infinite chain of tortoises, successively becoming tinier until they were soon imperceptible. The rank of the moment at which Achilles catches the big tortoise will now be denoted by the first ordinal number following the sequence: one times ω, two times ω, three times ω, four times ω, etc. This number will be ω "taken ω times." We write it $\omega \times \omega$ (ω multiplied by ω) or also ω^2 (ω squared).

And we may continue thus endlessly.

Does the transfinite number ω pose a problem for the mind? On the contrary, it is a comfort to the mind. When we try to imagine the succession of instants during which Achilles reduces his distance from the animal by half at each stage, we become fatigued with the effort almost immediately and we imagine at once the instant at which the runner reaches his goal; we think perhaps of 1, 2, 3, 4 separately, and then right away of ω.

This thought of ω, even though one does not name it, dominates all of mathematical analysis. The latter is founded on the notion of convergent sequence, that is, a succession of numerical stages analogous to those of the distance between Achilles and the tortoise he pursues, but with the difference that another Achilles is placed in front of the animal and strives to reach him by running backwards. During this action the tortoise will be immobilized. We shall also suppose that the stages observed will be sometimes those of the first runner and sometimes those of the second. In any case the convergent sequence denotes an infinity of stages indexed by the increasing sequence of natural numbers. But we shall only think clearly of a few of the simplest terms of the convergent sequence before we think of the result, which is only reached at the indefinite and inaccessible final stage of the operation. We think of what is called the limit of the sequence.

Here again, it is the rank ω, the immediate successor to all the finite ranks, that we hasten to consider, whereas we are discouraged or put off by integers that are somewhat large.

No idea is simpler or more restful to the mind than that of the transfinite. It enables us to pass over a tiresome, indefinitely long row of finite integers. It takes us just beyond this succession, as the flight of a bird or of an airplane eliminates a monotonous and interminable journey.

But the transfinite poses problems in the face of which the logic formerly used by mathematicians has proved helpless.

No matter what increasing sequence of transfinite numbers we have determined, no matter how this was done, we can always (at least it is so supposed, although falsely) imagine a transfinite number immediately succeeding all of these. The set of transfinite numbers must therefore be such that it remains unchanged by the addition of an element which was not present previously. It is not at all surprising that such a conception led to inexplicable logical difficulties. The famous French geometer Émile Borel strove often and well to eliminate the false belief that it is possible to conceive of a transfinite number of a nature different from that given to us by the example of the instant at which Achilles attains his goal in his pursuit of the tortoise. Mathematical logic had not experienced a crisis comparable to that caused by the discussion of the transfinite since the long dispute over the infinitesimal, which lasted for a century and a half.

This irresolution, this confusion in the reasoning of the science regarded as the most certain of all, thanks to its character of being in essence purely deductive, gives us living proof of the fragility of our least vulnerable certainties. Or rather the relativity of our logic, the impossibility of asserting that any method of proof examined and validated by it will be free of even a gross error, this inescapable, organic infirmity of our most rigorous thought, could give rise to a prejudicial suspicion regarding the certainty of any part of our knowledge.

Infinity, an infinity of purely human origin, turns out to be a stumbling block to our reason, which believed it had found in it a useful and submissive servant. The irrational numbers, popularly called "incommensurables," i.e., the continuum obtained by applying to the rational numbers (the fractions) the idea of an infinite sequence whose members change less and less; the infinitesimal introduced along with the differential and integral calculus by Leibniz and Newton; convergent infinite series; more recently, the transfinite numbers—all these diverse forms under which the notion of infinity has made its appearance again and again in mathematics—all cast the logic of their time into a state of bankruptcy, even that logic which the geometers of the time took to be an infallible mechanism capable of producing only unshakable truths.

And each time logic has had to laboriously and painfully seek the supplementary rules which it lacked by groping blindly after them for a long while. It is not by means of brilliant a priori insights, but rather by a careful analysis of the causes of its failures and successes that logic has modestly, humbly, learned the rules of exact reasoning applicable to these questions of a type completely new to it.

Reasoning, in mathematics as elsewhere, must be retempered periodically by empirical experience in order to acquire the strength which it cannot find within itself.

Science, even in mathematics, which appeared to be its asylum of certitude, no longer presents itself to our eyes as an unshakable structure erected on a foundation of bedrock, built of materials which defy time, constantly acquiring new, but always definitive, levels and annexes. Science is a ship voyaging on shifting seas. The wave which carries it is not rigid. It undulates and propagates itself, then cedes its place to its successor. But, what is essential, the ship floats and moves in the desired direction, tossing overboard as it goes the débris of obsolescent or controverted knowledge, constantly taking on a cargo of newly won information. And we must renounce the vain and ingenuous dream of ever reaching the shore of a single, primitive truth.

Science has fallen from the pedestal overlooking the universe to which we thought we had raised it, and has united itself with the universe, taking on the latter's characteristics of instability and perpetual evolution. Its ossified framework is the most sterile part of it; its most perishable elements are perhaps the most fecund. Daily, science sees itself betrayed by the principles which had appeared to be the most secure and it must ask the supreme oracle of Sibylline nature to reveal new guides in which it may place its confidence. Science now seeks its light in the darkness which confronts it, no longer in the illuminated regions it has passed by.

D. GROUP

Number, space and function are the three caryatids whose broad shoulders support classical mathematics. Certain of the more modern disciplines—the theory of sets, matrix and tensor analysis—are only indirectly involved in this book, the first in the articles of Fréchet, Eyraud and Denjoy, the others, of a less purely mathematical nature, in the article of Louis de Broglie devoted to the relations between mathematics and physics. Other aspects, scarcely suspected at the beginning of this century, are at least touched upon in the articles of Bourbaki, Dieudonné, Godement and Weil. In spite of its high degree of abstraction, we felt that it was impossible to pass over in silence, if not the theory of groups, at least the notion of group.

The great generality of this conception, a fruit of Galois's genius, in the first half of the 19th century, allows it to play a role in the most varied areas of mathematics, to relate their existence and mechanism to the structure of the human mind, perhaps even to the very architecture of the universe. The notion of group corresponds to a fundamental aspect of intellect: the aptitude for combining every new idea in every way possible with prior knowledge in such a way as to exhaust all the possibilities. But there is something autonomous and closed about the very nature of a group, and therefore in its function, which imparts to it, along with a concentrated efficacy, limits beyond which stretch new lands of Canaan that are no less marvelous. And is it not another of the powers of the human mind that it can escape the notion of group in order to gain new motions, foreign to those with which it is already familiar, with the goal of constructing "supergroups" (if we may be permitted this term) of ever wider import!

André Lentin is a young mathematician who is sensitive to all aspects of human culture. His chief interest is directed toward modern algebra, whose first and primary model has long been the theory of groups. His intelligent presentation will allow the reader to familiarize himself with the powerful instrument which has renovated mathematical thinking and whose revolutionary consequences have not yet lost their vitality.

F. LL.

17

THE NOTION OF GROUP
ITS POWER AND ITS LIMITATIONS

by André Lentin

For some sixty years the notion of group has had such extraordinary success in all branches of pure and applied mathematics that it is appropriate to devote a study to it.

We say that a finite or infinite set G of elements is a group if the following four conditions are satisfied:

1. There exists a *law of composition* which assigns to each pair x, y of elements (taken in this order) another element called the *product* of the first two.

We shall state and denote this product as is done in ordinary multiplication:

$$(1) \qquad x \cdot y = z;$$

2. The law is associative, i.e.:

$$(2) \qquad (x \cdot y) \cdot z = x \cdot (y \cdot z),$$

where the parentheses designate the multiplications to be carried out;

3. There exists an "identity element" e (or unity) such that

$$(3) \qquad e \cdot x = x \cdot e = x$$

for any x;

4. Every element x admits an inverse element (or reciprocal) x^{-1} such that

$$(4) \qquad x \cdot x^{-1} = x^{-1} \cdot x = e.$$

The reader may easily verify the fact that the set G_1 of positive real numbers (excluding zero) forms a group under the usual multiplication, and also that the set G_2 of all real numbers, positive, negative and zero, forms a group under addition. 0 is the identity in G_2; the inverse of a number x is its negative $(-x)$.

We are now going to define a group with a finite number of elements of a very different nature.

Let us consider three objects A, B, C. There exist six operations called *permutations* which simply consist in interchanging the order of the elements among themselves (either completely or in part). We shall give a name to each of these operations and characterize it by respectively writing to the right of each object A, B, C the object it becomes after the permutation.

$$I = \begin{pmatrix} AA \\ BB \\ CC \end{pmatrix}; \quad J = \begin{pmatrix} AB \\ BC \\ CA \end{pmatrix}; \quad K = \begin{pmatrix} AC \\ BA \\ CB \end{pmatrix}; \quad R = \begin{pmatrix} AA \\ BC \\ CB \end{pmatrix}; \quad S = \begin{pmatrix} AC \\ BB \\ CA \end{pmatrix}; \quad T = \begin{pmatrix} AB \\ BA \\ CC \end{pmatrix}.$$

The first operation causes no change; it is called the "identity operation."

Thus we define a set G_3 of six elements I, J, K, R, S, T.

By the product of two operations we understand the unique operation which has the same interchanging effect on the objects A, B, C as these two operations performed successively.[1]

For example, let us study the product $J \cdot R$.

A under J becomes B, which R changes into C;
B "   "   "   C,   "   "   "   " B;
C "   "   "   A,   "   "   "   " A;

therefore $J \cdot R = \begin{pmatrix} A & C \\ B & B \\ C & A \end{pmatrix} = S$. The reader will easily verify the fact that $R \cdot J = T$, and will readily convince himself that the product is not *commutative*; it depends on the order of the factors. This does not prevent the product of two operations x and y

(1) $$x \cdot y = z$$

---

[1] Cauchy was one of the first mathematicians to give this definition.

of G_3 from belonging to G_3. It follows from its very definition that the product is associative:

(2) $$x \cdot (y \cdot z) = (x \cdot y) \cdot z.$$

The operation I is the identity:

(3) $$I \cdot x = x \cdot I = x$$

for any x. Finally, any operation admits an inverse x' whose product on the *right* or on the *left* with x gives the identity back:

(4) $$x' \cdot x = x \cdot x' = I;$$

thus $J \cdot K = K \cdot J = I$, or $R \cdot R = K \cdot J = I$. R is its own inverse.

TABLE 1.

	I	J	K	R	S	T
I	I	J	K	R	S	T
J	J	K	I	S	T	R
K	K	I	J	T	R	S
R	R	T	S	I	K	J
S	S	R	T	J	I	K
T	T	S	K	R	J	I

Thus, we can form the appropriate multiplication table (Table 1), which we can read like a Pythagorean table, by agreeing to place the product of x by y at the intersection of row x and column y.[2] It will be noted that the elements I, J, K form a group *contained in the larger group*. It is called a "subgroup."

The rotations around a fixed axis in space form a group (commutative). The rotations around concurrent axes also form a group, since the product of two rotations whose axes intersect at a point O is still a rotation around an axis which passes through O. More generally, the *translations* in space, the transformations which map every figure into a congruent figure, form a group. These last two groups are not commutative, but it is easily seen that the law of composition is associative,

---

[2] Évariste Galois showed that the solution of an algebraic equation is linked to the study of a certain permutation group acting on its roots. The impossibility of solving the general equation of degree greater than 4 by radicals corresponds to a property of the total permutation group on n objects for $n > 4$ (Abel).

that there exists an identity transformation (one which consists in changing nothing) and finally that every translation admits an inverse translation.

Let us give a still broader category of examples which will show the importance of groups in geometry. Suppose that we have defined a space E and a group S of transformations which act on the points of this space. The properties of E left invariant by all the transformations of S form a geometry.[3] Thus, the geometry of Euclidean space is the set of properties invariant under Euclidean translations (see above).

Analytic expressions for these may be computed. We may also imagine how relations of neighborhood and a sort of continuity,[4] properties that are called *topological*, can be defined relating the rotations around concurrent axes.

However, more or less closely related groups lead us to projective geometry (related to perspective), to Lobachevsky's geometry, etc. Preserving only the notion of "continuity," we may arrive at the

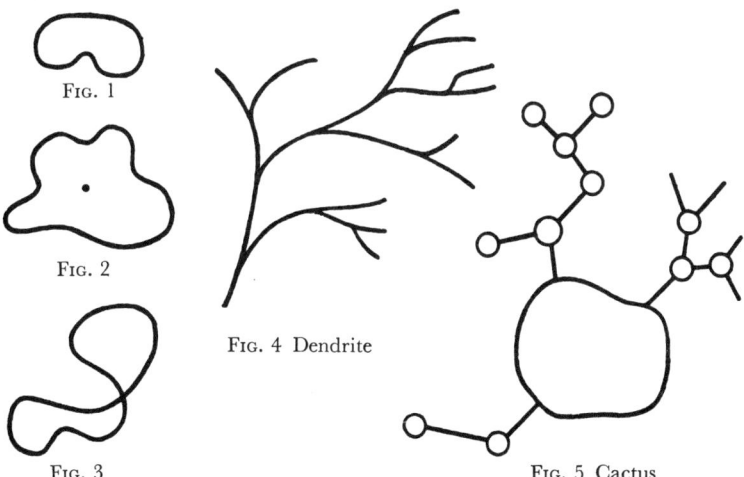

Fig. 1

Fig. 2

Fig. 3

Fig. 4 Dendrite

Fig. 5 Cactus

following notion:[5] Instead of drawing our figures on paper, let us draw them on a rubber sheet which may be stretched or released as we like without tearing. These transformations form a group (imagin-

---

[3] A notion made precise by Klein in his famous Erlangen Program, 1872.
[4] Results of Lie's researches. See the article by Élie Cartan in this collection.
[5] Too much technique would be required in order to make this precise.

ing the rubber to be ideal). The properties of a curve of being closed without double points (Fig. 1), of being a closed curve without double points surrounding a point (Fig. 2), of being a closed curve with a double point (Fig. 3), a dendrite (Fig. 4), a "cactus" (Fig. 5), etc., are invariant in this geometry[6] and suggest possible theorems.

After having extricated the idea of "group structure" from a certain number of examples of particular groups, mathematicians proceeded to develop the theory of this group structure in an autonomous fashion. We see in this an example of the constant effort made in mathematics to attain an ever higher level of abstraction and generality. When the first pioneers deduced the concept of natural number from the concrete notion of collection (of rocks, of apples, of horses or of men), their approach was not so different from those made in mathematics today.

This is not the place to discuss how mathematics leads us to define laws of composition very different from one another in their actual operation for objects of very varied nature: numbers, vectors, functions, ideals, equivalence classes, etc.

Classical geometry is not so much concerned with the nature of the entities with which it deals as with the relations that it establishes between them. What is a point? The idealized image of the mark left by the point of a sharp pencil. What is a line? The idealized image of a taut string. These definitions result from a cast of the mathematician's net into the sea of physics. Without them, geometry would never have arisen. They guide the creative activity of the mind. But are they ever brought into play in the course of a proof? No. On the contrary, a property purely involving *relation*, such as: "Two points determine a unique line," is in constant use.

Modern algebra, in which the theory of groups stands out, takes a further step in the direction of the abstract. It is no longer interested in either the objects on which a certain law of composition acts or the *concrete realization* of this law; it is solely interested in the *formal* properties of this law. Its history is one of progressive schematization and growing generality. We shall come back to the dialectic of this evolution.

Meanwhile, we expect that the reader will agree that two mathematical systems of extremely dissimilar appearance may, when their "secret structure" is examined, turn out to have identical and

[6] Very useful, for example, in the theory of analytic functions. See Riemann and Poincaré. It was then called *analysis situs*.

indistinguishable structures. Each of them is the concrete realization of the same abstract group G: we say that they are *isomorphic*.[7]

In a theory T_1 the elements a, b, c of the group G have as their concrete images the objects α_1, β_1, γ_1. In theory T_2 these same elements have other images α_2, β_2, γ_2. The law of composition manifests itself differently in T_1 and T_2. We may set up a sort of dictionary which allows direct translation from T_1 into T_2, and vice versa.

The multiplicative group G_1 of positive real numbers (excluding zero) and the additive group G_2 of all real numbers, positive and negative, *are isomorphic*.

The dictionary which translates the multiplicative theory into the additive theory is well known to all: it is the logarithm table.

Here is an extract from it:

multiplication	addition
division	subtraction
power	multiple
exponent	coefficient
1 (identity)	0 (identity)
2	0.30103...
etc.	etc.

As a result of this notion of isomorphism, when we study a group we may consider it *in abstracto*; or we may, on the other hand, study it "concretely" by means of a convenient representation.

Celebrated examples abound. Thus, Klein discovered[8] two isomorphic groups; one involving the regular polyhedron called the icosahedron, the other the equation of the fifth degree. Poincaré reduced the study of Fuchsian functions to that of the translation group in the Lobachevskian plane, for which he gives a Euclidean isomorphic image.

These two opposite methods complement each other. Here we see an aspect of the creative approach characteristic of mathematics. It is worthwhile to stop and consider this example for a time.

Today, certainly, the theory of groups has been sufficiently elaborated to allow of an axiomatic presentation in which formalism has been highly developed. We effortlessly attain the heights and ski elegantly on the paths traced out and climbed by the pioneers. But there is a

---

[7] Eupalinos: "This temple I constructed in the image of a girl of Corinth whom I loved happily," Paul Valéry, *Eupalinos, ou l'Architecte*.
[8] 1884.

great risk that we will form the most false ideas of the evolution of mathematics.

The investigator feels his way when in contact with the material of mathematics, and his tools, at first unsuitable, are fashioned and polished (method 1).[9] It then happens at a certain time that abstract reflection on these tools constitutes a real step in the direction of mathematical progress. Formalism is valuable: the abstraction is never carried so far from the concrete that it becomes impossible to go back and cause sheaves of new and varied results to spring up (method 2).

Past this optimum state, formalism can lead the investigator into a Scholastic desert. Then only a new supply of fresh material (for example, contact with physical theory) can save the mathematician from the schizophrenia which threatens him, can inspire him to pose meaningful problems, to create new tools (return to method 1).[10]

The history of the theory of groups illustrates this dialectic of the material and the tools.

The structure of a group is sufficiently simple and general to enable it to be discovered in many areas of the field of mathematics. This is what has assured and will continue to assure its success. The theory of groups coordinates and makes possible fruitful comparisons and syntheses.

We cited the older examples of Galois, Klein and Poincaré; we could cite numerous examples in more recent research.

Yet, group structure is only one particular structure. Important systems fail to possess this structure, and if we do construct "sham groups for the sake of symmetry" in them, only mediocre results are thus obtained.

The power of group theory is thus limited. It remains immense. It confers on its students a feeling—doubtless impossible to transmit—of the beauty of mathematics. All the abstract systems that men have slowly constructed on the basis of the concrete by "successive actions on the original sensation"[11] radiate the joy of a successful effort, but they also possess that indefinable grace always to be found in tools perfectly adapted to their task.

---

[9] Cf. Galois and the permutation groups that he discovered while studying the solution of algebraic equations.

[10] Relativity and quantum mechanics, for example.

[11] Cavaillès, *Méthode axiomatique et formalisme*.

E. PROBABILITY

The notion of group essentially arises from reflection by mathematics on its own structure, from an attempt to achieve more self-sufficiency and crystalline purity.

At the opposite pole of mathematics another revolutionary theory has found its source, on the other hand, in meditation on concrete phenomena, those of everyday life just as much as those of the physical sciences and even the biologic, economic, psychological and social sciences. We are speaking of the theory of probability.

The theory of probability arose in the 17th century, as did analytic geometry, combinatorial analysis and infinitesimal calculus. During the 18th century the principal snares were eliminated and it was given a theoretical form, although its applications were then only glimpsed. In the 19th century it was revealed to be an irreplaceable instrument of analysis in the study of certain physical phenomena, such as the kinetic theory of gases. 20th-century physics has given free rein to this surprising theory which would seem to have been created solely for its use. In the process of destroying the formulations of classical determinism, which had become too confining, and in replacing these by more flexible formulations, physics called on the theory of probability to serve as the foundation for the basic laws of the new mechanics, the only foundation that could be given from this point on. The two articles which follow will clarify this problem and will thus aid in the comprehension of certain articles to follow, notably the one by Kahan.

Aside from a few borderline incidents, we enjoy a relative calm in the other areas of mathematics. This is no longer the case when we come to consider the theory of probability.

On the very threshold of the theory we are faced with disturbing questions of its validity and even of its precise meaning and the definition of terms such as "probability" and "chance." It is worth the trouble to discuss the notions of probability and chance which lie at the base of the theory before we undertake the study of the theory itself in operation. We have called upon two well-qualified mathematicians to give this preliminary clarification.

Besides his studies in mechanics and articles on the theory of partial differential equations, Robert Fortet, a student of Fréchet, has made profound contributions to the theory of probability, notably in the area of probabilities of absorption. In his article he gives a felicitous presentation of the chief philosophical theses which confront one another concerning the crucial problem of the subjective or objective origins of the notion of probability. Here it is a question of an issue with which every cultured person of today should have an acquaintance. Basing itself on elementary examples, Fortet's clear exposition is a positive contribution to the scientific humanism which is the raison d'être of this book.

Pius Servien, the author of several appealing original works on esthetics, who is interested in formal problems and methods and highly informed about the movements of present-day art and science, and has written poetry whose sensitivity has been appreciated by Paul Valéry, has often concerned himself with questions of chance. His study is a highly personal contribution to the thesis of the partisans of subjective probability, and forcefully proposes a solution capable of eliminating a major ambiguity from this hazardous domain.

F. LL.

18

MODERN OPINIONS ON THE FOUNDATIONS OF THE THEORY OF PROBABILITY

by Robert Fortet
PROFESSOR AT THE FACULTÉ DES SCIENCES OF CAEN

1. Nothing is more commonplace than the use of the words probability, probable, chance, etc.; it must be agreed that the "man in the street" has a sufficiently clear idea of these notions for his current needs; this idea sufficed for the founding of the theory of probability, just as it sufficed for its initial development. (As we know, the honor of the founding and initial development goes to Pascal and Fermat; their independent labors were contemporary and almost equivalent. Let us also mention Huygens, who was the first to introduce mathematical expectation. It is only just to add that Pascal and Fermat had forerunners.)

But the variety and the importance of certain modern applications of probability, especially the essential role that it plays in quantum physics, its increasingly extensive and essential use in scientific investigation, as well as certain needs of mathematical rigor, have introduced new demands. A great number of philosophers and mathematicians have occupied themselves and still occupy themselves with the elucidation of the meaning of probabilistic statements relative to the human mind, and to the extent that they claim to explain or represent objective reality, their work partakes of philosophy as much as of mathematics, and has scarcely affected the theory of probability itself. However, we think that it is useful to devote several pages to an exposition of it, since we are dealing with questions that no educated man can fail to ask himself. Moreover, these authors, far from rallying round a common opinion, have more or less deep divergences, and it is interesting to note that these divergences cannot diminish interest in the theory of probability or alter our faith in the validity of its applications. Furthermore, since their discussions have had their fullest development in the course of the last few decades, if we ignored them we would not have a correct idea of the activity of the modern probability theorists. Thus, we propose to characterize the principal positions and theories presently held without giving a detailed discussion or criticism of them. However, there will be no discussion of problems connected with indeterminism or induction since these would carry us too far afield.[1]

We shall follow the customary usage in speaking of the "probability of an event E"; however, we might agree with J. M. Keynes that it would be more correct from the point of view of logic to speak of the probability of the statement or proposition asserting that the event E will occur (or has occurred), but this inaccuracy, if it is one, is convenient and will cause us no difficulty.

Subjective Probability

2. An event E in the future, for example, if it is well defined—we shall insist upon this restriction—will or will not happen; but we are frequently not in a position to decide about it in advance because we lack information or for some other reason. In such a situation, after studying the information we do possess, we can make a probabilistic statement, i.e., assert that the occurrence of E is *more or less probable*. The knowledge already possessed has an essential influence, so that we should not speak of the "probability of E" without adding or at least understanding: *relative to a certain fixed body of information*,[2] since the probability thus arrived at does not express an objective property of the event E, but only of the *opinion* that we infer from a certain collection of information.

If we wish to play heads or tails with a poorly made coin, it is clear that one side of the coin will be favored at the expense of the other,

---

[1] On this see Kahan's article in this collection. (F. LL.'s note.)

[2] It goes almost without saying that we include in the body of information "possessed" not only that which we actually have, but in addition, in certain cases, the information that we suppose that we possess; conversely, we should not include that which we actually possess, but which we decide not to use in the evaluation of the probability.

but it is certainly impossible for us to know which it will be without testing. We cannot anticipate that either of the two possibilities heads or tails will occur more often than the other, or wager on one with more confidence than on the other; in other words, we have to assume that they are equally probable. It is only after having made a sufficient number of tosses that we will know, if heads clearly turns up more often than tails, that the dissymmetry of the coin favors heads. We take this *new* information into account by making a new assertion, namely that the probability of heads is greater than that of tails. The initial probability, which expressed our ignorance, introduced nothing characteristic of the particular situation under study; the second probability corresponds to an objective reality—a greater occurrence of heads—but only because this objective reality had been included in the body of information at our disposal.

The name of *subjective* probability generally given to a probability of the type that we have just encountered finds its justification in these remarks.

3. From the fact, shown by the preceding examples, that there are equally probable events and events which are more probable than others, there arises the idea that the probability is a quantity that *can be represented by a (real) number*, a larger number corresponding to a greater probability. If we agree to take 0 as the minimum probability —that of an impossible event—and 1 as the maximum probability— that of an event certain to occur—we obtain the universally adopted scale 0–1 for the representation of probability.[3] The gain in precision achieved by replacing vague expressions such as "It is very probable, unlikely, etc.," by a number is evident, especially because only by such a substitution does the probability become amenable to mathematical manipulation. However, this substitution gives rise to some extremely ticklish problems.

First of all, is it certain that well-defined numerical value can always be assigned for the probability of an event E relative to a given body of information C? If that is the case, the role of the observer who determines the probability of E on the basis of C would be reduced to merely discovering the numerical value imposed by C completely

---

[3] The choice of the limits 0 and 1 for the subjective probability is purely conventional, but their use is clearly desirable, since this gives the simplest statement of the relations that exist between the subjective probability of an event and its future frequency.

independently of the observer (i.e., if two different observers arrived at different results for the probability, we would have to assume necessarily that at least one of them was mistaken). Our "subjective" probability would then equally well merit the name of objective in the sense of its complete independence of the observer. This is precisely the conception that Laplace had of probability; he qualified this only by saying that in many cases our talents were inadequate to allow us to actually carry out the determination of the value of the probability.

These problems cannot really be cleared up without a profound study of the way in which our mind evaluates a probability on the basis of a given body of information; this is the study undertaken by Keynes in a section—the most original—of his *Treatise on Probability*. He concludes that there exist situations in which the body of knowledge C does not impose a numerical value for the probability, and that in such cases the choice of a value is arbitrary and must be illegitimate.

Borel and Finetti have an entirely opposite opinion; according to them, in considering a real or fictitious wager, it is always possible to legitimately guide an observer to a numerical expression of his estimation of a probability.

In this regard, Borel compares the evaluation of a probability to that of a price; there are cases in which the body of information does not yield a well defined value for the probability, so that different observers may arrive at different values for the probability on the basis of personal factors without these divergences being necessarily inadmissible. In other cases, on the other hand, the body of information imposes the same value on all sane minds. This is the case, in particular, when we can arrange the situation in the form of "symmetric cases," i.e., events in which we cannot find a reason for supposing that one of them is more likely to occur than another. On the basis of a principle (sometimes called the "symmetry principle") which some feel lacks solid foundation, but which is admitted by Borel and many others, we can assert that these events are equally probable (thus the six faces of an honest die, or—if we are not aware of it—of a loaded die, or of a die that we know is loaded without knowing in what way, will all be considered to be equally probable).[4]

Finetti, who is an extreme "subjectivist," denies that the body of

---

[4] Let us take note here of the very interesting ideas of Pius Servien on "physical equality," which, however, do not appear to us to add anything essentially new relative to the "symmetry principle" as we present it here.

information can ever impose a value for the probability; according to him, its determination is a completely arbitrary act of the observer (who is constrained, however, to respect the axioms or principles to be discussed later on). However, it is obvious that there are cases in which observers of the most widely divergent views will arrive at the same value (if, for example, we obtain an average of seven heads to three tails after a large number of tosses of the same coin, we will all say that the probability of heads is approximately $7/(7 + 3) = 0.7$, and everyone will agree that if we continue to play using the same coin, the frequency of heads will continue to be near 0.7). We might see a decisive refutation of Finetti's doctrine in this accord, but he has shown that this accord, if it does occur, is due to the fact that the various observers have based their estimates of the probability on the same hypothesis, which is arbitrary, but so natural and, although precise, so purely qualitative, that they readily accept it unanimously, consciously or not (for instance, the hypothesis of the "equivalence" or of the "independence" of the tosses in the example of the game of heads or tails). This result, independently of the philosophical application that Finetti makes of it, has great importance for the mathematical theory of probability.

Objective Probability

4. However, many authors have interpreted in a completely different fashion the unquestioned accord of various observers in their estimates of a probability (at least in certain cases). Let us again take the example of the game of heads or tails. This is clearly representative of a large category, that of the cases in which the experiment (or, as is usually said in the theory of probability, the trial; a toss of the coin is a trial) can be repeated indefinitely under the same conditions. The main consequence of this is that the numerical value f of the frequency of heads observed in the *past* is universally adopted with confidence as the approximate value of the probability of heads, if a sufficiently large number of tosses was made. The larger the number of tosses, the more confidence we have that (f minus the probability) is small; in a somewhat converse sense, if we somehow know the probability, we expect that the *future* frequency of heads will be near that probability after a sufficient number of tosses, and nearer to it the larger the number of tosses. Clearly, this is possible only because the frequency of heads becomes "stable" after a sufficiently large number of tosses, i.e., because no matter how variable it is as the tosses considered vary, it is only slightly different—and with smaller variation as the number of tosses increases—from some central or average value (which is the probability, of course). Now, here we are dealing with a property which can be objectively observed, and has been effectively studied for games of chance, for various economic and social phenomena (crime, mortality, etc.), all of which can be conceived of as long sequences of repeated trials.

The first conclusion that we can draw from the foregoing observations is that there exist phenomena which objectively merit the name of random events; in other words, chance is not a mental construct, but is one of the modes in which nature acts. On this basis, according to Fréchet, for example, by means of an "inductive synthesis" the human mind was able to obtain a conception of probability as an objective property of random phenomena, a constant which is linked with them and whose existence we have to assume as an explanation of the observed stability of the frequency. The latter, a manifestation of the probability, at best gives a partial measure of it; again according to Fréchet, the fact that the frequency is not absolutely fixed, but only slightly variable, should not be any more disconcerting than the fact that when we repeatedly measure a length, the results always differ slightly because of errors in measurement.

The new kind of probability that we are thus led to conceive is usually called "objective"; we say that it is a new kind of probability, since the objective probability differs essentially from the subjective probability considered in the preceding sections; first of all, we cannot have objective probability unless there is a sequence of indefinitely repeated trials, or at least trials capable of being repeated indefinitely, while a subjective probability can be discussed without difficulty for a single trial. Let us go back to the example of page 212, in which the player has to play a game of heads or tails with a visibly asymmetric coin, but knows no more about it than that. We explained that from the "subjectivist" point of view the player is led to consider heads and tails as equally probable in spite of the asymmetry, which is absurd from the "objectivist" viewpoint. According to this view, the player is to consider the probability of heads to have a value which is objectively fixed, although of course unknown to him, but of which he will have a satisfactory estimate after having made a sufficient number of tosses; conversely, it is absurd from the "subjectivist" point of view to say that the probability of an event is well defined but *unknown*. A subjective probability is the expression of the body of information possessed *at present* and cannot be unknown (we have seen that we may find it difficult to express its exact numerical value, but that is something else entirely).

5. But it may appear that the objectivist point of view of Fréchet, while certainly sufficient and even convenient for justifying most of the practical applications of the theory of probability, turns out to be incomplete if we wish to deduce the foundations of even the notion of probability from it; the frequencies observed in the working out of a random phenomenon are, by the very nature of chance, only *approximately* stable. Therefore, there would exist frequencies that are sufficiently stable to be characterized as random, and frequencies which are not stable enough for that; where are we to place the dividing line? The frequencies derived from mortality tables (with certain other precautions taken) make possible the operation of life insurance companies, but we can easily imagine other applications which these frequencies would not allow.

It was partly in order to avoid objections of this type and to give a satisfactorily rigorous statement from the objectivist point of view that von Mises constructed his theory of "collectives"; indefinite sequences of trials are classified into two categories: the "collectives," in which chance plays a role, and the others, which, according to von Mises, cannot be considered in the theory of probability. He characterizes the collectives in the following way: For example, an indefinite sequence of heads-or-tails tosses $E_1, E_2, \ldots, E_n, \ldots$ is a collective if the frequency f_n of heads in the set of the first n tosses $E_1, E_2, \ldots, E_n$ approaches a limit p as n becomes indefinitely large, and if the same property (with the same value of p) is characteristic of all the partial sequences taken from the sequence of the E_n (for example, by taking every second or every third one among the E_n, etc.).[5] Then, *by definition*, the *limit* p *is the probability of heads*. Appropriately amended by Tornier, Wald, and others, the theory of von Mises is logically unassailable, and has won over many thinkers. However, it also lends itself to grave criticism; for example, the property possessed by a phenomenon of giving rise to a collective is certainly an objective one, but it cannot be verified, since it is not humanly possible to observe an infinite number of trials. Besides, if it is true that the principal

---

[5] Reichenback considers this requirement for the partial sequences to be excessive, and has constructed a theory in which it plays no essential role.

philosophical problem posed by the probability is to clarify its incontestable relations with the frequency, then the theory of collectives, the theory of Reichenbach, and more generally all those theories which not only base the notion of probability on that of frequency, but also go on to *identify* the two notions, evade this problem rather than solve it.

6. The preceding comments should suffice to convince the reader that the two notions of subjective probability and objective probability derive from two clearly separate conceptions; we may ask whether it is possible to adopt them simultaneously, or if it is necessary to reject one of them as erroneous, and which one. Cournot, who was the first, we believe, to clearly distinguish between the subjectivist and the objectivist points of view, saw no difficulty in accepting both of them; but it is unsatisfying to require the human mind to entertain two conceptions, the objective and the subjective, which are certainly distinct and practically opposed in certain respects, although sufficiently related so that we agree to give them both the same name of probability. Indeed, although Finetti, for example, is purely a subjectivist, the subjective conception is rejected as unfounded by the majority of modern "objectivists." Their position also has its difficulties: for instance, in the preceding example of the player of heads or tails who uses an as yet untried coin, according to what principles is he to decide his wagers on the first few tosses, for which he still has no useful estimate of the probability?[6] An analogous problem arises whenever we are discussing a single trial; Reichenbach thought that he had legitimately introduced a probability for a single trial, while retaining the idea of founding the notion of probability on that of frequency; but, as Borel showed, he did so by referring to a completely different frequency from the one which is involved in the objectivist point of view properly so-called.

---

[6] To say that it would be reasonable to abstain from wagering at the beginning is to forget that in life, where even abstention has its risks, we are always wagering in one way or another. The practical utility of a subjective probability 1/2 (introduced in conformity with the viewpoint expressed on page 213) may be contested, since it corresponds to a body of information insufficient for the needs of the player. However, this question of the practical utility is quite different from that of the foundation of the notion of probability. It cannot be assumed that there is a degree of knowledge such that for more knowledge the probability would be admissible, while it would be inadmissible for less. On this question see "La vérité concrète et les probabilités," a note by G. Darmois, Vol. IV, Fasc. III, p. 162 of *Traité de Calcul des Probabilités* by É. Borel, Gauthier-Villars, Paris, 1939.

The Sufficient Agreement of the Two Viewpoints in Practice

7. However, it turns out that these two very different conceptions of probability attach the same descriptive properties to it, both considering it as a real number, positive or zero and less than or equal to 1, assigned to certain events and constrained to satisfy certain conditions, called principles or axioms—the well-known axioms of total and composite probability, whose rigorous statement and justification give rise to some delicate problems, but on which, certain relatively secondary points aside, a universal agreement has been established. Since these axioms are the necessary and sufficient basis for the mathematical development of the theory of probability it cannot be contested. The theory of probability, after all, is nothing but the set of procedures or methods that allow the computation of the probability of certain complex events while taking account of certain axioms and making use of the probability of simpler events which are assumed to be known or at least to be accessible.

Clearly, it is important that the latter be legitimately introduced and correctly evaluated; in order that the computation and, in general, the application of the theory of probability to a "concrete phenomenon" be possible and verifiable by experiment, it is sufficient to accept a single principle, which we shall call, following Borel, the "single law of chance." It may be stated as follows:

"An event to which a sufficiently small probability has been assigned never occurs (or more precisely: we should act as if it could not occur)."

This principle suffices by itself because it is in fact always possible to link the phenomenon studied to an event of arbitrarily small probability; if after some trials this event does occur, then we must assume, in virtue of the single law of chance, that the given probabilities which were the starting point for the calculation were incorrectly estimated, or even that the phenomenon in question is not random and cannot be considered within the framework of the theory of probability.

Now, this law is evidently of a metamathematical nature, and has been given different interpretations by the subjectivists and the objectivists, but all accept it as definitive, and one will easily understand it if one makes use, as does Borel, of the celebrated example of the typist: A typist who is not acquainted with German is left alone with her machine and hits letters at random. After three months, she finds that she has typed out, without any variants, the German text of the complete works of Goethe; such a "miracle" has an extremely small probability, but is not strictly speaking impossible; however, any sensible person will consider it to be "impossible in practice," and if someone told us that it had occurred, we would firmly believe it to be a fraud.

Naturally, we need to know what value the probability of an event must not surpass if it is to be considered as "impossible in practice"; this depends, of course, on the range of phenomena with which we are dealing. However, for a given range Borel has shown that we can estimate the order of magnitude of the limiting value with greater precision than might have been thought on first consideration. Thus, in the human range, i.e., for the situations of everyday life, a probability less than 10^{-6} is considered negligible (10^{-6} is almost exactly our probability of being run over when we go out into the streets of Paris, a risk that we do in fact neglect); in the cosmic or supercosmic range to which a scientific theory must refer, the limiting value must be lowered to 10^{-100} or 10^{-200}.

The general acceptance of the "single law of chance" and of a well-defined system of axioms explains the fact that, even if there remains a certain lack of agreement on the deeper significance of probability and if perhaps there even is still something obscure about this notion, neither the theory of probability nor the validity of its applications can be deposed or doubted because of it.

19

CHANCE AND MATHEMATICS

by Pius Servien
ACADÉMIE DES SCIENCES OF RUMANIA

It has become more and more apparent in our time that no matter where we begin to dig in the soil of science we finally encounter probability.

Of course, the theory of probability, the basis of all science, is itself only based on an evident defect of thought that many have striven without success to remove. No one has succeeded in defining probability,[1] the very material from which the edifice is constructed.

Mathematicians would not be content to allow such a fundamental theory to remain without a solid foundation unless they were quite weary of the many failures to supply one, from Laplace to Poincaré and their successors. But this resignation has been more tolerable because of a state of mind which goes back to the very beginnings of this unusual science born of dice and card games. This science has become accustomed to being different from the others. It has remained the science of a certain mysterious chance, full of philosophical obscurity. This poorly defined nature even caused its founder, Pascal, to believe that it would be useful in theology. Laplace and Poisson applied it to legal disputes.[2] It still seems to be the best meeting ground for scientists and philosophers at scientific congresses; it is now even charged with the task of settling the philosophical problem of determinism and freedom.

---

[1] "It seems close to impossible to give a satisfactory definition of probability. The complete definition of probability is therefore a sort of vicious circle," H. Poincaré, *Calcul des probabilités*.

[2] Laplace, *Essai philosophique sur les probabilités*. Poisson, *Recherches sur la probabilité des jugements*.

We think that this insoluble problem, the definition of probability, depends on another which should be attacked first: the problem of determining the nature of this mysterious chance.

No one has even touched on this essential problem. Doubtless, people have at times spoken of chance and of the various senses that this word seems to possess.[3] But it is appropriate to distinguish between such a philological description of the meaning of a word as used in everyday speech, and the only way a word ought to be defined in science, which consists of using a label such as "chance" to represent a unique, well-defined object. We are not concerned with listing the various meanings which may be attached to a word like "energy"; we must attach the label "energy" to a unique, well-defined mathematical or physical concept. Once this has been done, we then leave aside all other meanings that the word may have in ordinary speech.

We say for example—and such problems form the subject matter of geometrical probability:[4] "Choose a chord of a circle at random (i.e., by chance). What is the probability that it is less than the side of the equilateral triangle inscribed in the circle?" We have used two words in this statement one of which, probability, is admitted to be undefinable and the other of which, chance, no one has sought to define. We might just as well replace them by "integrity" and "grace," respectively, and then suppose that we had stated a mathematical problem, and hand it over to someone to solve. "Choose by chance." You are willing to accept the meaning that I am free to attach to these sounds, without giving me any way of verifying that my meaning is

---

[3] For example, Hadamard, *Cours d'analyse professé a l'École polytechnique*, Hermann, 1930, Vol. II, pp. 537–538. "We usually consider that a phenomenon is itself determined when all the circumstances which can influence it are determined. At least this is what is assumed in all purely scientific research, since wherever factors involving human psychology are involved, the preceding hypothesis gives rise to a question of metaphysics, that of free will.

"The idea of probability is involved when the circumstances in question are only incompletely determined, and when as a consequence of this the phenomenon itself is indeterminate to an appreciable extent, either the indeterminacy involved in the initial conditions is itself important, or (in conformity with what has just been said) we may even suppose that determinacy is made impossible by the intervention of human freedom, or (and this is especially likely) extremely small variations in the circumstances may be capable of producing significant variations in the result. If this is the situation, this is what we express when we say that chance plays a role in the phenomenon in question."

[4] See, for example, the works on probability by Bertrand, Poincaré, Borel and others.

the same as your meaning, or even that we know what we are saying when we make these ritual sounds. If you say to me, "Pick out the multiples of three among these numbers," or, "Pick out the cherries from among these fruits," then I have a clear idea of what I am to do, and we will both obtain the same results.

Therefore, we have to decide what is meant by the special kind of chance used in the theory of probability, to the exclusion of all the other meanings that this syllable may have in everyday language.

Let us consider a balance; one of its pans carries a bronze weight, the other pan some flour; they balance. I say that there are two equal weights, two equal magnitudes. I clear the pans, then put the same objects back into the same places; I again find the same balance. But here it is a question of gross weight, the same as at the grocer's. With a precision balance I would observe a different phenomenon; the two magnitudes when compared would always turn out to be unequal. But after several weighings, if I had not observed any systematic inequality, I would again say, as I just did: "These magnitudes are equal."

But if I analyze this statement exactly, I see that I have set them equal by decree; observation alone would never allow me to discover this. Let us stick with the observations; for instance, after twenty observations, I observe that A exceeds B ten times and B exceeds A ten times. One further observation would be a choice by chance between A and B. I have compared two magnitudes in the case in which there is no systematic difference between them, and I see that I have just discovered the precise definition of chance; one that is general and unique. For example, if I had called the left pan "heads" and the right pan "tails," I would see that I was playing a game of heads or tails and that the case of a coin thrown in the air would be equivalent to the same general scheme, well defined and unique. The same would be true for cannon shots, photons traversing a polariscope, etc.

Now, armed with this criterion, let us take up again the preceding problem of geometrical probability. Either "chance" does not have in this area the unique meaning that we have just discovered, and thus such a problem has no place in the theory of probability; or, by convention, we attach that unique and well-defined meaning to such a problem, which we then have to formulate. But then the chapters on geometrical probability found in all the treatises have no right to appear there as distinct sections; they are incorrectly classified as separate branches.

Among the numerous clarifications that we shall owe to our discovery of the nature and unity of chance, the first will surely concern that classical and insoluble problem: the definition of probability. Let us recall another, analogous series of failures: the attempts to square the circle. Finally people tired of it, just as they tired of defining probability. One day Lindemann showed that this problem, still open at the time, was actually impossible. Next we will consider an attempt to weld together mathematics, which is based on equality, and the notion of chance—an attempt which must of necessity fail. In this regard, the recent exposition[5] of probability based on the "axiom of chance" appears to be even more profoundly illusory than the "circular reasoning" of the classical theorists of probability.[6]

Let us at least learn caution from so many failures. We must never again confound mathematical equality with the unusual situation that we have just considered, the absence of systematic inequality.

We must recognize that the relation between these two things, one mathematical, the other observed, is essentially the same as the relation between celestial mechanics, on the one hand, and the observations of stars, recorded exactly as observed, on the other. It is a question of a mathematical model which is proposed in order to account for a certain category of observations. Here we are dealing with what we might call, rather improperly, physical theory.

We thus have admitted that the model "mathematical equality" has simply been superimposed in some way on the class of observations in which we have recognized the unique notion of chance.

It is here, at this point fundamental for the entire structure of science, that the insurmountable difficulties encountered by Laplace, Poincaré, and others, entered; or rather, this is their deepest root. It is therefore the point to which we must pay closest attention.

When A is always greater than B, I have a faultless mathematical model; I write $A > B$. The same is true when I always have $A < B$. Let us consider (leaving aside a mixed intermediate case) the case in which $A > B$ is observed n times and $A < B$ is observed n times.

I propose the mathematical model $A = B$ in order to account for this. We have just noted that this is only a model which has been

---

[5] R. von Mises, *Théorie des probabilités*, Annales de l'Institut Henri Poincaré, 1932.

[6] Sometimes people have tried to see a new fundamental notion in the notion of probability, analogous to that of "point" or "point particle." See for example Darmois, *Statistique et applications*, 1934, p. 19.

superimposed by us on the observed situation. Now we make a second remark: this model is not forced on us. If I find "heads" 478 times and "tails" 522 times, this does not force me to use the "physical theory" 478/1000 in order to represent the observed phenomenon, to the exclusion of the physical theory 1/2.

Finally, we make a third essential remark: the model in question doubtless has its advantages, but it is not without disadvantages. It is useful as long as we lack a better one. Let us examine it more closely.

My quantity A, such as I have observed it, has not shown itself to be systematically greater or systematically smaller than my quantity B. A is neither greater than B nor less than B. If I write $A = B$, this mathematical model gives a good account of this aspect of the observation.

But there are other aspects of the observation that the model $A = B$ does not account for at all. It asserts that if I consider the two given quantities at a given instant, I will find them equal. But this never happens.

There is therefore an irremediable contradiction between this situation, in which we recognize the unity of chance, and a mathematical model which is based on equal quantities.

Every definition which attempts to bring together chance and mathematics must fail because of this contradiction.

This remark opens our eyes by obliging us to consider the impossible as impossible and to observe more carefully what is going on when we throw dice. We have to distinguish between two phases: the dice in the air and the dice after they have fallen.

The theory of probability treats only the dice in the air. As for choosing by chance, it remains outside its scope; the theory can never traverse this uncrossable boundary.

We have thus explained the universal nature of the theory of probability: it is a bridge set up between two universal domains, the mathematical domain and the domain of unique chance, just as we have erected one in considering the measurement of two magnitudes. The separation of these two domains is thus definitively obtained and there is an end to the physicomathematical hybrids that have been produced by attempts to define under the name of probability a certain mixture of chance and mathematics.

Such a hybrid, no less universal and also the subject of incessant failures—for instance, Gauss' law—is manifested in this same viewpoint: "Everyone believes it," Lippmann told me one day, "since the experimenters think that it is a mathematical theorem and the mathematicians think that it is an experimental fact."[7]

In the same way a deeper unity is given to the theory of probability, whose problems group themselves around three centers, the interconnections between which seem unclear: a theory of games, a theory of errors and a theory of statistics. Our analysis clearly points up the link joining them by showing that everything can be traced back to a single thing: the application of a theory to a particular physical situation, which act, as we have just seen, can be indifferently regarded either as a measurement of "physical equalities" or as a choice by pure chance.

Moreover, the purely arithmetic nature of the theory is now revealed. The definition of probability is simply based on equal things and not on "equally probable" things. We have likewise shown that the two theorems—sometimes regarded as axioms or principles—of total and composite probability are two very simple propositions of a purely arithmetic nature which concern what are properly called fractions. It would be vain to look for chance in this domain.[8]

[7] Poincaré, *Calcul des Probabilités*. The situation has not changed since then. See Mineur, *Éléments de statistique mathématique applicables à l'Astronomie stellaire*, p. 14.

[8] See our works published in *Actualités scientifiques* (Hermann); *Le langage des sciences*, 1931 (2nd ed., 1938); *Le choix au hasard* (Feb. 1941); *Base physique et base mathématique de la théorie des probabilités* (March 1942); *Probabilités et physique*, 1945.

PART II
The Mathematical Epic

"The essence of Mathematics is its freedom."

G. CANTOR

"We are of a race divine and possess the power to create."

R. DEDEKIND

The historical point of view is certainly missing neither in the first nor the third part of this anthology. However, it is particularly as a function of the present that we have oriented the description of the Temple of Mathematics and have undertaken the evaluation of the Influences (Part III). There remained at the center of this volume place for an interlude: "The Mathematical Epic." It will usefully complement the preceding part and at times prepare the way for the following one.

It is not bombastic to describe as epic this prodigious intellectual adventure—mathematics, at the same time both Iliad and Odyssey. It possesses some of their characteristics: loftiness of theme, the supernaturalness of certain turns of fortune and that close connection with the general progress of civilization which is found in all great collective works.

We have attempted to restrict ourselves and have called to this tribune only a small number of subjects. In choosing them, we have endeavored to elucidate points of view presented in the first part, while relaxing the rigor of presentation which was less avoidable in the preceding studies.

The division of this second part into three books—Past, Present and Future—follows quite naturally from its aim.

Book One

THE PAST

Paul Germain's article will be received with a hearty welcome by readers who may have experienced difficulties in this work. By putting in their historical perspective the disciplines already discussed in a systematic—and for that very reason more abstract and difficult—fashion he will clarify them and facilitate their comprehension. We shall appreciate the set purpose of simplification of this study which concentrates its spotlight on a small number of great moments: the Greek miracle, the 17th century and finally modern times. Enriched by the additions certain articles make to it (those, especially, of Dubreil, Valiron, Desanti, Brunet and Chapelon, all filled with historical essences), it empties into Godeaux's study as a river into the sea.

It is especially toward mathematical ideas, that is number, space, function, group and probability, that we have oriented the articles of Part I, Book Two, "Disciplines," pp. 67–226. We shall profit from this second part by attuning a privileged audience to mathematical methods. The most important place, from this point of view, goes to that infinitesimal analysis which has proved to be the ideal tool for probing the mystery of function. Indeed, we wrest from the idea of function many of its secrets by relating it to number and space and by considering these two principal characters as engaged in variations which make values evolve from it and which correspond to each other. And how can we mention infinitesimal analysis without evoking the great figures of its three principal inventors: Fermat, Leibniz and finally Newton,[1] to whom Pierre Brunet devotes an article in which he takes up again the details of an episode which Germain could only outline. Brunet, who is Secretary of the History of Science Division at the International Center for Synthesis, is particularly well informed on the movement of

---

[1] We know that the scientific world preferred to wait for the end of World War II to celebrate with more pomp the third centenary of Newton's birth.

ideas in the 17th and 18th centuries. We enjoy in his works that great intellectual integrity which seems to us the most valuable quality of the historian. He has been able to elucidate the roots of the Newtonian intuition which go down deep into physical reality and lay the foundations of pure mathematics on the idea of time borrowed from practical mechanics.

Elie Cartan has revived modern mathematics in at least three broad directions. First of all, in higher geometry, as much by an admirable use of the method of the movable frames of reference, to which he has been able to give a very great generality, as by the discovery of torsion spaces which came just in time to procure for the general theory of relativity the groundwork necessary for its development. Next, by the application he has made of Sophus Lie's theories to the study of hypercomplex systems with rational coordinates. Lastly, if the history of group theory had to be summed up in three names, certainly those of Evariste Galois, Sophus Lie and Elie Cartan would have to be the ones mentioned. E. T. Bell has been able to say of Cartan's 1894 thesis that it was "so exhaustive in its treatment of certain timely problems that it actually eliminated for many years the interest shown in continuous finite groups."[2] The article which Cartan devotes to the life and work of Sophus Lie, whom he knew personally, felicitously reinforces that of Lentin.

The prominent women mathematicians discussed by Mme. Dubreil-Jacotin will remind us that mathematics, like literature and painting, has its Sévigné and its Berthe Morizot. Herself a distinguished mathematician, author of a thesis on the theory of waves in hydrodynamics and of works on the theory of equivalences in modern algebra, Mme. Dubreil-Jacotin has been able to unite artfully in her article an appealing narrative and scientific competence. Incidentally, with regard to Sophie Germain, this bouquet of monographs will help the reader to further clear up certain points connected with number theory and the study of curved surfaces; with Sophie Kovalevski, he will deal again with the important theory of differential equations and with equations with partial derivatives. Finally, on the subject of Emmy Noether, he will have a new echo of that body of doctrine called modern algebra (or also, in a closely related sense, abstract algebra) which illumines numerous articles of this work, notably those by Bourbaki, Dieudonné, Weil and Godement.

F. LL.

---

[2] Numerous connections are established among these various questions and impart to them a kind of unity. See, for example, the article by Godeaux, pages 284–286.

20

A GENERAL VIEW OF THE EVOLUTION OF MATHEMATICS

by Paul Germain

We can take many points of view when we study the history of the sciences, as Pierre Boutroux has said.[1] There is the historian who discovers and studies original texts and documents which acquaint him with the thought and methods of scholars of a more or less remote period; there is the historian who investigates with precision the authorship of this or that great discovery; nevertheless, we cannot take either of these two points of view. We shall follow rather the new and apparently very fruitful direction Boutroux indicated when he declared: "The history whose outline we shall try to sketch will concern itself very little, as we have said, with isolated discoveries, detached from their background; it will have as its principal aim the study of the great currents of mathematical thought, assigning to each fact the position which it deserves—not in science as it is today, but rather in the science of the scholars who especially studied this fact and who ascribed to it an important role." This is also the same perspective that Pelseneer takes in his book *Esquisse du progrès de la pensée mathématique*[2] [Sketch of the Development of Mathematical Thought].

Péguy declared that in the general history of humanity one must distinguish between "periods" and "epochs." "Periods" correspond to a slow, but progressive and regular evolution of humanity; "epochs," on the contrary, are the turnings and tackings of this humanity. "Epochs" are the moments of strenuous life, bringing new conceptions to the world, "periods" only the long development of the germinal ideas

---

[1] *L'Idéal scientifique des mathématiciens, dans l'antiquité et dans les temps modernes.* Pierre Boutroux, Félix Alcan, 1920.

[2] J. Pelseneer: *Esquisse du progrès de la pensée, mathématique des primitifs au IX*ᵉ *Congrès international des Mathématiciens,* Hermann et Cie Éditeurs, 1935.

discovered in the "epochs." The same is true in the history of science, and in particular in that of mathematics. It also has its "epochs" of fertile creations and its "periods" of development and of refinement. Now, if we are looking for the great epochs in mathematics, "ages," using Pelseneer's term, we are led to distinguish the Greek epoch, the Cartesian epoch and the modern epoch. It is chiefly these "epochs" that we shall concern ourselves with, intentionally overlooking even the very rich periods, which are sometimes indispensable to the organic development of science but much less interesting for the person casting a rapid glance over general history. Nevertheless we shall make an exception in the case of the early period, poorly known and surely ill-defined. But it does seem indispensable in a study such as this one, in which the problem of the genesis of mathematical ideas plays an important role, not to pass over without notice those long centuries when the human mind advanced slowly towards an unexpected scientific concept; it is in this confusion, this vagueness, this empiricism that the ideas of number and forms were born. Why deny these origins? This is the very rhythm of life; it searches and feels its way for a long time before finding its new direction.

Primitive Mathematical Thought

It is often said that the world of mathematics is a world apart, sufficient unto itself; according to certain people, it could have developed independently of the history of nations. This is very debatable, but in any case we can say with certainty that from its beginning mathematics has been a tributary of contemporary conditions of life. Let us take for example the idea of number. Among primitive peoples —as certain recent sociological findings prove—the idea of number is that of correspondence.[3] Primitive man needs number, the lowest integers at the very least (some tribes do not count beyond 20) in order to count this or that category of objects. He needs them to verify the count of his modest herd or to effect his rudimentary business exchanges. Geometry is non-existent as yet; what good would it be among these nomadic tribes? It can come into the world only among a people who measure land or who build.

---

[3] It is interesting to compare this remark with the various articles in this work concerning set theory, in particular those of Fréchet and Eyraud.

The Earliest Ancient Civilizations

It is with the Egyptians that we can speak of mathematical science. They were compelled by nature to survey their lands and were mindful of their buildings; they were thus led to work with numbers and lines. Are we to see among them the first mathematicians in history? This is not Boutroux's opinion: "The Egyptians were acquainted with mathematical facts, and knew how to work with numbers and reason about geometrical figures; but, as far as we can judge at the present time, the ends they pursued were utilitarian and practical. They do not appear to have held a distinct concept of theoretical science, a scientific ideal." This debatable question on which people like Tannery and Gaston Milhaud on the one hand, and Emile Picard or Brunschvicg on the other, have different opinions, is still far from being resolved. Examination of Egyptian mathematical papyri nevertheless reveals interesting results to us. We find there arithmetical problems already requiring a certain reasoning ability; the exact calculation of the volume of a truncated pyramid with square base, given by the formula

$$V = r \frac{h}{3} (a^2 + ab + b^2),$$

the area of a hemisphere by means of a formula equivalent to that now used, with the excellent approximation for π of $(\frac{16}{9})^2 = 3.160\ldots$, and solutions of equations of the first degree. True, the problems considered were only special cases. But on the other hand, we possess extremely few documents. Perhaps these are only random notes, and not the most general account of the best-trained people. It seems, moreover, that a minimum of theory and general ideas must have guided the development of these complicated formulas, which, under the contrary hypothesis, would be nothing more than recipes. Pelseneer's hypothesis is not improbable: "Only calculating tables have come down to us from the Egyptians—and this is a question of probabilities—but didn't they have a Fermat or a Pascal?"

A similar problem arises with the Babylonians. Documents show us the solution of problems of the second degree, but here again, there is no interest in systematization. And yet do we not have here the vestiges of a Babylonian science?[4]

---

[4] This is Picard's opinion: "It is not possible that it did not have its theoretical aspects," he states in his book *La Science Moderne* (Flammarion).

(Recent works by Neugebauer and his school have just established that the

The Greek Miracle

Despite all the interest presented by the study of these isolated mathematical facts, we cannot in the present state of our knowledge confidently speak with regard to them of science, scientific thought or a scientific ideal. The situation is quite the reverse when we come to "the Greek epoch." Historians all speak of "the Greek miracle" whatever their opinions concerning the contribution of the East to Greek mathematics. Indeed, it is certain that these Greeks did not extract everything from their own genius, and that they were influenced by Egypt and the Orient. The Greeks were great travelers. Pythagoras, Plato and many others were reported to have gone to the Orient. Some, like Democritus, were even brought up by wise men of the East entrusted with their instruction and their education. Need we recall the geographical position of Miletus or Samos which were the centers of the schools of Thales and Pythagoras? There was a definite influence there, difficult to estimate, but far from explaining everything. Between a utilitarian technique which was doubtless given to the Greeks and that systematic conception which they handed down to us there is a discontinuity, a void to be filled. The miracle is that they made human thought take this decisive and important step. With them ideas appeared which until then had most probably remained in the shadows, without quite being uncovered: abstraction, generalization, analysis and synthesis. All these general processes of thought, which were lying dormant in man, but of which man had not yet become aware, would from now on be at the service of human reason. It would be very interesting to study from texts this human progress from experience to reason. But such documents are scarce. How many manuscripts and books—and surely the most interesting—have been lost and are known to us only through quotations, appreciations or references!

However, let us try to fathom this "miracle." *For the Greeks, mathematical ideas are purely abstract*. The fundamental notions are those of number and form. The integer, despite the impossibility of having a logical definition, is a conception of the mind, an "idea" in the Platonic sense of the term. As to form, we are aware with what insistence Plato, Aristotle and many others stressed the purely rational nature of

---

Babylonians did have more advanced knowledge and, in particular, solved equations of the third degree. Note by F. LL.)

this idea which until then had been held to be empirical. For the Greek geometer, a triangle or circle existed only in the mind; the visible figure is only a very imperfect representation of it and the relation between these two terms is the same as that between an idea and the word which expresses it. Furthermore, these "ideas" of number and forms represented for the Greeks a thing of such purity that he often used them in his mystic theology. On the subject of proof we encounter a perhaps more remarkable point of view: *No statement without a logical proof.* We are far removed from the recipes presented to us in Egyptian and Assyrian manuscripts. Geometers became aware that there were a certain number of results which indeed had to be accepted, especially those which concerned the possibility and uniqueness of certain constructions. They had uncovered the idea of postulates. With these elementary logical rules and some carefully stated definitions, they possessed the essentials for the development of geometry. They were superb logicians—perhaps even too much so, as we shall see. The steps of a proof, analysis, synthesis, proofs by *reductio ad absurdum*—all these ideas were familiar to them.

If we penetrate still further into the ideal and thought of the Greeks, if we examine the range of their studies, the aim and ends of their science, the methods which they championed, we shall be led to discover the principal faults of these mathematicians, as well as the "seeds of death," according to Boutroux's expression, which Hellenic mathematical thought contained in itself. Boutroux has admirably analyzed the role of "construction" in Greek geometry. What ideas should be included in science? Which ones should be chosen among all those which can arise in the thinker's fertile imagination? At its very first appearance, is science going to take upon itself the study of all phenomena? Is there not the danger of its overextending itself? For all these reasons the Greeks confined themselves to the study of ideas and problems which at the present time we call "of the second degree," that is, those which admit of a construction—ideal, naturally—with ruler and compass. This restriction, despite all its inconveniences, is a marvel of intuition. Almost without realizing it, the Greeks reserved for themselves a homogeneous field of study. This concept of construction, according to Plutarch, goes back to Plato; in any case it is certain that Euclid made use only of intersections of lines and circles. But this famous construction was to fetter the Greeks a good deal before very long. What would be its equivalent in three-dimensional space? Thus the Greeks arrived at this phase only very late, and even then they limited themselves to elementary ideas such as those of the sphere, the cone and the cylinder (of revolution).

Another fundamental idea, this one relating to the object of science, was to impede progress; as has long been observed,[5] the Greeks cultivated in mathematics as in everything else what was simple, what was beautiful, what was harmonious. But what is beautiful for the Greek cannot be the proof, the remarkable procedures for simplifying an exposition, or the discovery of unexpected results. Beauty is found "in ideas and not in what man adds to ideas." From this arose in arithmetic considerations on perfect numbers,[6] on amicable numbers, Speusippus's discourse on the properties of the number 10. It is also in reference to this ideal of beauty and simplicity that we can understand the aversion of the Greeks to irrationals. In geometry we again find the same preoccupations. The sphere, the equilateral triangle, the regular tetrahedron are thought of as having a divine essence. Greek geometry does not pursue difficulty. Knowledge is an intellectual vision. In order to discover, it suffices for our mind to observe; the thinker does not create a fact, he ascertains it. This ideal of purity kept the mathematician distant from utilitarian ends. No practical application was considered. Plato objected with force against the very term geometry. But in refusing this contact with experience and nascent physics, they also shut out a whole category of problems; and they remained shut up in their somewhat artificial ivory towers, contemplating the simple and clear ideas discovered by reason.

All these rigorous requirements are in their exposition. Nowhere, except perhaps with Archimedes, as we shall see later, is there any concern to retrace for the pupil the development of thought in discovery. The rigorous proof was sought, but not the natural one. This is why these expositions seem so artificial to us, and why we find it difficult to reconstruct Greek thought in this field. This mode of exposition was to weigh heavily on mathematics for a long time.

But already, as early as the Greek epoch, certain mathematicians felt themselves ill at ease in the midst of all these requirements. They soon discovered the relation of geometry to kinematics. Hence the appearance of "kinematic" or "mass" concepts alongside "geometric" ones. Thus were introduced examples going beyond the limits of traditional geometry, such as the conchoid of Nicomedes or the cissoid

---

[5] See G. Milhaud: *La Géometrie grecque, œuvre du génie grec*.

[6] Euclid's theorem on perfect numbers is presented as the consummate achievement of his three books on arithmetic.

of Diocles. But with Archimedes a still more remarkable stage was reached. This author, unlike his predecessors, indicates to us in his "method treating of mechanical problems" the path that led him to his discoveries. Although he himself declared this mode of exposition imperfect and afterward returned to the traditional method of proof, we have here a stage of thought worthy of interest. It is to Archimedes that we likewise owe the first processes for evaluating areas and volumes. These processes go beyond the Greek preoccupations and methods. And we can see in them the first beginnings of the integral calculus. Must we regard the case of this scholar as an exception to the Greek ideal? No, he is linked with the classical conception which we have set forth. Like the Platonists, he studied "the essential properties of figures, but those unnoticed by the men who had cultivated geometry before him."[7] Archimedes overcame that "distrust of infinity" which, according to Picard, was the weak point of the Greeks, but he ventured there only with regret. We can see in him an ingenious attempt to broaden certain narrow concepts, but despite this, the unity of Greek science remains unimpaired.

The last big question which broke through the framework of Greek science was that of the relationships of numbers and magnitudes. It was at first with great satisfaction that the Greeks discovered certain relationships: proportions, figures whose sides have some simple relation; all this satisfied for a long time their ideal of beauty and simplicity. The Pythagorean school took a deep interest in these matters: "All things are numbers." But precisely the famous Pythagorean theorem (which was considered a calamity) proved that geometry quickly introduced the irrationals, the "incommensurables" as the Greeks called them, to testify to their impotence when confronted with this vague idea. The unity of Greek science was shaken by this discovery and, fundamentally, the difficulty was never resolved. Indeed, examples are found of equations of the third and fourth degree solved by geometrical constructions, but we must wait a long time before witnessing the birth of an independent study of these algebraic ideas, as well as a use of calculation for geometry.

It is now easy for us to come to a conclusion. Greek science owes much to the ideal of beauty and harmony of the Greek spirit: its ingenious ideas, its taste for pure and theoretical science, its concern for rigor, but also its arbitrarily limited field of study, its artificial

---

[7] Quoted in Boutroux, p. 59.

methods of explanation, that "distrust of infinity" reacting against the methods of Archimedes or the incommensurables which geometry made it discover.

The Greeks loved clarity, order, precision. But a science in the process of developing is subject to the same laws as life. Life, we have already pointed out, feels its way, searches, advances and retreats, before finding its way and taking a new step forward. Disciplined by their ideal, the Greeks never wanted to enter the troublesome paths which opened up before them. Thus science, however perfect it may have been at the close of the Greek epoch, had to wait several centuries before attaining a new level of development.

The Romans, the Christian and Moslem Middle Ages, the Renaissance

The Greeks had brought mathematics a great step forward by achieving for it the first degree of abstraction: that of considering general ideas and studying mathematical objects. The second degree of abstraction was attained more than ten centuries later in the great Cartesian epoch. It was then that the laws of algebra and their multiple applications were evolved. But what was taking place during the early centuries of our era? A period of great maturation. The Moslems, who at that time enjoyed great power, deserve the credit for synthesizing the knowledge of the Greek scholars and the Hindu calculators. It was in this way that the Arabs were surely the initiators of algebra. What were the aims of these first algebraists? They were utilitarian ends above all; they applied themselves to solving practical problems and wished to obtain results quickly. They very often neglected rigor and thus quickly found out that to be successful it was not necessary to keep constantly in mind the meaning of the objects upon which one was operating.

We understand now that the Greeks were unable to be good algebraists, with their somewhat narrow conception of science. Those among them who were precursors in the subject, such as Diophantus, were "calculators," much despised by the Greek scientists of the period. When, beginning with the 14th and 15th centuries, the Christian world awoke to this new discipline, it still remained strongly influenced by this Greek thought, of which it was the heir. The mathematicians of these times did not have the boldness of the Arabs and Hindus, for their theoretical scruples were too great. Thus competent algebraists like François Vieta were unable to reason about magnitudes without depicting them; in the 17th century, Gregory de Saint-Vincent, who was a great scientist in his day, wrote his papers without the slightest algebraic notation, staunch disciple of Euclid that he was. During the entire latter part of the Middle Ages and the beginning of the modern epoch, mathematicians in the Latin-using countries always returned to the geometric considerations of the ancients in their attempt to establish algebra and demonstrate its rules.

The Great Syntheses of the 17th Century

In order for algebra to attain its independence, it was necessary to build its foundations anew, and it is here that Descartes's work takes its place. Most certainly, at first sight this seems strange. Is not Descartes known for his conception of geometry, and is not this word "geometry" the very title of the book in which his algebra is expounded? Does geometry represent for him too the essential object of mathematics? This would be a serious misunderstanding; as Boutroux states, for Descartes algebra takes logical precedence over the other branches of mathematics and is not in the least influenced by the nature of the problems to which it is applied. One of the treatises which Descartes was unable to finish, and of which we have only a rough draft, shows this mathematician trying to treat algebra as an independent discipline, without basing its development on external considerations. Besides, the methods of reasoning and proof which he embodied in his work are new. Essentially synthesistic, they consist in studying simple elements and building up progressively from them by combinations more and more complicated objects. We see, then, the decisive step taken by Descartes, and it is deservedly that his name serves to baptize this epoch of which he was the innovator. Let us state a little more precisely Descartes's views; we shall see more clearly their originality and at the same time their systematic character. *For him algebra is not a science; it is a method*, which teaches us to reason about abstract and undetermined quantities. For him, the object of [pure] mathematics is valueless, for it has no part in the explanation of the universe. Mathematics becomes mechanical, easy, demanding no great effort of the mind. Mathematical thinking is mechanized, automated; it is enough to combine elements ceaselessly among themselves. Does this say that Descartes had no admiration for algebra? Not at all, but for him method is paramount. Now, algebra is the method of "universal science." The explanation of the universe by Descartes is essentially geometrical and mechanical. He himself applied algebraic calculation to geometry with success, and he foresaw the considerable role of algebra and calculation in mechanics. Caring little for the beauty and harmony of the ideas he studied, disregarding the disinterested object of mathematics, and giving his attention only to an abstract method of logical combinations of undetermined elements, Descartes makes a clean and open break with the Greek ideal, and by this very means opens new perspectives to the mathematicians of the future. After the ideal of contemplative science came that of constructive science. On this point, however, Descartes was not always followed by his contemporaries. Fermat, for example, who had employed coordinates in his geometrical studies long before Descartes[8] (certain historians ascribe to him—incorrectly, in our opinion—the authorship of analytic geometry), remains faithful to the Greek tradition and, fundamentally, does nothing but apply the principles which we already find in Apollonius. Newton[9] likewise, despite his contributions to the infinitesimal calculus, was attached rather to the Greek spirit than to the Cartesian. And most happily, the ideal of beauty and harmony of intuitive and impartial science was preserved by many of the mathematicians of the 17th century.

The second sensational discovery which appeared in the course of the Cartesian epoch was the birth of the infinitesimal calculus. Must we consider with this event a new epoch in mathematics? Are we in the presence of a new spirit? This is a very delicate historical problem which remains the subject of numerous controversies. The first reaction is to see in this attainment an originality and a progress so considerable that we are tempted to mark with Newton and Leibniz the beginning of a new epoch. The "great step to infinity" is indeed accomplished. Moreover, as Emile Picard notes, ideas were introduced in science which furthered the birth of those new branches of mathematics—kinematics and dynamics. Thus we find the beginnings of that long series of discoveries in analysis which are the fruit of mathematical endeavors in mechanical and physical problems. Certainly we would not have the derivative and the integral, nor the ideas dependent upon them, had there not been an immediate area of application. Let us bear in mind that the foundations of rational

---

[8] *Ad locos planos et solidos isagoge*, Œuvres de Fermat, Vol. I, p. 91.
[9] See in Pelseneer, p. 101, a very significant letter from Newton on this subject.

mechanics were laid down in the 18th century. There is an entire set of problems in differential equations raised in this way (in particular we find, thanks to mechanical principles, existence theorems accepted which were not to be proved until a century later). Moreover, d'Alembert wrote the equation $\Delta \phi = 0$, and recognized the importance of this equation in potential theory; he thus opened the way for the solution of a large number of problems in mathematical physics.

However we shall not consider this infinitesimal analysis as arising from a spirit different from Cartesian algebra. This new calculus contains no principle contrary to those developed by Descartes. To realize this, we have only to recall rapidly the principal stages in the beginnings of the infinitesimal calculus; we shall have occasion to ascertain a certain parallelism with the history of the birth of algebra. Like algebra, it is geometric in origin. In 1635, Father Cavalieri published his *Géométrie des Indivisibles*. Like every mathematical innovation, it met with opposition from certain minds too enamoured of rigor, the algebraists especially, on account of the vagueness of its still badly suited language. Pascal upheld the new points of view and exerted himself to strengthen the new theory by showing its direct connection with the geometry of the ancients. About the same time, the differential calculus made its appearance. Fermat and Roberval were the first to make an attack on the problem of tangents. The algebraists, even Descartes himself joined the attack. But it is doubtless the lack of general methods that was objected to. It suffices to transcribe these dawning theories into the finite terminology of derivatives to realize that the calculation of derivatives is only a branch of algebra and the problem of tangents to plane curves is merely the second chapter of Cartesian geometry. Moreover, classical science quickly absorbed these ideas which appeared as the direct extension of algebraic calculation.[10] There remains a last province of the infinitesimal calculus which, a priori, seems to present a more original innovation: the calculus of series, which made rapid progress thanks to the labors of Newton and Leibniz. But, as Boutroux has shown, the step needed to pass from elementary algebraic functions to functions which can be expanded into series (and more particularly into Taylor's series) is the same as the one in number theory separating the rationals from the irrationals: to take this step it is necessary to

---

[10] We might think that these notions presupposed the new idea of continuity, but this is not, in our opinion, a serious objection; see below, note 13, p. 242.

introduce the notions of approximation and of convergence, which had already been formulated in a more or less precise fashion as early as the 16th century.[11] With series expansions, we come to a new chapter in mathematical analysis, an enlarging of algebra (since the trigonometric, exponential and logarithmic functions are related in many details to the algebraic functions). But in this new acquisition we do not see the occurrence of any very new idea able to justify the introduction of a new epoch. Moreover, this was indeed the point of view of Newton, who said himself that these discoveries were the extension of finite algebra. Leibniz's opinion at first sight seems quite different. This thinker, who dreamed, when still quite young, of a "universal combinatorial system" [*combinatoire universelle*], was a man of system, like Descartes. He very quickly realized the power of the new calculus in whose development he had played a very effective part, and at the same time, he was quick to disparage Cartesian algebra. In reality, here must be seen one example of those failures to understand in which the lack of elapsed time, the vagueness of the meaning of certain terms,[12] and even a certain rivalry sometimes play a large part. What put these two great thinkers in opposition was principally their philosophic views. But this has nothing directly to do with the principles of their mathematical structures. We can say, in concluding, that this Cartesian epoch contained two great events: the birth of algebra systematized by Descartes and the birth of analysis systematized by Leibniz; they were derived from a single synthetic spirit, but one performs upon finite combinations what the other performs on infinite combinations.[13,14]

---

[11] For example, the formula $\pi = \sqrt{\frac{1}{2} + \frac{1}{2}\sqrt{\frac{1}{2} + \sqrt{\frac{1}{2}\ldots}}}$ was given by Viète. See the article by Dubreil, p. 101.

[12] Like those of analysis and synthesis in the 18th century.

[13] One could object that the theorems of analysis presuppose the notion of continuity, and could insist upon this point of view, to take a position different from ours. But this notion of continuity in the 18th and 17th centuries remains very vague; it corresponds basically only to geometric interpretation and it is regarded as a leading idea. It is precisely the thorough investigation of this notion of continuity by Cauchy, Weierstrass and others which was to become one of the principal signs of the dawning of a new epoch.

[14] The 17th century likewise saw the birth of the calculus of probability (see the articles of Fortet, p. 211, and Kahan, Vol. II, p. 105) and of combinatorial analysis (see our article on "Beauty in Mathematics", note 10). Desargues had some intuition then of projective geometry and Leibniz of analysis situs and logistic curves. Let us point out only as a reminder the extraordinary arithmetic work of Fermat who, according to the sound views of Germain, reflect the Greek ideal. How can we explain this sudden and powerful blossoming of such

The 18th Century

Before continuing, let us note well that science is built up progressively; especially in mathematics, nothing is ever abandoned. If one epoch, characterized by this or that bent of mind, finds itself replaced by another, the point of view which dominated it is never forsaken. But it is new elements and new problems which proceed to enrich classical science, which proceed to amalgamate with it, and which extend at the same time both its areas of research and study, and its possibilities of action on this new material to be elaborated.[15] It would be particularly wrong, as we shall have occasion to point out, to believe that algebra ceased to interest mathematicians beginning with this or that date, and that they systematically abandoned synthetic methods which proceed from the simple to the complex. The "algebraic-logical synthesis" survived and it is still producing new results today. The Greek spirit, which operates more intuitively, preferring a direct contact with the object studied, also carries on in this or that branch of modern mathematics. In a word, a new epoch marks the enrichment of points of view of the human mind.

It is evident, indeed, that the perfecting of means of calculation is always a fundamental problem for mathematicians. Thus we see theories develop which clearly go beyond the concepts of elementary algebra, but which originate directly in the Cartesian spirit: the theory of determinants, the calculus of matrices, the vector calculus which simplifies the calculations of analytic geometry. It is again to the algebraic-logical synthesis that we must credit the developments in classical differential geometry; Cartesian geometry is the application of algebra to geometric concepts. Differential geometry is the application of analysis to curves and surfaces. It had to wait until the beginning of the 20th century for its nearly definitive development by Gaston Darboux. Nevertheless, despite the many studies which had to be made in this field, mathematicians of the end of the 18th century showed signs of a certain weariness. The mathematical horizon

---

diverse disciplines? We think it is proper to consider the 17th century in the West as a *classic* century in the arts and literature, but only as the true *renaissance* in mathematics. This renaissance began after the 16th century, for reasons which we cannot go into here. See Vol. II, p. 209. (Note by F. LL.)

[15] In the 17th century, for example, elementary geometry made much progress with Pascal, Lahire, Desargues and others. We have not spoken of it because, according to our point of view, we are studying principally only facts attesting a new spirit.

seemed to them obstructed; little by little they had approached the study of overly complicated questions which were wanting somewhat in scope. Mathematicians felt strongly that it was necessary to find something else: to study new ideas, to employ new methods.[16] It is indeed the appearance of new elements which will mark the beginning of the modern epoch; but we shall not find ourselves here in the presence of a unique and sensational fact comparable to the birth of algebra. Precisely what did the thorough investigation of this notion of continuity by Cauchy, Weierstrass and others consist in, this research which was to be one of the principal signs of the dawn of a new epoch? This is difficult to say in a single sentence. We shall understand better at the end of this section the essential, very complex points which give it its originality. Let us simply say for the moment that it is characterized, roughly, by a profound and critical analysis of every idea, by a broadening of the mathematician's view which often goes beyond the simple ideas of line and of number, and by methods in new fields of study. The richer and more complex, the more difficult to analyze; we shall content ourselves with studying a few examples.

Modern Mathematics

The transition between two epochs is difficult to determine; there was no abrupt change valid for the totality of mathematics, but a transition for each branch,[17] these transitions taking place at times occasionally separated by some sixty years.[18] The first problems

[16] Here are passages from a letter of Lagrange to d'Alembert, dated 1781: "It seems to me that the mine is already almost too deep and that unless new lodes are discovered, sooner or later it will have to be abandoned: it is not impossible that the chairs in geometry in the academies may become what the chairs in Arabic in the universities are today." Quoted by Pelseneer.

[17] In many respects one could choose Gauss, perhaps the greatest mathematician of all time, as the "hinge." To the spirit of the 18th century, in the virtuosity of which he partook, he brought the penetrating views of the 20th century. In addition to fostering, under the congruence sign, the theory of numbers, in a body of organized doctrine, and his works on curvilinear coordinates which prepared the ground for the Einsteinian "octopus-coordinates," was he not the first to imagine non-Euclidean geometries before Bolyai, Lobachevsky and Riemann, quaternions before Hamilton, functions of complex variables before Cauchy, elliptic functions before Abel and Jacobi, the application of topology to the theory of functions before Riemann? We have by no means tried to exhaust the list of his principal works in mathematics, nor to point out his contributions to physics and astronomy. (Note by F. LL.)

[18] One of the latest transitions is that from differential geometry to "direct infinitesimal geometry" brought about by Bouligand on the basis of remarks by Henri Lebesgue.

marking the beginning of this new spirit are those of continuous functions and of existence theorems. We have already had occasion to show that the problem of continuity had not been studied; the idea was still very vague. It was thought in particular that every continuous function was differentiable at every point, apart perhaps from a finite number or a denumerable infinity of exceptional points.[19] Cauchy was the first, in the middle of the 19th century, to apply himself to giving a precise definition of what is called a continuous function. He resolved this question and then took up with success the very fecund study of continuity. It is indeed a critical mind which presides over the elaboration of these very rich ideas; there is here a point of view very different from that which animated the mathematicians of the preceding century. It is no longer a question of putting expressions together, of forging new means of calculation, but of analyzing ideas until then held to be intuitive. Let us now take the example of existence theorems. We have said that nascent mechanics had promoted the development of the study of differential equations. But the fact that a differential equation of the first order has one and only one integral curve passing through a given point was always regarded as intuitive. Some mathematicians, Cauchy and Lipschitz in particular, exercised their critical intelligence on this point; they gave different proofs of this theorem, while making very precise the conditions for its validity. Similarly Riemann[20] proposed a proof of the famous Dirichlet principle, which states that a harmonic function is completely determined if we are given the values of the function on its boundary. This theorem is also an existence and uniqueness theorem for a partial differential equation of the second order: the equation $\nabla \phi = 0$.[21] It was to play a fundamental role in potential theory and in mathematical physics. Mathematicians launched out on these new paths and today these two matters of continuity and existence theorems are still of moment. It is not so long since continuity was analyzed in a very precise and seemingly definitive way by Baire in his *Leçons sur les théories générales de*

[19] Weierstrass was the first to construct functions continuous at each point, yet having derivatives at no point. See the article by Valiron, p. 161.

[20] Cauchy and Riemann are in our opinion the two great innovators in mathematics of the modern epoch.

[21] Weierstrass was the first to point out that the proof was, at the very least, incomplete. Hadamard proved by an example that it was false. Hilbert took up Riemann's method again and made it rigorous. This indecision over the validity of a proof is almost sufficient to prove the originality of the point of view envisaged by Riemann. On this subject, see the articles by Valiron, p. 161, Montel, p. 174 and Dieudonné, p. 907. (Note by F. L.)

l'analyse, and since Picard gave his famous method of successive approximations which was to prove a great step in the chapter of existence theorems relating to differential equations, partial differential equations and even more complex questions.

The originality of the modern epoch is not only in its having studied with a critical spirit a certain number of notions already known,[22] but also in having introduced into science new ideas which were very quickly to prove of great value. We cannot undertake here the study of all the new points which came up in just one century in mathematics. We shall continue to proceed by examples. Let us take that of set theory.[23] Up to this point, mathematics had been based essentially on the notions of number and line (the latter had already been more or less reduced to that of number in analytic geometry). The modern epoch was to introduce that of set. Set is a basic concept denoting a collection of certain objects, finite or infinite in number, regardless of the nature of these objects. This notion logically precedes all earlier notions (and, in this sense, one can say that the modern epoch marks the third degree of abstraction in the history of mathematics). It appears naturally in the theory of continuous functions: certain functions such as polynomials are defined for any value of the variable, but others like $n! = 1 \cdot 2 \cdots n$ are defined only for integral values of the variable. Thus we are led to consider the "set" of values that the variable can assume. We introduce similarly "the set" of values taken on by the function; the function is then the correspondence established between these two sets. Clearly it is advantageous for the pursuit of the study of functions to develop the theory of sets; and this was all the more productive since theorems on sets, considering their high degree of generality, could be applied to other branches of mathematics, such as arithmetic and geometry. The field of set theory is very extensive at the present time. It includes "the algebra of sets," whose spirit is that of the algebraic-logical synthesis; it studies the algebraic structure of sets, that is, the different operations (described axiomatically) possible in the interior of a set on its elements;[24] the

---

[22] One could also cite in this connection Hilbert's study on the *Foundations of Geometry*. This problem had never been completely attacked in two thousand years of geometry.

[23] See the articles by Fréchet, p. 70, Eyraud, p. 109 and Denjoy, p. 191. (Note by F. LL.)

[24] In this manner we distinguish groupoids, semi-groups, groups, rings, fields. (For groups, see especially the article by Lentin, p. 202. Note by F. LL.)

external operations (those concerning the mode of combination of the elements of a set with other elements outside the set);[25] combinations of sets and the operational properties of these combinations. In set theory we also consider what Bourbaki calls "general topology," or the study of the structure of sets (independently of the algebraic point of view mentioned above). A deeper analysis called forth the ideas of openness and closure, of dense sets and of sets nowhere dense, along with those of compact sets and complete sets, so often employed in questions of limits and of convergence. Another point of view is that of the measure of set, a problem intimately tied in with that of integration, which was to lead to the consideration of new definitions of the definite integral, broader and more satisfactory, the most famous of which is due to Henri Lebesgue.[26] From set theory, the labors of Fréchet (1906) brought us to the theory of abstract spaces, in which it is enough to define certain elements (functions, for example) as points in a fictitious space. Without stressing this further, let us point out that this notion has been very conducive to the development of the theory of operators, which finds application in many fields and is indispensable in certain problems of wave mechanics.[27] There is doubtless no point in laying further stress on the essential newness of these considerations.

Not only the field of study is broadened, but also the methods utilized. When the modern mathematician meets with a problem not solvable by classical calculations, far from giving it up, he accosts it by other methods. He skirts the difficulty completely by attacking the problem from other angles. There has been much stress on the nature of modern research, in which the scholar seems to struggle against a rebellious and resisting nature. Among the Greeks, constructions of the mind were studied; in the 17th and 18th centuries elements were combined with one another. At the present time we study questions raised, which absolutely must be resolved. In a certain sense we can say with Evariste Galois that mathematical method in modern times approaches that of the experimental sciences. We have already mentioned the method of successive approximations; the very term would have made the Greeks and the Cartesians shudder! In the theory of

---

[25] This is a central consideration in the theory of hypercomplex systems.
[26] Set theory has raised serious problems which have sometimes divided mathematicians. The most famous is that of the axiom of Zermelo, or the axiom of choice, which was the subject of very many discussions.
[27] See de Broglie: *La Physique nouvelle et la Théorie des quanta*, Flammarion.

integration, it was not possible to evaluate by direct calculation expressions of the form:

$$u = \int_\alpha \frac{dz}{\sqrt{az^4 + bz^3 + cz^2 + dz + 2}}.$$

By studying the properties of this expression directly, it was shown that the function $u = f(z)$ is (in the complex plane) uniform and doubly periodic; we have here the starting point for the study of elliptic functions. Certain differential equations were not integrable; mathematicians began to study the "singular points" and to deduce from them very interesting properties on the behavior of the solutions. In the theory of algebraic equations, it was known how to solve those of the third and fourth degree. No one had ever gone further. Leaving aside the direct problem of the solution, Evariste Galois studied the groups of roots of algebraic equations, and showed in particular the impossibility of a solution beyond the fourth degree.[28] What further characterizes the modern epoch is not only the multiplicity of studies, the considerable number of branches into which mathematics has ramified, but also the profound and frequent contacts among all these domains. The most beautiful results at the present time often arise from ingenious comparisons or conjunctions.

Having arrived at this point, the reader will perhaps ask the question: Are we on the eve of a new "epoch"? Do mathematicians feel nowadays that weariness which their predecessors experienced at the end of the 18th century? It is not my task to provide an answer to this question; it can be sought in other articles of this work. Are we perhaps passing through a "period"? In any case it does not matter very much; however it may be, we must always work with earnestness. This is the indispensable way of preparing in a more or less immediate future new points of view, which will again betoken new progress for the human mind; their import and breadth will doubtless surpass what we are able to imagine today.

[28] And he availed himself of this opportunity by constructing group theory. On this subject, see the articles by Lentin, p. 204, note 2, and Cartan, p. 264. (Note by F. LL.)

21

VIEWS ON NEWTON'S MATHEMATICAL THOUGHT

by Pierre Brunet

DIRECTOR OF THE SECTION ON THE HISTORY OF SCIENCES
AT THE INTERNATIONAL CENTER OF SYNTHESIS

IF Newton's work has a just claim to be considered a culminating point of the soaring of mathematical thought which marked the end of the 17th century, it is no less true that the mathematical labors of this thinker of genius "will always be," according to the remark by Arago, "only his second claim to immortality." It might be said that he had a presentiment of this, in not publishing them until long after his astronomical discoveries." And yet Newton seemed to be interested from the very first in mathematical questions. It seems to us that there is a deeper reason for the apparent anomaly in the reversal of chronological order of discovery in the systematic exposition, than the hypothetical influence of an obscure presentiment offered tentatively by Arago. This is that Newton really conceded to mathematics only a kind of instrumental value. To expound for their own sake certain new methods of calculation, before their fecundity and their necessity in certain problems brought up by the machinery of the universe had been made apparent, might therefore have seemed to him to a certain extent inopportune, if not altogether useless. We see from this that Newton's attitude on this point was not dictated to him either, as has been occasionally suggested, by the desire to astonish the readers of the *Principia*. At the most it could be said, with Clairaut, that he presumed too much on their sagacity, in leaving to them the necessity of supplementing, in many cases, the insufficiencies of his own explanations.

Be that as it may, even though Newton had been in possession of his method of fluxions (which we shall have above all to examine) as early as 1665–1666, and had nearly perfected his theory of gravitation some ten years later, it was only with the publication of his great work, in 1687, that he acquired his fame as a physicist and astronomer, and equal fame as a mathematician. He had been a professor at Trinity College, Cambridge, since 1669; but doubtless he was often absorbed in what he himself called his "mystic reveries" (he wrote a commentary on the *Apocalypse*), and it is very likely that a natural timidity (perhaps related to the delicacy of his constitution) also caused his reluctance to make his reflections known. He had attracted the attention of his colleagues only by his independence of mind; this was to merit his being chosen by them in 1688, after his fame was established, to represent them in Parliament, where he performed his duties until 1695. However, as early as 1671, the bishop of Salisbury had obtained his nomination to the Royal Society of London, and in 1675 he had received from the king the necessary dispensation to continue teaching at Trinity College without taking orders. Thus it was within the bounds of this college that a large part of Newton's life was passed, since, previous to his replacing Barrow as professor there he had first been his student, as of June 1660; he was at the time a little less than 18 years old (having been born on December 25, 1642, at Woolsthorpe, a hamlet of the parish of Colsterworth, in Lincolnshire), and his mother, widowed and remarried, had just decided, after an unfruitful attempt to initiate him into country life, to let him follow his inclination for study.

It is likely, moreover, that despite the fame of the *Principia* in the scientific circles of all of Europe, Newton would have lived out his life at Cambridge, if his career had not been suddenly changed by the advent to power (as Chancellor of the Exchequer) of the future Lord Halifax, who, a former student at Cambridge University, had remained a friend of the Trinity College professor. The support of this official personage procured for Newton the post, in 1695, of Warden of the Mint, then, in 1699, Master of the Mint. He then discontinued the professorship, although remaining incumbent of the chair until 1703, the date when he was chosen as president by the London Royal Society. He was to be subsequently reelected to this position each year until his death in 1727. It was shortly after his arrival in London that there broke out between Leibniz and him, or rather between the followers of these two scholars, the famous dispute over the priority of one or the other in the discovery of the infinitesimal calculus. We shall lay stress later on on this idea, by determining Newton's contribution in this affair. We shall content ourselves here, without giving the details of this dispute, with pointing out that, without the bitterness shown on both sides, it would have been possible to realize that there had been no plagiarism in the affair, and that the two mathematicians had arrived simultaneously, and by different paths, at analogous results. According to what one of his biographers (Brewster) says, Newton willingly compared himself to a child playing with his shells on the seashore, while the ocean of truth remained hidden from his eyes; very modest language, which has appeared to some to contradict his attitude in the quarrel of his partisans with those of Leibniz. But we must not be astonished to see a shy man show a great deal of irritability. Moreover, even in these circumstances, he deviated as little as possible from his horror of quarrels and love of tranquility. His disinclination to publish his works would be sufficient, in the absence of other testimony, to reveal this essential facet of his character.

Even if we isolated from the rest of his work its more strictly mathematical part, we could not propose considering here its various aspects, and we are obliged to restrict ourselves to pointing out its basic framework and its characteristic principles. What we have referred to above as the conception of mathematics as a tool deserves first of all to be brought to the foreground. Calculations ought above all to be of use in the solution of concrete problems; and, far from being an end in themselves, they are only a practical means of expressing in precise and concise formulas the data on which the mind works, of thus making them more tractable and leading to more distinct results. Does this mean that the mathematician should, as a consequence, consider himself as absolutely free to use this or that type of calculation in an entirely arbitrary manner? In following Newton in his work as closely as possible, it is seen, on the contrary, that it is never by chance that he decides to use a particular proof. In each case, what seems to have been a primary consideration for him is precisely the choice of method, at the same time as those considerations which will make it plausible and rational. Hence the necessity of at least an implicit criterion, intuitively discovered in the greater number of cases, independent, up to a certain point, of all fixed method, like every a priori deduction. Newton paid so much attention to the concrete character of problems that it is precisely the nature of each of them which was to decide the method to be applied to solve them. Concern for reality, in the physical meaning of the term, caused Newton to adopt an attitude which could, if one were willing not to see a gross anachronism there, be characterized as pragmatic. What matters after all is the success of the solution, even if the coherence of the mathematic structure suffers somewhat from it in its appearance. As with a perfected tool, which it is, we shall demand of the mathematical machinery, in every circumstance, to serve us as well as possible. However complicated, a problem must always be attacked by the method which will lead to its solution with the least difficulty. "Having several different methods at hand," Newton wrote in his *Method of Fluxions*, "we can always take the easiest at the same time as the simplest."

Still there must be a clear understanding about this simplicity which is always to be sought: we ought not to judge it from the abstract point of view alone, thinking only of the formulas involved. In a significant passage from the *Universal Arithmetic*, we read: "Constructions made by the parabola are the simplest of all, considering only analytic simplicity, then come those done by the hyperbola, and in the last place constructions by the ellipse. But, if you regard only practical simplicity, reverse this order." This practical simplicity is what Newton calls elsewhere "simplicity of description," that is, simplicity of graphical construction. From this point of view, we should, in certain problems, prefer curves expressed by more complicated formulas to the various conics.[1] We see how the concern for reality and experience ties in with the desire not to give up geometrical considerations.

Thus, for Newton, there is more a hierarchy of methods than there is, properly speaking, a hierarchy of problems. And if, depending upon the conditions of their use, a hierarchical classification of the methods of calculation is shown to be impossible, it is no less so if we start from historical considerations of the order of appearance of these various procedures. They occurred, each in its turn, at the exact time when the set of those effectively attained at a given moment proved

---

[1] It is known that the three curves considered in Newton's text are classed together under the heading of conics (conic sections are those obtained by the intersection of a cone by a plane). The constructions with which he is concerned are the intersections of curves useful in the solution of certain problems. (See the article by F. Le Lionnais, "Beauty in Mathematics, "Vol. II, Book Four, pp. 121–158.)

insufficient. Though they are successive deepenings of the resources of the mind, and as it were a progressive realization of all its potentialities for research, they do not imply, in each transition from one to the other, a discontinuity in the progress of thought. If Newton, on this point as on so many others, did not express himself very precisely, it seems hardly doubtful that the way in which he regarded his own discoveries implies that he saw them more as a kind of extension of earlier methods than as a break with traditional procedures. As Pierre Boutroux very properly remarked, he "does not seem to have had the feeling that he was going to change the face of mathematics." The enthusiasm of his followers did not permit them, in general, to maintain the same reserve, and it was they above all (followed in this by certain historians of science) who presented the creation of a new mathematics as a veritable revolution.

It is not astonishing, consequently, to recognize that, indeed, Newton, while finding himself at the origin of the brilliant movement of mathematical thought constituted by the development of analysis,[2] did not on this account repudiate resorting to geometry, practiced not only in the Cartesian form of analytic geometry (the application of algebra to geometry), but also in the fashion of the ancients. Pemberton tells us, moreover, that Newton claimed to have learned as much from reading Euclid as from Descartes. Thus Michel Chasles has been able to write, contrary to those who after having applied in analysis the geometric methods employed in the *Principia*, paid honor to that one of all the discoveries which that work contains: "A gratuitous supposition, which shows a lack of understanding of the rich nature of geometry and the extreme facility of its natural, and often even intuitive, deductions, in matters where it can be applied first. But it is enough for us to recall that, in order to attribute Newton's discoveries to the analytic method, we are obliged to admit that this geometer, in following this path, would have made use of the calculus of variations, the invention of which is due to the illustrious Lagrange." There is, without doubt, much to be retained from these remarks. Let us say, from the very first, so as not to return to this point, that it seems quite plausible to us not to make of Newton, in the absence of sufficient proof, a true precursor of the calculus of variations. But especially,

---

[2] We use the term here in its present sense, which is that of infinitesimal analysis. In Newton's time, the word analysis had a much broader meaning, coextensive with the idea of algebra.

apart from the scientific interest which Chasles might have had in keeping Newton in the camp of the geometers, up to a certain point,[3] this knowledgeable historian of geometric methods has very opportunely drawn attention to an important peculiarity of composition of the *Principia*. For, if it is indeed true that the whole work is, in Bertrand's forceful expression, "penetrated and dominated" by the calculus of fluxions, it is no less true that geometry maintains a large place therein.[4]

To better understand the range of Newton's mathematical discoveries and their precise meaning, a brief historical survey seems called for here.[5] Not to go back further, let us recall that Cavalieri, in his *Géométrie des indivisibles* (1635), considered planes as formed (not actually but abstractly) by an infinity of lines and solids by an infinity of planes, and denoted by the term indivisibles the infinitely small elements thus assumed. This method was to seem to Newton "not very geometric," and the hypothesis upon which it was based "too difficult to admit"; but it is seen that his thinking had been directed along these lines. Earlier, Pascal, without concealing from himself the element of arbitrariness possibly inherent in such hypotheses, had employed analogous, but more rigorous ones, in the solution of certain problems concerning the cycloid (the curve described by the rolling of a circle on a line). Thus the difficulties encountered in some mathematical questions, particularly in the quadrature of curves,[6] had led to a glimpse, as it were, of the calculus called integral, that is, that in which one determines, by the operation known as integration, the limit approached by a sum of terms whose number increases indefinitely,

---

[3] For we can say that at the time of Chasles, mathematicians were divided into analysts and geometers, and that the researches of the latter (not very numerous, by the way) were regarded by the former with some disdain.

[4] It is even in part to this that is due the apparent obscurity which certain passages present to us, and also the insufficiency of some explanations on various points. Indeed, instead of supposing with Clairaut and then Delambre that, in the theory of the irregularities of the moon, Newton presented as rigorous theorems results simply deduced from the examination of observations, it seems more likely to us to admit that instead of persisting in solving difficult problems of analysis, he found it more expedient to look for the solutions wanted by a geometric method which does not appear in the publication.

[5] One should compare what follows with the corresponding passage in the article by Germain, Part II, Book One, pp. 240–242. (Note by F. LL.)

[6] We understand by quadrature of a curve the finding of the area of this curve.

while the magnitude of each approaches zero. But we can say that this calculus did not exist before Newton.[7]

Quite as much as problems of quadrature, the problem of tangents had baffled the sagacity of geometers up to the moment when, almost simultaneously, Fermat (1636) and Descartes (1637) each proposed a method intended to solve it. The solution was sought for by Fermat in his method of maxima and minima, based on the principle that in the fluctuations of a quantity, those in the neighborhood of a maximum or minimum are virtually imperceptible. This brief statement is enough to indicate that the method tends to introduce the notion of the derivative.[8] Nevertheless, if it was, in a certain sense, a prefiguration of the calculus dealing with infinitesimal quantities, it still did not call for reasoning about the infinitely small. It is only in Newton's work that there truly develops the differential calculus, in which, as its name indicates, we have the occurrence of differentials, that is, infinitely small increments through which a variable passes in taking on successive values.

Still it was necessary in order to have the complete infinitesimal calculus—the combination of differential calculus and integral calculus—for the connection between these two new branches of mathematics to be explicitly stated (the integral calculus henceforth was going to be more and more off-handedly defined as integration of differentials). And it is again to Newton that this progress is due. Certainly, there had been a presentiment of the relation between the two branches of the calculus, since people had already designated as "the inverse problem of tangents" the one which consists in passing from the tangent to the curve, and which is in fact a problem of integration. But they had not succeeded in a generalization which

[7] We imply here, as below, the more precise addition: "and Leibniz" (so that we take into account our earlier conclusion on the question of priority of one or the other of these two mathematicians).

[8] The quantity denoted by this term (which appears only later) is in effect the limit of the ratio of the increments of two variables depending upon one another; that is, in the language of infinitesimal analysis (according to the Leibnizian notation which, as we shall see, prevailed), the ratio of infinitely small increments called differentials. Let us add that the word "derivative" is used as an abbreviation for the expression "derived function" (in contrast with the "primitive function" from which it is derived); and let us recall that one quantity is said to be a function of another quantity x when it varies with it and takes on a fixed value for each value given to the variable x.

We cannot sum up here in a few words Descartes's procedure, which, by the method called the method of indeterminates, involves considerations very close to those appearing in infinitesimal analysis.

saw that problems of quadrature also depended upon a like mathematical procedure, infinitesimal analysis. Now, after speaking of the method of tangents, Newton added: "That is rather a particular corollary of a general method which can be extended, without cumbersome calculation, not only to obtain the tangents to any curve whatever ... but also to solve other kinds of very difficult problems involving curvature, quadratures, rectifications, centers of gravity of curves."[9]

It remains for us to see how the infinitesimal calculus occurs in Newton's work. It assumes two forms there, almost equivalent, moreover, and of which one can be considered as coming from the other, the method which he named "of first and last ratios"*—quantities which arise and vanish—and which is none other than that of limits (the term usually employed by d'Alembert and which is also already found in Newton), and the method called "the method of fluxions." In the first of these methods, depending upon whether increasing or decreasing variables are being considered, upper and lower limits are sought—limits approached by the increments or decrements; the last ratio of "vanishing quantities" (quantities which approach zero) being "that which diminishing quantities have to one another, not before they vanish, nor after they have vanished, but that which they have at the very moment when they do vanish." "In the same way," Newton continues in his *Principia*, "the first ratio of nascent quantities is that which increasing quantities have when they begin to be and the first or last sum of these quantities is that which corresponds to what they had at the beginning or at the end of their existence, that is, at the moment when they begin to be augmented or cease to be diminished."

As to the methods of fluxions, the principle of it is thus summed up by Newton himself in his *Treatise on the Quadrature of Curves*: "I do not consider mathematical quantities as composed of extremely small parts, but as generated by a continual motion. Lines are described and generated, not by any apposition of parts, but by a continual motion of points. Surfaces are generated by the motion of lines; solids by the motion of surfaces; angles by the rotation of their sides; time, by a continual flux, and similarly for the rest. Therefore, considering that quantities increasing in equal times are greater or less according

---

[9] On the notion of quadrature, see page 254, footnote 6. To rectify a curve is to calculate its length, or better, to construct a straight line of equal length.

* [Also referred to in English as prime and ultimate ratios. TRANS.]

as they increase with a greater or smaller velocity, I sought a method for determining the quantities from the velocities of their motions or increments, by which they are generated; and by calling the velocities of the motions or of the increments by the name of fluxions, and the generated quantities fluents, I did light upon the method of fluxions." In the *Principa*, after making clear that by moments he means the momentary increments or decrements of the variables, Newton adds, "It will be the same thing, if, instead of moments, we use the velocities of the increments and decrements, which may also be called the motions, mutations, and fluxions of quantities." And it is again Newton who, in his *Treatise on the Quadrature of Curves*, remarks that "the fluxions of lines, whether straight or curved, in all cases whatever, as also the fluxions of surfaces, angles, and other quantities, can be obtained in the same manner by means of the method of First and Last Ratios." Indeed, the fluxions are the first and last ratios of the spaces traversed and of the time consumed in traversing them. And this is how the method of fluxions derives naturally from that of the first and last ratios.

It is quite certain that to give the method its full efficacy, it was necessary to create an adequate system of notation. To denote the fluxions of the variables x, y, z, Newton used the same letters with a dot over each, while he represented the fluxions of the fluxions by the same letters still, with double dots over each; he used one or two indices also placed over the letters x, y, z to indicate the fluents and the fluents of the fluents, respectively. It is true that we do not find in his writings a symbol for integration comparable, above all from the point of view of convenience, to the one which mathematicians continue to borrow from Leibniz (namely, $\int$) and which, with the characteristic d, also due to Newton's rival, constitutes the essential algorithm of the infinitesimal calculus.

In the text from the *Principia* devoted to the method of fluxions, Newton states precisely that he "intermingled" this method with that of infinite series; and he had already made the same remark in a letter to Collins, dated December 10, 1672. The calculus of series, as practised by Newton, found its antecedents in certain works of English mathematicians, such as Wallis (*Arithmetica universalis*, 1665), Mercator (*Logarithmotechnia*, 1668) and Barrow (*Lectiones geometricae*, 1670). The idea on which it is based is that of approximation. It was seen, in fact, that the theory of expansion in series allowed the representation, by convergent algebraic expressions involving a single variable x, even of transcendental functions[10]—in the same way that it made possible from the very first the representation of an irrational number by a convergent arithmetic expression. Even before Newton, such processes of the calculus had been applied in the solution of problems of quadrature, as well as the problem of tangents; and it was moreover partly by this path that Newton arrived at the discovery of the infinitesimal calculus.

But, still more than by the way in which, for Newton, it was related to this calculus, the method of expansion in series was to prove itself fertile, in the hands of this scholar, by the unlimited field which it would open, thanks to him, in the general study of functions. In this order of ideas, his discovery of the formula for the expansion of the binomial $(1 + x)^m$ for any value whatever of the exponent (positive or negative, rational or irrational) marked a considerable progress which, with others of the same kind, largely contributed to the growth in power of the new calculus.

The way in which Newton proceeded, in certain cases, in the course of his study of series expansions, enables us to say that he sometimes employed arguments by analogy. If, for example, the first terms of a series exhibit a law of regular succession, it can be assumed that the regularity will continue to be manifested; similarly, one is correct in supposing that an approximate result obtained by replacing infinitesimal terms with quantities that are only very small will assume an exact value in the limit. But analogy plays a no less characteristic role when, applied to formulas, it lets us make interesting comparisons between certain ideas which geometry sets apart. It was thus that Newton explicitly noted that certain problems concerning the

---

[10] We have recalled above (p. 255, note 8) what is meant by functions. When the quantitative relations connecting variable quantities involve only the fundamental operations of algebra, the functions are called algebraic, while they are transcendental when other operations are involved (examples: $\log x$, $\sin x$). Let us note that we give the name of series to the sum (or an alternation of sums and differences) of an infinite sequence of terms (numbers) formed according to some specified law; and that a series is convergent when the sum of its terms approaches a certain limit indefinitely as we take more and more terms. This limit is the sum of the series.

As examples of the results obtained by Newton in this area, we can give:

$$\arctan x = x - \frac{x^3}{3} + \frac{x^5}{5} - \frac{x^7}{7} + \cdots$$

$$\sin x = x - \frac{x^3}{1 \cdot 2 \cdot 3} + \frac{x^5}{1 \cdot 2 \cdot 3 \cdot 4 \cdot 5} - \cdots$$

circle and the hyperbola could be treated analytically in a similar manner, although geometrically these curves seem quite different.[11]

But there is another method used by Newton of which he himself liked to say that the problem solved by it was "perhaps the most beautiful of all those whose solutions he desired"; this was interpolation. This ingenious procedure, indeed, permits solving by approximation certain problems involving curves whose direct solution would present too many difficulties. The idea on which this method is essentially based is that a curve, however complicated it may be, can always be intercalated between two simpler curves, exact knowledge of which allows us to calculate approximately the desired curve; clearly, the greater number of common points the greater will be the degree of approximation. We see how, more than in analogy, interpolation assumes continuity, and in what sense we can consequently consider this idea, also implied directly by the infinitesimal calculus, as essential of Newtonian thought.

The interest shown by Newton in geometry and Cartesian analysis is especially revealed in his *Enumeration of Curves of the Third Degree*. We know, and Newton expressly remarked this, that geometric curves[12] are completely classified in orders according to the degree of their equation,[13] or, alternately, according to the number of points in which they can be cut by a line. But he notes explicitly that, if consequently a curve of the third degree can be cut by a line in three points, it sometimes happens that two of these points coincide, and even, in certain cases, that the double point (= two coincident points of intersection) is carried off to infinity (the curve is then effectively intersected in only one point). Now, for all curves having a double point, Newton established that they can be constructed when seven points, including the double point, are known. In the matter of the number of curves of the third order, he distinguished seventy-two kinds of them, reducible, moreover, to four cases. Despite the ingenuity of these insights, interest in this work would remain less than that in Newton's other studies, if it did not contain this remarkable

---

[11] The reader should compare this passage with the article by Deltheil, "Analogy in Mathematics," Part I, Book One, pp. 37–43. (Note by F. LL.)

[12] In the 17th century curves which we now call algebraic (according to the criterion indicated in an earlier note) were referred to as geometric. In contradistinction, curves presently designated as transcendental were then called mechanical.

[13] Descartes had already very clearly explained this. Practically, we have come to speak normally of curves of this or that degree.

assertion, given, it is true, without proof, but afterward confirmed by Clairaut, Nicole and other mathematicians: "Just as the circle, being exposed at a luminous point, gives all the curves of the second degree by its shadow, so do the five divergent parabolas give by their shadow all the other curves of the third degree."[14]

In the field of algebra, the notes furnished by Newton in the theory of equations concerned not only the means of calculating the roots,[15] having the advantage of being applicable as well in the case of transcendental equations as in that of algebraic equations, but also certain rules for finding the maximum number of roots of this kind or that. What he established, in this order of ideas, is particularly interesting when imaginary roots, which he called "impossible" (Descartes had already called them "false") are involved. It is to be remarked, moreover (the expression is significant), that Newton did not yet have, any more than his predecessors, the idea of employing calculations on imaginaries; the determination of such roots assumes, for him, its entire usefulness from the fact that it permits the recognition of cases not capable of concrete interpretation. "It is necessary," he wrote, "for equations to have impossible roots, without which, in applications, certain impossible cases would become possible." We see once more by this example that Newton never lost sight of the reality of experience; and we understand so much the more why precisely the imaginaries could have no value of their own, in his eyes. If he did not dwell upon this longer, it is not due to an oversight on his part, but a very deliberate attitude, dictated again by the conception of the instrumental nature of mathematics.

[14] The equation representing the parabolas in question is of the type:

$$y^2 = ax^3 + bx^2 + cx + d.$$

[referred to today as the semi-cubical parabola. TRANS.]

Let us note that the proofs given later by various mathematicians of the 18th century are all imbued with analytic considerations. It is, on the contrary, by geometric proofs (very likely more consistent with Newtonian thought on this point) that Chasles came to the same conclusion, and besides established another way of generating all curves of the third degree by the projection of five of them.

[15] Let us recall that the roots are values to be given to the unknowns in order to satisfy the equation; we can therefore say that they are the solutions of it. These roots (whose number depends upon the degree of the equation) can be real (positive or negative) or imaginary (if they involve in algebraic symbolism the extraction of the square root of a negative number, which is naturally an impossible operation).

This remark, which brings us back to our point of departure, invites us besides to consider that, if mathematics appeared to Newton as an instrument of expression, of research and sometimes even of discovery, it was also, for him, in a certain sense, an instrument of control. In fact, it permitted the scientist, in many cases, to assure himself that his explanation of phenomena did not involve him in unsafe hypotheses, incompatible with the spirit of true experimental science. We must not forget that the formula which became famous, *hypotheses non fingo* (I frame no hypotheses), corresponds to a profound tendency of the Newtonian mind; it is not fitting, in astronomy any more than in physics, to devise explanations more or less tinged with metaphysics or depending upon arbitrary and gratuitous suppositions. And precisely every impossibility of translation into formulas must be interpreted as an unfavorable presumption with regard to this or that idea, however ingenious it may seem in other regards; while conversely we have the right to feel satisfied with every conception which, expressed as a formula, allows us to attack reality in an efficacious manner. Subject to this criterion, the theory of universal gravitation takes on its full meaning and its full value.

22

A CENTENARY: SOPHUS LIE

by Élie Cartan

MEMBER OF THE ACADEMIE DES SCIENCES AND OF THE
BUREAU OF LONGITUDES

THE year 1942 marked the centenary of the birth of an illustrious Norwegian mathematician, Sophus Lie. The cultivated public knows the great novelists, the great dramatists, the great musicians of Norway; it is less well acquainted with the contributions which that nation has made to the other branches of intellectual activity. Science owes it in particular two very great mathematicians: the first, Niels Henrik Abel, who was born in 1802 and died at the age of 27, left behind a work of wonderful profundity; the second is Sophus Lie (1842–1898).

Marius Sophus Lie, born on December 17, 1842, was the son of a pastor of the little village of Eid, at the base of the Eidesfjord, one of the branches of the Nordfjord, which plunges deeply into the land 200 kilometers north of Bergen and some 120 kilometers southwest of Andalsnes, whose name was often mentioned in the tragic hours of May 1940. Pastor Lie having been called in 1851 to carry on his ministry in the small town of Moss, located on the shore of Oslo Fjord, not far from the capital, which was then known as Christiania, little Sophus attended the town's elementary school; at 15 he was a pupil at a private school in Christiania, and two years later began courses at the University of Christiania. He was a good pupil in all subjects, nothing about him foreshadowing a particular aptitude for one field rather than another. He thus had none of that mathematical precocity which made Pascal and Clairaut famous early in their lives and which Abel's mathematics professor had noticed in him when he was quite young.

Armed with his bachelor of science diploma by 1865, Sophus Lie hesitated for quite some time over the path he should follow, uncertain of his calling and suffering from this uncertainty. Suddenly, in 1868, his reading of the works of the French geometer Poncelet and the German geometer Plücker revealed to him his own potentiality and he felt irresistibly the call of mathematics. He retained throughout his whole life a great admiration for these two masters, whom he did not know personally, since Poncelet and Plücker had just died, the first in 1867, the second in 1868; but they both exerted a profound influence on the general orientation of his work.

The first articles by Sophus Lie procured for him in 1869 a government scholarship which enabled him to enjoy a lengthy sojourn abroad. He spent the winter of 1869–1870 in Berlin, where he made the acquaintance of a pupil of Plücker, Felix Klein; he became friends with him; they worked together and exchanged ideas, and there is no doubt that this exchange was as fruitful for one as for the other. In the summer of 1870 they went together to Paris, where they made the acquaintance of Camille Jordan and Gaston Darboux, the latter the same age as Lie, the former his senior by four years; these are two of the glories of French science. It was in Paris that Sophus Lie made one of his most beautiful discoveries, the famous transformation which bears his name and which establishes an unforeseen relation between lines and spheres in space on the one hand and between asymptotic lines and lines of curvature of surfaces on the other. It elicited the admiration of all geometers and it would be sufficient by itself to immortalize his name. Upon the declaration of the Franco-Prussian war, Klein returned to Berlin; Sophus Lie remained in France and conceived the idea of going on foot to Italy, crossing all of France. He was unfortunately arrested almost at the beginning, while passing through the forest of Fontainebleau; perhaps his absent-minded mathematician's ways—perhaps simply his physical appearance, which revealed him without possible doubt as a foreigner—led to suspicion of espionage, and he had to spend four weeks in the Fontainebleau prison, from which Darboux's pressing intervention succeeded in having him released. Sophus Lie does not seem to have held a grudge against the French for this little mishap, which, after all, perhaps provided him with an asylum where he could meditate and work in complete tranquility. From Italy, where he arrived without further hindrance, he returned to Christiania through Switzerland and Germany.

Having received his doctorate of science in 1871 with a thesis in which he developed the numerous consequences of the discovery he had made in Paris, Sophus Lie taught for a year at the University of Lund in Sweden (at this time Sweden and Norway were one country); but, thanks to the intervention of his friends and the recommendation of foreign scholars who appreciated the value of his work, the Norwegian Parliament created a chair for him at the University of Christiania. He was then able to devote himself with complete tranquility of spirit to the development of his ideas. It was at this time that he married a distant relative of his famous compatriot Abel; it was a happy household, blessed by two daughters and a son.

This period of his life is marked by great intellectual activity. Papers followed papers, papers in geometry, papers in analysis on the integration of partial differential equations. Naturally I cannot here analyze these works, in which was made felt, even though in an indirect way, the influence of the great French geometer Gaspard Monge, one of the founders of the Ecole Polytechnique, who had shown in his work the mutual support analysis and geometry could give instead of being opposed to one another. But Sophus Lie owes his glory above all to a theory of which he is the uncontested creator, the theory of transformation groups, of which he had early felt the necessity and fruitfulness. In this sense—and he proudly proclaimed it himself—he is the continuator of the French mathematician Evariste Galois, whose short and tormented career (he died at 20) left an immortal work, the repercussions of which have not yet perhaps been entirely felt.

Galois had shown that the idea of the group of permutations furnishes the key to the entire theory of the solution of algebraic equations, and gave in particular the underlying reason for the impossibility of the solution by radicals of equations of degree higher than four, an impossibility which had been shown for equations of the fifth degree by Abel. Lie had already shown the role that the idea of transformation groups can play in theories of integration; as early as 1872, his friend Felix Klein, in the famous *Erlangen Program*, had shown in a masterly fashion the fundamental role which it plays in geometry. We can now say that it has invaded almost every area of mathematics and physics. But the theory of transformation groups itself, its technique, had not been created, and nothing indicated the path to be followed for that creation. Sophus Lie devoted himself to this work from 1873 and by intense labor rapidly managed to construct the fundamental theorems from which he quickly deduced very many consequences.

In 1882, upon reading a paper of the French mathematician Halphen, Sophus Lie realized that his earlier researches enabled him to see in perspective the problem considered by Halphen. He felt the necessity of expounding in one great didactic work the results of his earlier researches, particularly those dealing with group theory. Thanks to the devoted collaboration of a young German mathematician, Friedrich Engel, the projected work was written and published after nine years' labor; it appeared successively in three large volumes between 1888 and 1893.

In 1886 Lie was summoned to the University of Leipzig to succeed Klein, who had been appointed professor at the University of Göttingen. He was then able to gather pupils around him; he was especially happy and proud to see coming to Leipzig a succession of young students of the Ecole Normale Supérieure, sent by Jules Tannery, the director of the school, to study group theory at its very fountainhead. I shall content myself by mentioning among them Ernest Vessiot, honorary director of the Ecole Normale, who greatly contributed by his writings to making known and extending Lie's work.

In 1892 Sophus Lie went to Paris to spend six months, interesting himself with a great good will in the research which young French mathematicians were devoting to group theory. He could often be seen with them around a table at the Café de la Source, on the Boulevard Saint-Michel; it was not unusual for the white marble table top to be covered with formulas in pencil which the master had written to illustrate the exposition of his ideas. The indifference with which his first work had been greeted, one could even say ignored, by the majority of mathematicians, had now given way to admiration; most of the great academies, except, however, that of Berlin, had been anxious to consider him one of their members. During his stay in Paris, on June 7, 1892, the Paris Academy of Sciences accepted him as a corresponding member in the geometry section.

It was the following year, in 1893, that the third volume of his great work appeared; it was dedicated to the Ecole Normale Supérieure. This acknowledgment was a mark of respect for a school which had earlier had Evariste Galois as a pupil and whose director had understood the importance of the work of the great Norwegian scholar. Sophus Lie was to return to Paris a short time later in 1895; indeed, he was anxious to attend the celebration of the centenary of the Ecole Normale and, in the book published on this occasion and dedicated to the glories of this school, he had felt the obligation to write an article of his own, one full of deep insight into the mathematical work of the "archicube" (third-year student) Galois.

In 1897 the Mathematical Physics Society of Kazan was to award for the first time the Lobachevsky prize. Due to a foundation created in honor of the Russian mathematician who was the first to make a systematic study of non-Euclidean geometry based on the denial of Euclid's postulate, this prize of considerable value was to be conferred every five years to the best book published in geometry, in particular non-Euclidean geometry. Upon a recommendation—very interesting in itself—by Felix Klein, the prize was awarded to Sophus Lie. A large part of the third volume of his great work had indeed been devoted to a profound discussion of the hypotheses or axioms needed as a basis for geometry in order to deduce from them either the classical geometry of Euclid or the non-Euclidean geometries of Lobachevsky and Riemann, depending upon whether or not one admits Euclid's postulate. These studies formed a continuation of earlier works by the great physicist Helmholtz, correcting them on some points, making them more precise and carrying them on to a conclusion.

It was shortly after the conferring of the Lobachevsky prize that Sophus Lie left Leipzig to occupy at Christiania a chair in group theory which his country had just established for him. But he arrived there in a very precarious state of health. Pernicious anemia slowly consumed his strength and he passed away quietly on February 18, 1898 at the age of 56.

In 1921, Engel, with the help of Heegaard, professor at the University of Christiania, began the publication of Lie's *Complete Works*; they comprised six large volumes, filled with many notes and with very interesting extracts from Lie's correspondence with various foreign mathematicians. This beautiful monument to Sophus Lie's glory had hardly been completed when in 1936 an international congress, made up of a large number of mathematicians from every country of the world, assembled at Oslo. At one of the meetings of this congress, in the presence of the most eminent Norwegian personages, in the great hall of the university, there was unveiled a bust of Sophus Lie, and one of the French delegates paid the homage of the Ecole Normale to his memory. Since 1939 a replica of this bust, a gift of

Herman Lie, son of the great geometer, graces the scientific library of this school.

Sophus Lie was of tall stature and had the classic Nordic appearance. A full blond beard framed his face and his gray-blue eyes sparkled behind his eyeglasses. He gave the impression of unusual physical strength. One always immediately felt at ease with him, certain beforehand of his sincerity and his loyalty. He was not afraid to admit his ignorance of branches of mathematics unfamiliar to him, which nevertheless did not keep him from being aware of his own worth. He was not always, so it seemed, easy to get along with, above all after a severe attack of neurasthenia which had deprived him of all sleep and any possibility of work. He had found it necessary to make a long sojourn in a sanitarium. He had gradually recovered his health and resumed his activity. But if we are to believe his collaborator Engel, from whom I got these details, his personality was changed; particularly, he had become extremely irritable. He had a tendency, despite his own richness of thought, to believe that his pupils were stealing his ideas. But what are these petty eccentricities in comparison with the greatness of his work? Posterity will see in him only the genius who created the theory of transformation groups, and we French shall never be able to forget the ties which bind us to him and which make his memory dear to us.

23

WOMEN MATHEMATICIANS

by Mme. Marie-Louise Dubreil-Jacotin
PROFESSOR, SCIENCE FACULTY, POITIERS

WOMEN who have left behind a name in mathematics are very few in number—three or four, perhaps five. Does this say, as a common prejudice would tend to persuade us, that mathematics, so very abstract, is not congenial to the feminine disposition? This would be to ignore the true character of mathematics, to forget, as Henry Poincaré said, "the feeling of mathematical beauty, of the harmony of numbers and of forms, of geometric elegance. It is a genuinely esthetic feeling, which all mathematicians know. And this is sensitivity." And should not this very sensitivity, contrary to the prejudice, make mathematics a feminine domain? Moreover, if we consider the other sciences, the great inventions whose applications have staggered the world, we do not find any more women's names associated with these areas. Must it then be said, as Maurice d'Ocagne seems to conclude in his *Etudes sur les Femmes de Science et sur les Mathématiciennes*, that woman is generally destitute of inventive spirit and creative genius? The growth of female education, the overthrow of prejudices, the profound changes in the kind of life and in the role assigned to woman during the last few years will doubtless bring about a revision of her position in science. Then we shall see in what measure she can, as the equal of man, emerge from the role of the excellent pupil or the perfect collaborator, and join those of our scientists whose work has opened new paths and bears the mark of genius.

Yet, it is as early as Greek antiquity that the first woman who can be considered as a mathematician makes her appearance: Hypatia, born at Alexandria in the year 370 of our era. In a school which she opened in her native city, she expounded at one and the same time

Plato, Aristotle, the works of Diophantus and the conic sections of Apollonius of Perga. It is also believed that a commentary on the tables of Ptolemy, which has come down to us under Theon's name, is due to her. Hypatia acquired a great reputation as much for her science as for her eloquence and her beauty. She died tragically at the age of 45; she was thrown from her carriage and stoned by a crowd inflamed by monks at the instigation, it appears, of the patriarch St. Cyril.

Then, from the 16th century on, while the lords devoted themselves to politics and war, the ladies of high society gave themselves to the joys of the mind, and mathematics, by its very abstraction and beauty, had a preferential place there which did nothing but expand in the two following centuries, when a certain snobbishness seems, however, to have been mingled with the charm found by certain women in algebra and geometry. And it is perhaps this very epoch that saw the fullest appearance of those truly feminine qualities which make some great lady not a scientist, but the best of pupils or the fine and clear-sighted collaborator of some great scientist. There was Vieta, the creator of algebra, who had as his best pupil Catherine de Parthenay, Princess of Rohan-Soubise. Then Descartes, who had as his disciple Christine, Queen of Sweden, and Elizabeth of Bohemia, Palatine Princess; and Leibniz with Sophie, Electress of Hanover, and her daughter Sophie-Charlotte, Queen of Prussia and mother of the great Frederick. Euler wrote his *Lettres à une princesse d'Allemagne* to teach science to the Princess Anhalt-Dessau. Newton was studied and understood by Caroline of Brandenburg Anspach, and Emilie de Breteuil, Marquise du Châtelet, noted for her liaison with Voltaire, translated his principles and added certain personal notes to her translation—notes with which it may be Clairaut was not entirely unfamiliar. The latter, moreover, used the long calculations made for him by Mme. Lepaute, the charming wife of the celebrated watch-maker of Louis XV, without even mentioning her, "to oblige," says Lalande, "a woman jealous of the attainments of Mme. Lepaute and who had pretensions without any kind of knowledge. She succeeded in making a wise but weak scientist, whom she had captivated, commit this injustice."

But let us abandon these brilliant women, more or less amateurs, and the admirable astronomers' assistants such as Mme. Yvon Villarceau, Mme. Lefrançais de Lalande, and others, in order to come to those who, if they cannot all be set among the greatest names, can at least cut an honorable figure among the great mathematicians.

First we have the Italian Maria Gaetana Agnesi (1718–1799), the first woman professor of mathematics on a faculty; indeed, she was appointed professor at the University of Bologna by Pope Benedict XIV. Maria Gaetana Agnesi was a very gifted and precocious child. Her father, Don Pietro Agnesi, feudatory of Monteveglia, actively encouraged his daughter in her studies and, proud of her, exhibited her in the kinds of academic meetings fashionable at that time. But, of a self-effacing nature, she went into the background again after the death of her father and finally went into retreat among the nuns of the "Azure" order, where she is said to have given of herself unsparingly and to have been an object of great edification. She left behind, under the name of *Instituzioni analitiche*, a "remarkable account of ordinary algebra, with the solution of several solved and unsolved geometric problems"; a second volume, entirely devoted to infinitesimal analysis, a science then quite new, was declared "the most complete and the best done in this field" by the commissioners of the Academy of Sciences of Paris, who were assigned to examine this work at their meeting of December 6, 1749. This book was translated into French and published "under the license of the Royal Academy of Sciences" by a decision of August 30, 1775, signed by the permanent secretary of the Academy, upon the advice of the commission composed of d'Alembert, the Marquis of Condorcet and Vandermonde. The contribution of Agnesi to the study of curves of the third degree was great enough that one of them still bears her name: "the Witch of Agnesi."

Far from being encouraged by their families, it was only by main force against their parents that the two following women succeeded, almost simultaneously, in becoming brilliant mathematicians. First, Sophie Germain, born in Paris in 1776, the daughter of a rich middle-class silk merchant. As a child, Sophie received a very careful education and upbringing. At the age of 13, she found by chance in her father's library Montucla's history of mathematics. She thus read that Archimedes was killed by Roman soldiers because, entirely absorbed in the study of a problem, he was unaware of the capture of Syracuse and did not answer their questions. That one could be so thoroughly absorbed by a mathematical question as to forget everything, even the threat of death, filled her with such admiration that she wished at any cost to plunge herself into the study of this science. It was then that she encountered paternal hostility; nothing daunted her—she worked at night wrapped in a blanket, because they had taken her clothing away to keep her from getting up; they took away her heat, her light—all this only hardened her resolve and increased her fervor, so that her father finally gave in and she was at last able to devote herself to mathematics.

Her knowledge was already quite extensive in 1794 when the Ecole Polytechnique was founded; she was then able to obtain the notes of Lagrange's course and studied them with profit. It was from their study that her first personal observations took form. She wished to submit them to her master Lagrange and decided to write him; but, as she confessed much later, fearing the "ridicule attached to the name of *femme savante*," she signed her letter "Le Blanc, pupil at the Ecole Polytechnique." But Lagrange wanted to meet the Polytechnic student with such interesting observations. Astonished and charmed, Lagrange became a valued adviser for Sophie Germain and introduced her to all the French scientists of the time. She was quickly appreciated in scientific circles, as much for her learning as for the charm of her conversation. Nevertheless, later on, when she wanted to write Gauss, after the publication of his *Disquisitiones arithmeticae* in 1801, to discuss with him results she had obtained in the theory of numbers, she once more concealed herself under the pseudonym of Le Blanc, Polytechnic student. But Gauss, too, learned the true identity of Le Blanc. It was during the time of the German campaign and French troops were entering Brunswick, Gauss's city; Sophie Germain, haunted by the memory of Archimedes' death, began to fear for the scholar and wrote to a friend of her father, General Pernety, who was at the very moment in Brunswick, to commend her master to him and to entreat him to watch out for his safety. The latter hastened to reassure Sophie Germain and . . . to show her letter to the party concerned, who naturally continued with the young feminine mathematician the epistolary contact begun with the self-styled male Polytechnic student.

Sophie Germain died at the age of 55, after two years of terrible suffering which she bore with an admirable courage and stoicism. Her moral worth was a match for her beautiful intelligence; she loved virtue, it was said, like a geometric truth. She had the great honor in 1816 of receiving from the Paris Academy of Sciences the Grand Prize of the Mathematical Sciences, for a paper on the vibrations of thin elastic plates, a question put up for competition since 1811. It was in the year of her death that her important work on the curvature of surfaces appeared, in which she introduced for the first time the notion—classic today—of mean curvature. Her arithmetic work is no less important. Attacking the proof of Fermat's last theorem with the help of Legendre's formulas, she supplied an important theorem and its application to the proof of Fermat's theorem up to the hundredth degree.[1] Finally, aside from her mathematical work, she left a certain number of articles on the history and philosophy of the sciences, articles of a genuine value, which Auguste Comte quoted with praise in his course on positive philosophy.

Her contemporary, Mary Fairfax, born in 1780, met from her father, a Scotch admiral, the same hostility toward her mathematical studies; and, despite her precocious propensities, it was only after a short widowhood and then remarriage to her cousin Somerville, that she succeeded in asserting herself as a mathematician. A long life—she died in Naples at 92—made it possible for her to leave a respectable output under the name of Mary Somerville, as well as the memory of a devoted wife and a good mother. Mary Somerville's principal work consisted of translating and thus making known to her contemporaries the celestial mechanics of Laplace and of adding to it personal notes of real value. Mary Somerville also left a goodly number of papers in mathematics and physics; she was pensioned by Queen Victoria for her scientific work.

Mary Somerville had also had the honor of introducing to mathematics the only daughter of Lord Byron, Ada Byron—Countess Lovelace—born in 1815, died in 1852. Raised far from all paternal influence, Ada Byron was early attracted to mathematics, and distinguished herself therein; she left original works which she signed A. L. L., a pseudonym whose true meaning was disclosed only thirty years later by General Menabrea, a correspondent of the Paris Academy of Sciences, and Italian ambassador to France.

Here finally are the last two, the two most striking—the Russian Sophie Kovalevski (1850–1891) and the German Emmy Noether (1882–1936)—so different from one another.

The first, Sophie Kovalevski, née Korvina-Krukovski, was a direct descendant of Mathias Corvinus, King of Hungary; her grandfather, by his marriage to a Gypsy, had lost his hereditary title of prince; on

---

[1] On this subject, see the article by Got, Part I, Book Two, pp. 81–91. (Note by F. LL.)

her maternal side, her grandfather was a stern scholar and to these contrary heritages Sophie attached great importance, explaining by them the sudden changes of her passionate character, so essentially feminine. "This morning," she wrote to her friend, Mme. Anne-Charlotte Leffler, when, as professor at Stockholm, she was spending a few days of vacation at Berlin, "I awakened with the greatest desire of having fun; all of a sudden who came in but my maternal grandfather, the German pedant, that is, the astronomer. He looked at the scholarly dissertations which I had promised myself to study during the Easter recess, and reproached me most seriously for unworthily wasting my time. His severe words put to flight my poor Gypsy grandmother."

It is impossible to recount here the childhood of Sophie in Russia, her life in the great family house, that winter spent in Saint Petersburg, where at thirteen she was infatuated with Dostoevski and jealous of her sister, whom he came to see frequently; one should read all this recounted by herself with so much talent in her *Recollections of Childhood*, which has merited this appreciation: "The Russian and Scandinavian literary critics have been unanimous in declaring that Sophie Kovalevski was the equal of the best writers of Russian literature, in style as well as in subject matter." Her premature death prevented her from realizing all her literary projects; in particular she had wanted to write "The Razhevski Sisters during the Commune," a reminiscence of her trip to Paris in 1871. She loved enormously to write and found in literature an escape and a respite from the excessive labor to which she applied herself during the periods of her life when mathematical research engrossed her. She wrote, moreover, in her memoirs: "At twelve years old I was thoroughly convinced I was born a poet." But at the same age, she listened eagerly to her uncle, who had just bought several books on mathematics, speaking to her about the quadrature of the circle, and, she says: "If the meaning of his words was unintelligible to me, they struck my imagination and inspired me with a kind of veneration for mathematics, as for a superior, mysterious science, opening to its initiates a new and marvelous world, inaccessible to the ordinary mortal." As to her subsequent ease in understanding analysis, she attributed that to the deep impression left in her mind by the hours she spent as a child contemplating the unusual wall covering in her room, in the large country mansion; not having any wallpaper, her father had used the lithographed pages of Ostrogradsky's course on integral calculus!

But let us come back to the life of Sophie Kovalevski. At seventeen she shared the thirst for science of all the intelligent Russian youth of the period, its desire for freedom and its political aspirations. Being a girl, she did not have the necessary independence to pursue her studies, so she decided with her sister and a girlfriend that one of them would make a fictitious marriage which would permit the three of them to attend a foreign university. Kovalevski, a very gifted young student, consented to help them, but on the condition that Sophie marry him. The young couple were installed at Heidelberg and spent a year wholly devoted to work—he studied geology, she mathematics. According to the report of her contemporaries, she seemed perfectly happy, with very great good fortune, of which she nevertheless later spoke with a certain bitterness. Small, slender, her hair short and curly, she captivated everybody with an unconscious charm. The coming into her home of her sister and her friend put an end to this intellectual intimacy with her husband; the latter left for Jena and then Munich, and if he came to see her frequently, Sophie nonetheless suffered from this separation and from the impression that her husband, totally absorbed in his studies, suffered from it less than she did. "He loves me only when near me," she said, at the same time refusing to end their irregular state of affairs. She always had, it seemed, a certain predilection for strained relations; she was a tortured being; she needed to be encouraged and admired. Too impassioned, she was not satisfied with abstract research. She had to be understood, encouraged with each new idea which came to her; a woman's weakness, Mittag-Leffler said deservedly—but how complex and winning a personality.

In 1870 she decided to go to Berlin to see Weierstrass, who, at the age of 55, was then in all his glory, and asked him for his advice. "To conceal her emotion, Sophie had put on a large hat with a broad brim, so that Weierstrass could see nothing of those marvelous eyes whose eloquence no one could resist when she wanted something." Weierstrass was won over nevertheless, and not being able to have Sophie admitted to his courses at the faculty—the university then being closed to women—he decided to give her private lessons. Sophie went to see him every Sunday afternoon, and Weierstrass came to see her one day a week. For four years, Sophie worked thus furiously, having no diversion, entirely absorbed in her studies, which were directly inspired by Weierstrass' work, of which they were either appendages or developments.

In 1874 she received from the University of Göttingen the title of doctor *in absentia* for her thesis *On the Theory of Partial Difference Equations*. But she was then completely exhausted by this excessive labor. She also lived in lamentable material circumstances, for, thoroughly unpractical and totally absorbed in mathematics, she was unaware of the insufficiency of her diet, the lack of comfort in her room and even, so they said, of whether her clothing was torn. Excessively worn out, surfeited with science to the point of no longer taking pleasure in research or discovery, she returned to Russia; she enjoyed herself, read novels, played cards and, upon the death of her father, feeling herself all alone, finally allowed herself to be loved by her husband and became his wife. The couple settled down in Saint Petersburg and led a brilliant life. The joy of living carried her away in a whirl of pleasures and parties; she was surrounded, admired, flattered. She forgot mathematics and her teacher Weierstrass; she no longer wrote to him and did not answer his letters, not even the one in which he urgently asked her to deny the rumor current in Berlin that she had become a society woman and had abandoned mathematics.

In 1878, Sophie brought into the world a little girl, Foufie; the enforced rest gave her back some desire for work and she wrote Weierstrass for a research project. The latter, forgetting her former ingratitude, always ready to help her, answered her; but it was only two years later that Sophie Kovalevski truly returned to mathematics. The immediate cause was a financial disaster: the brilliant life of the young couple required an income greater than theirs, and they had gambled on speculations which had good success only in the beginning. In this misfortune, Sophie found herself again, the born mathematician, able to live only for science. Her husband, on the contrary, could not decide to return to the simple life of the scholar and persisted in continuing his collaboration with the speculator who had involved them; it was the break-up of their marriage. Sophie went away again to work abroad. She stopped off first at Berlin, and upon the advice of Weierstrass, attacked the problem of the propagation of light in a crystalline medium.

It was at Paris in 1883 that she learned of her husband's death, and it was a terrible shock for her. Several months later, Sophie Kovalevski became Privatdocent of Professor Mittag-Leffler at Stockholm; moreover, she was to become professor for life on this same faculty. Weierstrass had finally succeeded in finding for his pupil, despite the difficulties caused by her sex, a position worthy of her. Sophie, completely happy, blossomed in the pleasant life of Stockholm. However, she quickly became weary of it—her Gypsy grandmother, she said—and soon considered Stockholm as an exile and went to Berlin or Paris as often as possible. The sentimental element then played an important part in her life. Indeed, she had met a man who finally seemed bound to give her perfect happiness. She was prey to a terrible struggle between her feminine aspirations and her scholarly ambitions. She wrote to Mme. Leffler from Paris, in the midst of her work: "Yesterday was a rough day for me, for big M... left in the evening. If he had stayed here I do not know how I would have been able to work. He is so tall, so powerfully built, that he manages to take up a great deal of room, not only on a sofa, but also in my thoughts and I would never have been able in his presence to think of anything but him." It was exactly at this moment of her life that she produced her master work. Her reputation as a mathematician was also at stake; all her scientific friends knew that she was competing for the Bordin Prize. She succeeded in presenting in time her paper entitled *On the Rotation of a Solid Body about a Fixed Point*, and was awarded the prize.

The equations for the problem had been set up in definitive form by Euler, who had himself given an integrable case of it (a case named after Euler and Poinsot); later Lagrange and Poisson discovered a second one, and Jacobi gave the general solution of the problem in terms of elliptic functions. But in her paper, Sophie Kovalevski found a new integrable case, that is to say a new case in which it is possible to find for the equations of the notion of a heavy solid moving about a fixed point a third algebraic "first integral" distinct from that of kinetic energy and from the integral for areas with arbitrary initial conditions. Husson, moreover, proved later that these three cases are the only ones. Sophie Kovalevski's paper was of such exceptional worth that the Commission of the Academy of Sciences, to attest its importance, increased the value of the prize from 3000 to 5000 francs. Her teacher Weierstrass expressed his joy to her in these terms: "I do not need tell you how much your success has rejoiced my heart and those of my sisters, as well as all your friends here. I have particularly experienced a real satisfaction; competent judges have now given their verdict that my faithful pupil, my 'weakness,' is not a frivolous marionette."

Sophie experienced a real triumph in Paris on Christmas Eve, 1888, at the time of the conferring of her prize. Feted, invited everywhere, she was in full glory, she seemed completely happy. Her friend had come to rejoin her. But this happiness was of short duration; with her demands and her tyrannical and jealous love, she asked too much of him; she fancied that she found in him an admiration rather than a love comparable to hers. Besides, she was unwilling to give up her career and become simply the wife of this man whom she loved and admired, and so it was that they separated often, full of bitterness, becoming reconciled only to part violently once more. Unable to do without him any more than to live with him, exhausted, torn by this incessant strife, she became ill and died in 1891 at the age of 41 after a brief attack of influenza. Her death was universally mourned; every journal published articles in her praise. Her funeral was magnificent; from every country people sent a profusion of flowers and the women of Russia had a monument erected to her.

Emmy Noether was born in Germany in 1882 in the little university town of Erlangen where her father, Max Noether, had been appointed professor in 1875 and where he remained until his death in 1921. Max Noether, a celebrated mathematician who played a considerable part in the development of the theory of algebraic functions, had two children: Emmy and her brother Fritz, two years younger than she. Both inherited mathematical ability from their father. Fritz was attracted to applied mathematics—Emmy, on the contrary, "thought only in concepts," calculation and intuition being equally strange to her! Emmy Noether developed, then, in a mathematical environment, with her father and his friendly colleagues, Gordan in particular. She received a carefully supervised education and a solid foundation, but nothing seemed to foreshadow the great mathematician she would become. Nothing seemed to impel her particularly toward research. She was a girl of good family, cultivated, taking part in household duties and in social activities. Admittance to German universities and to scientific careers becoming possible for women, she naturally directed her steps in this direction, always ready to accept life as she found it. Hers was not a tormented nature, still less rebellious. Devoid of feminine charms, she had neither women's trickeries nor their false-heartedness. Hers was a simple and good soul, without ambition, full of courage and life, a faithful friend. "There remained about her to the very end," said Weyl, "something child like, as if an entire part of her being, overwhelmed by her mathematical genius, had not been developed." How far she was from Sophie Kovalevski, a prey to the torments of her ambition struggling against her love; a less complex personality, she was without doubt also happier. Quite naturally then she first worked with Gordan and did her thesis, *On Complete Systems of Invariants for Ternary Biquadratic Forms*, with him—she upheld it in 1907. But she quickly turned away from Gordan's influence, which so ill suited her nature—Gordan, who turned out pages of calculation with hardly a word of text, and who said one day about an abstract proof of one of his results, given by Hilbert simultaneously with a broad generalization, "This is no longer mathematics, but theology." Yet it was probably the influence of the Erlangen environment which made an algebraist of her. She learned much from Fisher before settling at Göttingen under the influence of Klein and Hilbert. There Emmy Noether found a long period of work, and results which were interesting, but in which her personality had not yet appeared. She was still somewhat dependent upon her teachers; her creative power, so original, even brilliant, revealed itself only later. Weyl made it begin in 1920 with a work published in collaboration with Schmeidler on modules of non-commutative domains. It was starting from this moment that she began to change the aspect of algebra. She did it in a series of papers, but also in her teaching and by her effect on her pupils. She had, in fact, been "qualified" at Göttingen in 1919, thanks to the new regulations put into force by the German Republic. During the war Hilbert had already tried to obtain her qualification, but in vain; he ran afoul principally of the hostility of the philologists and historians. There was an anecdote often told at Göttingen that to defend her he thought it clever to say to the council of the university: "I do not see why the sex of the candidate should be an argument against her appointment as Privatdocent; after all, we are not a bath-house . . ." Meanwhile she was authorized to give lectures at Göttingen, then in 1922 appointed "nichtbeamteter ausserordentlicher Professor" [unofficial professor extraordinary], a simple title with no obligations and no salary. She was entrusted, nevertheless, with an algebra course with modest pay—a mediocre position which she kept until 1933 and which was unworthy of her personal ability and the wide influence of her classes, which were attended by many foreigners attracted by her methods. She was nevertheless not a good lecturer; she even seemed devoid of pedagogical qualities. Van der Waerden relates that "the touching trouble she took to explain her statements, even before she had completely uttered them, by corollaries rattled off at top speed, had rather the opposite effect." And what work to follow her! But what profit for a small group of those who succeeded in doing so; each lesson was a syllabus! She did not teach completed, polished things, but rather theories in the process of becoming, distributing her ideas generously to her pupils, happy when they profited by them and pursued her investigation. Round about her—I was going to say "him," for they more often called her *der* Noether, not so much because of her face with its masculine features and her corpulence as for the unfeminine authority which her mathematical superiority conferred upon her—they formed a little family. She loved them, was interested in their personal affairs, was kind and motherly to them, always ready to help them but also always an implacable judge. Of those on whom she had the greatest influence, let us mention among the best-known: Krull, Grell, Koethe, Deuring, Fitting and F.-K. Schmidt. Van der Waerden came from Holland to follow her courses, and learned from her the notions which enabled him to formulate his own ideas and solve the problems which he had set himself. A part of his remarkable book *Modern Algebra* is only the work of Emmy Noether, but clarified and set in order by him. She also usefully collaborated with Hasse and Richard Brauer. With Alexandroff she was linked by a deep friendship; moreover, she spent a semester at Moscow around 1930 and Alexandroff often came to Göttingen as a guest.

It was in the midst of her scientific activity and influence that the racial laws introduced by the revolution of 1933 interrupted her career at Göttingen and led finally to her departure along with Born, Courant, Landau, Neugebauer and many others. In the tragic months which preceded their departure she was a comfort to all with her courage and her indifference to her own fate. After a short stay at the Institute for Advanced Studies at Princeton, she became a professor at the nearby women's college Bryn Mawr. With her happy nature, she adapted herself there admirably and "her Bryn Mawr girls became as dear to her as her Göttingen boys." There and at nearby Princeton she once more formed about her a nucleus of researchers. In America as in Germany she pursued her research, the source of her greatest joy. Emmy Noether was at full maturity, she had just published her major works on non-commutative algebras—the principal subject of her research, once she had completed her axiomatic theory of ideals[2]—when her accidental death interrupted this beautiful, hard-working life. She had just undergone an operation which seemed to have been completely successful, when in a few hours she died, on April 14, 1935, of an unforeseen complication.

She left a considerable body of work, published both alone and in collaboration. Her most important personal papers are *Idealtheorie in Ringbereichen*, 1921; *Abstrakter Aufbau der Idealtheorie im algebraischen Zahlkörper*, 1923; *Nichtkommutative Algebra*, 1933.[3]

Dedekind had introduced ideals in algebraic number fields and established the decomposition of an ideal into the product of prime ideals. Lasker had proven that for ideals of polynomials there is in general only one decomposition into the lowest common multiple of primary ideals. In her first paper, Emmy Noether introduced her famous axiom of divisor chains, which permitted her to establish the theorem of the decomposition of an ideal into l.c.m. of primary ideals in every ring satisfying this axiom, and in particular in polynomial rings according to Hilbert's finite base theorem. In the second paper she exhibited the axioms which allowed the final result of decomposition into the product of prime ideals.

Her works on hypercomplex systems, the theory of representation and, more generally, non-commutative algebra are especially characterized by the important role played in them by the notions of module, ideal and automorphism, and by the fact that her results are valid whatever the fundamental field, whereas in the works of Frobenius and his immediate successors, this field was that of complex numbers or that of real numbers. By theorems such as that of the "cross product," developed by herself alone or in collaboration with Hasse and Brauer, Emmy Noether attained results in algebra and arithmetic of great depth, where hypercomplex methods are brilliantly applied to difficult problems in the theory of class fields.

This work not only puts Emmy Noether in the first rank of women mathematicians, but places her among the very great mathematicians.

---

[2] See Got's article, Part I, Book Two, pp. 87–90, on the subject of Kummer ideals, which preceded those of Dedekind and of modern algebra. (Note of F. LL.)

[3] Theory of Ideals in Rings; Abstract Construction of Ideal Theory in the Domain of Algebraic Number Fields; Non-Commutative Algebra.

Book Two

THE PRESENT

We had intended to include within the framework of this collection an examination into the progress of mathematics in the last ten or twenty years. It was worth the trouble, to give more weight to this self-examination, to wait for the end of the Second World War and to consult the best mathematicians of the times, but this would have delayed the completion of the book too much. It seemed worthwhile to us, while awaiting the accomplishment of this purpose in a second collection, to state the position of modern mathematics, by going back from the last years of the 19th century up to the period between the two wars. Lucien Godeaux, of the Royal Academy of Belgium, is the great stimulator of mathematics in his country. His work in algebraic geometry, along with that of the French school (founded by Emile Picard) and the Italian school (led by Cremona and Severi), is rightly well-known. Armand Colin has published his charming book on Geometries. *The evaluation he has been kind enough to write for our work most especially brings out the relations between geometry and analysis. He continues the historical study given by Germain and provides us with a useful transition in the direction of the most recent aspects of mathematics such as are touched upon in the articles of Bourbaki, Godement and Weil.*

The Second World War has seen the disappearance of one of the greatest French mathematicians, Henri Lebesgue. His discoveries, the result of a keen intelligence applied to intuitions of striking originality, have given a new impetus to classical analysis. After the master stroke of Cauchy, who carried over into the complex field the classical conceptions of Euler, Lagrange and Laplace, the extension of the notion of the integral has been one of the great preoccupations of the 19th and 20th centuries. Riemann was the first to increase the efficacy of this tool. After him, Lebesgue was able to extend its dominion further and to annex to it the boundless domain of discontinuous functions. This endeavor has been extended in our time in the notion of totalization, due to Denjoy; from this point of view, the clear and engaging article which Louis Perrin has devoted to Henri Lebesgue provides a bridge between the explanation of Desanti and the discovery of Denjoy.[1]

It was against the wind and tide that Cantor had to forge that extraordinary theory of sets which, with group theory, constitutes one of the most valuable and most incontestable contributions of the 19th century to modern mathematical thought. It overcame the initial prejudices by regenerating the theory of functions with a success which soon became impossible to ignore. But analysis did not exhaust the incomparable renovating force of set theory. Digging down to the very rock bottom of the foundations of mathematics, this theory has called into question principles which mathematicians had never dared to meddle with for fear of desecrating truth itself. Jean Cavaillès, in whom mathematics and philosophy are one, as they are united by set theory, had entrusted to me a valuable work on the modern aspects of set theory. He was investigating the revolution which it has brought about in traditional logic and was reviewing its applications to problems of axiomatics and the non-contradiction of arithmetic. It is known how Jean Cavaillès was assassinated in January 1944 by the Nazis after actively participating in the internal Resistance. Another philosopher, a common friend, Pierre Kaan, whom I met at this time at Montluçon and Néris, served as our intermediary. Pierre Kaan, also, paid with his life for his refusal to accept Hitler's oppression. Originally intended for the Revue Philosophique, *Cavaillès's article (which he had delivered to me on the occasion of a trip to Clermont-Ferrand) surpassed in length the scope of this work. We had agreed to meet with one another to discuss necessary revisions; circumstances did not permit that. We have had the pleasure of knowing that so profound a book will not be lost. It will appear (published by Hermann) under the title:* Infinity and Continuity.

The prince of mathematicians during these last thirty years, a German who never bowed down before the Nazis, David Hilbert, died in 1943. His work assigns him a choice place in the mathematical pantheon, alongside the greatest. Such a genius personifies in himself an entire era. The thought of David Hilbert has profoundly marked our own era, and, without doubt more than anyone else's, has contributed to separate it from the preceding period. It has seemed quite appropriate to us, in order to bring it to life again, to address ourselves to one of

---

[1] The same question is also investigated in Godeaux's article.

the leaders of that "Bourbaki School" whose members profoundly refer to the Hilbert ideal. Jean Dieudonné, moreover, is the author of remarkable works on polynomials, topology and topological algebra. This profound familiarity with areas of most interest to modern mathematics is accompanied by a personal insight of great interest into the foundations of mathematics. That is why his article, in which a fine enthusiasm for the queen of the sciences is made manifest, is not limited to a dry enumeration. It succeeds in making comprehensible the unbelievable vigor and the innovating power of Hilbertian thought. The merit of his essay is augmented by the desire for simplicity from which it never deviates.

But however great a man may be, he would not be able to embody an era all by himself, even when, like Hilbert, he dominates and characterizes it. It seemed worthwhile to retrace that collective activity which is tending more and more to develop on the intellectual plane, just as society itself—from which it emanates—is tending to become collective on the economic and social planes. We have asked the great Genevan mathematician, Rolin Wavre, to be good enough to trace the history of those congresses which spur the interaction and osmosis of ideas with such fruitful results. Wavre's mathematical work presents the example of a beautiful conjunction of practical problems and a penetrating analysis. Starting from geodesy, he came to the theory of Newtonian potentials; the study of their polydromes led him to partial differential equations; recently he has applied himself to the theory of operators. Wavre, who has personally and actively contributed to so many international gatherings of mathematicians, has been able to review all the manifestations of any importance which have taken place since the first international congress which met at Zurich in 1897, just 50 years ago.

F. LL.

24

MATHEMATICS AT THE BEGINNING OF THE 20TH CENTURY

by Lucien Godeaux

MEMBER OF THE ACADÉMIE ROYALE OF BELGIUM,
PROFESSOR AT THE UNIVERSITÉ DE LIÈGE

INTRODUCTION: FROM ALGEBRA TO GROUPS

Algebra. As early as 2000 B.C. men had to solve problems leading to equations of the second degree in one unknown. This solution is classic today. The solutions of equations of the third and fourth degrees were the work of Italian algebraists of the 16th century (Tartaglia, Cardano, Bombelli). The roots of these equations are expressed in terms of radicals involving the coefficients. As soon as we pass on to equations of degree higher than the fourth, this is no longer the case. Abel and Ruffini proved that it is impossible to solve the general equation of the fifth degree by radicals; or, in other words, that in order for an equation of the fifth degree to be solvable by radicals, certain relations among the coefficients must exist. It was left to Galois to show the fundamental reason for this fact. At the same time, Galois introduced the notion of group, which little by little was going to penetrate every field of mathematics.

FROM GROUPS TO GEOMETRY[1]

The concept of a continuous group of transformations led Felix Klein and Henri Poincaré to a fruitful interpretation of geometry. Let us consider a plane and the group G of movements in this plane.

[1] This section is an extension of the articles by Lentin and Cartan. (Note by F. LL.)

Every movement preserves distances and angles; further, two equal figures are figures such that one can be brought into coincidence with the other by subjecting the first to a movement. This is essentially the same definition of equality given by Euclid. Euclidean geometry can therefore be considered as the study of properties of the plane which are not altered by movements, that is, which are invariant with respect to the group of movements. In general, the geometry of a variety V with respect to a continuous group G is the study of properties which remain invariant when the transformations of the group G are applied to the figures drawn on the variety.

However, we can conceive of geometries not contained in this framework. Let us imagine a variety V, continuous, whose points are determined by n coordinates. We can define the distance between two points of the variety in the following way: let $x_1, x_2, \ldots, x_n$ be the coordinates of a point, and $x_1 + dx_1, x_2 + dx_2, \ldots, x_n + dx_n$ those of a point infinitely close to the first. By definition, the distance between these two points will be a certain expression:

$$ds = E(dx_1, dx_2, \ldots, dx_n),$$

where E is a given function, homogeneous and of the first degree in $dx_1, dx_2, \ldots, dx_n$, whose coefficients are functions of the coordinates of the first point. Let us next consider two points A, B and a certain curve, traced on V, beginning at A and ending at B. The distance from A to B along this curve will be expressed by $\int_a^b ds$ taken along this curve, from A to B. This distance will obviously depend upon the curve chosen, and one of the problems to be solved will be to determine the curve giving the minimum distance from A to B. We shall thus obtain the geodesics of the variety V. Usually we take for E the square root of a homogeneous polynomial of the second degree in $dx_1, dx_2, \ldots, dx_n$; this is what Riemann did. The case where E is any function whatever has been studied by Finsler.

Elie Cartan succeeded in bringing the geometry of the variety V based on the idea of distance into the framework of Klein and Poincaré—only, however, by enlarging this framework somewhat. For greater simplicity, let us suppose that V is a surface. For each point of this surface, let us draw the tangent plane and let us establish in this plane a geometry of the group G. We shall suppose, in order to be specific, that G is the group of movements. This leads to establishing in the tangent plane α at the point A, two reference axes Ax, Ay, rectangular for instance; we know that the movements are expressed, with respect to these axes, by formulas analogous to the transformations of coordinates. Let us draw on the surface a path starting from A and ending at a point B. In the tangent plane β at this point, axes Bx, By are determined. Let us make the plane α roll on the surface in such a way that it constantly remains tangent to the latter, the point of contact traversing the fixed path AB. When α comes into coincidence with β, the axes Ax, Ay will have fallen into a certain position A$'x'$, A$'y'$, distinct in general from Bx, By. It is understood that the position A$'x'$, A$'y'$ depends upon the choice of the path AB. Let us now imagine that the point B coincides with A; we therefore return to A after traversing a certain closed path, and this time the axes A$'x'$, A$'y'$ are situated in plane α. We pass from A$'x'y'$ to Axy by means of a transformation of coordinates, that is, by a certain movement T. Let us repeat the operation for all possible closed paths, drawn on the surface, starting from A and returning to it; we shall obtain a family of movements analogous to T and forming a group G'. All the transformations of G' belong to G, but the inverse is not necessarily true. G' is called the group of holonomies of the geometry considered. It is understood that geometries can be constructed by starting with a variety V, a continuous group G fixed in a space associated with each point of V, and from a law joining the groups G of the spaces associated with two neighboring points of V. (In the preceding example, this law was determined by rolling the tangent plane on the surface.) Thus we obtain the non-holonomic geometries of Elie Cartan, who was led to them by the study of problems presented to geometers by the theory of relativity.

Let us return to the Klein-Poincaré concept. We have supposed that group G is continuous, but algebraic geometry leads to the consideration of groups whose transformations can be determined only if we are given a certain number of quantities which can vary continuously, and certain integers. A discontinuous group G is thus obtained and this shows that geometers have been led to a dual study: first, that of varieties V, which gives rise to topology; second, that of the most general continuous and discontinuous groups.

There is still material here for numerous investigations.

Uniformization of Algebraic Functions and Algebraic Geometry[2]

Let us consider a curve, for example an ellipse with equation

$$b^2x^2 + a^2y^2 = a^2b^2.$$

If we set $x = a \cos u$, $y = b \sin u$, we shall obtain, for any u, a point on the ellipse. Thus we have a parametric representation of the ellipse.

Can anything analogous be obtained when we are dealing with any algebraic curve of the form $f(x,y) = 0$? It is then a matter of finding two functions $x = x(u), y = y(u)$, such that when they are substituted into the equation $f = 0$, an identity is obtained. First of all, we must define exactly the nature of the functions of u being considered. We shall suppose that these are uniform functions of the complex variable $\zeta + i\eta$; that is, to a point ζ, η of the (ζ, η)-plane, there corresponds one and only one value of each of the functions $x(u), y(u)$. The problem, trivial for rational curves, was solved first for certain curves called elliptic, among which are found for example plane cubics without a double point. For $x(u), y(u)$ doubly periodic functions are obtained. To tell the truth, the latter were not encountered on the occasion of the solution of the problem in question. When we try to express the length of an arc of an ellipse, we are led to the integration of a function involving the square root of a polynomial of the fourth degree; this integral cannot be evaluated by elementary means. Let us write it in the form $u = \int F(x) \, dx$. Abel and Jacobi considered this equation as defining x as a function of u. We then find that x is a doubly periodic function of u, that is, that there exist two numbers ω, ω', whose ratio moreover is imaginary, called periods of the function, such that

$$x(u + \omega) = x(u), \qquad x(u + \omega') = x(u).$$

These doubly periodic functions were named, for the reason just indicated, elliptic functions, whence the name of elliptic curves given to curves the coordinates of whose points are expressed as doubly periodic functions of a parameter, the periods being the same for the two functions.

---

[2] This section is in accord with the end of Valiron's article, which it completes and extends. (Note by F. LL.)

It was left to Poincaré to give, in the general case, the solution of the problem raised, by proving that the coordinates of the points of an algebraic curve are expressed as Fuschsian functions of a variable, with these functions corresponding to the same group of substitutions. We say that the curve is uniformized by means of Fuchsian functions.

Let us suppose that a given algebraic curve C has been uniformized. Under what conditions will the uniformization of another algebraic curve C' be reduced to that of C? The solution of this problem leads to the division of algebraic curves into classes, two curves belonging to the same class if it is possible to pass from the points of one to those of the other by a biunique and algebraic transformation. More precisely, the coordinates x, y of the points of curve C are expressible as rational functions of the coordinates x', y' of the points of curve C', taking account of the equations of the curve, and vice versa. We say that the curves are related by a birational transformation. It is a question of finding criteria permitting us to recognize whether two curves belong to the same class, that is, of looking for the characteristics of a curve which are invariant with respect to birational transformations. This problem has been attacked by different methods, and the results obtained are important. One of the methods used consists in the study of systems of groups of points belonging to an algebraic curve; operations which are not altered by birational transformations of the curve are applied to these systems.

The extension of the preceding considerations to algebraic surfaces creates very difficult problems, which are far from being solved.

Let us divide algebraic surfaces into classes analogous to the classes of curves. Two surfaces belong to the same class if we can pass from one to the other by a birational transformation, that is, if there exists a biunique and algebraic correspondence between the points of these surfaces. Just as with the curves, it is a question of determining when two surfaces belong or do not belong to the same class. This problem has been attacked by transcendental means—and here the work of Picard is of great importance—and by geometric means. It was especially the Italian geometers Castelnuovo, Enriques and Severi who employed these means. It was by the consideration of systems of curves drawn on the algebraic surfaces and of operations invariant with respect to birational transformations applied to these systems, that these geometers were able to construct an extremely elegant theory. Recently, Severi has likewise introduced the consideration of systems of groups of points. Thus certain integers (the genus and the plurigenus of a surface) which are the same for two surfaces belonging to the same class can be introduced, but other conditions remain to be discovered before we can characterize the surfaces of a class.

Next it is a question of uniformizing an algebraic surface representative of a class, that is, of finding the expression for the coordinates of a point of an algebraic surface $f(x, y, z) = 0$ in terms of uniform functions of two parameters. The solution of this problem is known only in simple cases (rational surfaces, ruled surfaces, hyperelliptic surfaces). Picard has constructed functions (hyperabelian functions, hyperfuchsian functions) which lead to particular algebraic surfaces, study of which, incidentally, remains to be made. The problem of the uniformization of algebraic surfaces still seems very far from a definitive solution.

Functions of a Real Variable[3]

It is probably the representation of a curve by an equation, in Cartesian geometry, which gave rise to the general idea of function. This notion is based on that of correspondence. If to each value of a variable x (or to each system of values of n variables $x_1, x_2, \ldots, x_n$) corresponds one value of a variable y, we say that y is a function of x (or of $x_1, x_2, \ldots, x_n$). From the beginning of the study of functions, one is also led to separate functions of real variables from functions of complex variables.

The modern theory of functions of a real variable originated in the works of Baire, Borel and Lebesgue; it is based essentially on the idea of set, due to Cantor. A function of a real variable can be visualized by considering, in a plane referred to two axes Ox, Oy, a continuous curve met in one point by every parallel to Oy; assuming that this curve has a tangent at every point, then the function $y = f(x)$ which represents the curve is continuous and has a derivative at every point. More precisely, we say that a function $f(x)$ is:

1. defined in an interval (a, b) when to every value of x between a and b there corresponds a value of the function;

2. continuous at a point x if to every arbitrarily chosen positive number k, as small as desired, there corresponds a number h such

---

[3] This section takes up again certain ideas already put forth in the articles by Desanti and Perrin. (Note by F. LL.)

that if $x' - x$ lies between $-h$ and $+h$, then $f(x') - f(x)$ lies between $-k$ and $+k$, where x' is a point close to x;

3. differentiable at a point x if, when x' approaches x, the ratio of $f(x') - f(x)$ to $x' - x$ approaches a well determined limit, called the derivative of $f(x)$ at the point x.

If the function $f(x)$ is continuous at each point of the interval (a, b), we say that it is continuous in this interval. It is called differentiable in (a, b) if it is differentiable at each point. A continuous function is not necessarily differentiable, and there exist continuous curves without tangents. On the contrary, a differentiable function is continuous.

The idea of integral has played a great role in the development of the theory of functions. Riemann defined the integral in the following manner: Let us consider a function $f(x)$ defined in the interval $a \leq x \leq b$. Let us divide this interval (a, b) into small intervals by the points $x_1, x_2, \ldots, x_n$ arranged in increasing order, and let ξ_i be a value of x between x_i and x_{i+1}. Let us form the sum $\sum (x_{i+1} - x_i) f(\xi_i)$, taken over all the intervals. If this sum has a limit when the number of intervals increases beyond all limit, each interval approaching zero, then this limit is the integral, in the Riemann sense, of the function (x) in (a, b). It is represented by

$$\int_a^b f(x) \, dx.$$

Lebesgue has given another definition of the integral. Let us suppose that in the interval (a, b) the values of the function lie between the numbers m and M. Let us divide the interval (m, M) into a certain number of intervals by the numbers $m_1, m_2, \ldots, m_n$, arranged, let us say, in increasing order. The points x which give to $f(x)$ values lying between m_i and m_{i+1} form a certain set. Let us assume, for greater simplicity, that this set consists of points of a finite number of intervals and let e_i be the sum of the lengths of these intervals. Furthermore, let μ_i be a number between m_i and m_{i+1}. Let us form the sum $\sum e_i \mu_i$ taken over all the intervals. If this sum approaches a limit when the number of intervals increases indefinitely, each interval tending towards zero, the limit is the integral, in the Lebesgue sense, of $f(x)$ in the interval (a, b). It is also represented by the symbol

$$\int_a^b f(x) \, dx.$$

In reality, the definition of the Lebesgue integral is much more general; the set of values of x giving to $f(x)$ values lying between m_i and m_{i+1} is not in general formed by the points of a finite number of intervals, and it is necessary to introduce the notion of the measure of a set due to Borel. This would take us too far afield.

A function which is Riemann integrable is Lebesgue integrable, but the converse is not true. The procedure of Lebesgue is therefore more general.

Functions of a Complex Variable[4]

Cauchy, Riemann, Weierstrass and Méray created the theory of functions of a complex variable. Let us consider a uniform function $Z = f(z)$ of a complex variable $z = x + iy$. If we wish to define the derivative of such a function at a point z_0, we must consider the limit of the ratio of $f(z) - f(z_0)$ to $z - z_0$ when z approaches z_0. Now in the (x, y)-plane, the point z can approach z_0 along a fixed path and the desired limit (if it exists) depends in general upon the path chosen. A function such that the limit in question does not depend upon the path chosen is called monogenic, and this limit is called the derivative $f'(z)$ of $f(z)$ at the point z_0.

An example of a monogenic function is given by a polynomial integral in z, this point z belonging to a certain region located at a finite distance. Such a function always remains finite and furnishes in addition an example of a holomorphic function, that is, a function which always remains finite in the region in which it is defined. A holomorphic function can be expanded in a Taylor's series, in integral powers of $z - z_0$, converging in the neighborhood of z_0.

Instead of a polynomial, let us consider a rational function of z, that is, the quotient of two polynomials in z, in a certain area A. The function is in general monogenic, but here there occurs a new fact: There exist points z at which the denominator vanishes and where the function therefore becomes infinite; these are the poles of the function. Generally, we call a point z_0 a pole of the function $f(z)$ if and when z approaches z_0, $f(z)$ grows beyond all bounds, but $(z - z_0)^m f(z)$ still remains finite, m being a positive integer (the order of the pole). A point at which $f(z)$ is infinite and which is not a pole is called an

---

[4] This section treats further some of the questions considered in Valiron's article. (Note by F. LL.)

essential singular point; such for example is the point $z = 0$ for the function log z. A function which, in a certain region, possesses only poles is called meromorphic.

The singular points (poles, essential singularities) of a function may form sets of very different natures; they can even form continuous lines. One of the questions which arises is the study of the values of a function in the neighborhood of one of its singular points, a question mastered by the famous Picard theorems.

We can define a holomorphic function by its Taylor's expansion about the point z_0, an expansion convergent in a certain circle with centre at z_0, and then investigate in which region of the plane this function remains holomorphic. This is what is called seeking the analytic extension of the function; this is the point of view taken by Weierstrass and Méray, which leads to the notion of analytic function.

If $Z = f(z)$ is an analytic function of z defined in a certain domain A, to every point $x + iy$ of this domain corresponds a point $Z = X + iY$ of a certain domain A'. We thus obtain a correspondence between the areas A and A', which preserves angles and which is called a conformal representation. The study of these representations forms an extremely fruitful chapter of the theory of functions.

Montel has considered families of functions satisfying certain conditions and which he has named normal families of functions. We have here an instrument of very great scope in the study of analytic functions.

Along with analytic functions, functions known as quasi-analytic functions have recently been under examination. A function is defined in a region by the values of its derivatives at each point; the Taylor expansion of the function can then be written. If this expansion is convergent, the function is analytic; if not, the function is quasianalytic. These functions have been studied by Borel, Carleman, de La Vallée Poussin and Denjoy.

The study of multiform functions, that is, functions $f(z)$ such that to one value of z correspond several values of the function, as is the case for example for the function u of z defined by a polynomial $F(u, z) = 0$ of degree at least equal to 2 in u, can be reduced to the study of uniform functions by the introduction of Riemann surfaces. This method consists in replacing the (x, y)-plane by superimposed sheets, equal in number to that of the values of the function at a point z, these sheets being joined two by two, crosswise, along certain lines, in such a way as to allow passage from one sheet to another.

All these questions have been extended, in a certain measure, to functions of several variables.

Differential Equations

The applications of mathematics lead to differential equations, that is, to relations between one or several functions of a variable and their derivatives with respect to that variable. It is then a matter of determining this or these functions.

Analytically, a differential equation is obtained by starting with a relation $F(x, y, C_1, C_2, \ldots, C_n) = 0$ and by eliminating the constants C from between this relation and those derived from it by differentiation with respect to x. A relation is thus obtained between x, y and the first n derivatives of y with respect to x. This shows that if, inversely, we start from this last relation, the integral of the differential equation, that is, the function $y(x)$ which satisfies it, will depend in general upon n arbitrary constants.

The first investigations of geometers had as their objective the reduction of the integration of differential equations to quadratures and, finally, the expression of the solutions in terms of known functions. It is indeed obvious that only very special differential equations could thus be obtained.

With Cauchy, the aspect of the problem changed. Cauchy considered a differential equation, for example, of the first order, $y' = f(x, y)$, and raised two questions: first, the existence of a solution; second, the uniqueness of the solution.

As in the theory of functions, it is convenient to treat separately the cases in which the variables are real or imaginary. In the second case, Cauchy proved that if $f(x, y)$ is holomorphic in a certain domain, there exists one and only one holomorphic function $y(x)$ satisfying the equation and reducing to y_0 for $x = x_0$, where the point x_0, y_0 belongs to the domain considered. The method of proof is interesting. Cauchy begins by constructing a series of powers of $x - x_0$ which formally satisfy the equation. He next constructs a second series whose convergence involves that of the first. Now, this new series satisfies a differential equation which can be integrated by quadratures. The solution obtained by this method is holomorphic under conditions easily determined, it coincides with the second series, therefore the latter is convergent and the same is so for the first series. The theorem follows. To construct the second series, use is made of what is called a "majorizing function"; this is a function of x, y which is greater, at the point x_0, y_0, than the modulus of $f(x, y)$, and all of whose derivatives are greater, at the same point, than the moduli of the corresponding derivatives of $f(x, y)$. This method "by comparison" has been called by Cauchy "the method of limits"; it can be extended without difficulty to differential systems and, in general, to partial differential equations.

When the variables are real, the continuity of $f(x, y)$ involves the existence of an infinity of solutions $y(x)$ reducing to y_0 for $x = x_0$. For the solution to be unique, it is necessary that the function f satisfy another condition in addition. It is generally assumed that f satisfies the Lipschitz condition, that is, that the absolute value of the difference $f(x, y_1) - f(x, y_2)$ is less than the absolute value of the difference $y_1 - y_2$, multiplied by a constant factor, called the Lipschitz constant. This condition is sufficient for the uniqueness of the solution, but it is not necessary. If the function f is differentiable with respect to y, it satisfies the Lipschitz condition besides, but the converse is not true. Cauchy has proven the existence and the uniqueness of the integral by a method which recalls the one used to arrive at the concept of the integral in the Riemann sense. Later, Picard came to the same result by using a method of successive approximations. By quadratures, he constructed a sequence of functions approaching more and more the desired solution, which is the limit of the sequence. This method can also be extended to systems of differential equations.

The preceding investigations led to the study of solutions of differential equations in the neighborhood of an ordinary point (or of a holomorph point) x_0, y_0; they were to be followed by the study of solutions in the neighborhood of a singular point. This study, which is very difficult, is not very advanced; the works of Poincaré, Picard and Painlevé are fundamental in this type of question.

Poincaré considered the integration of differential equations from another point of view. He proposed to construct curves defined by a differential equation

$$P(x, y)dx + Q(x, y)dy = 0$$

in the (x, y)-plane. This is thus no longer the study of the solution in the neighborhood of a point, but in the whole plane. This viewpoint supposes in the first place the qualitative study of the solution, that is, the investigation of the geometric form of the curve; it has to be followed by the numerical determination of the values of the solution.

Differential Equations and Partial Differential Equations

Let us consider a partial differential equation $F(x, y, z, p, q) = 0$, in which z is the function of x, y to be determined, and where p and q are the derivatives of z with respect to x and y. Let us interpret x, y, z as coordinates of the points of the space. A solution of the equation, that is, a function z of x, y which when substituted into the equation reduces it to an identity, will be represented by a surface, called an integral surface. Cauchy's problem consists in constructing an integral surface passing through a curve given a priori. It has a unique solution under certain very broad conditions imposed on the data of the problem.

If the equation under consideration contains besides the second derivatives of z with respect to x and y, Cauchy's problem consists in determining an integral surface passing through a curve given a priori and touching a given surface along this curve. The problem has a unique solution under certain conditions imposed on the data. However, the problem can be impossible for certain curves called characteristic curves; it is proven that every integral surface is in general produced in two ways by the characteristics.

These two examples show the direction of research in partial differential equations.

Functional Analysis

We have encountered above the idea of a geodesic of a surface, that is, the curve of minimum length drawn on a surface. Lines are the geodesics of the plane, great circles those of the sphere. The determination of geodesics arises in the calculus of variations, which was created by Lagrange. To make comprehensible the nature of the problems which come up in this chapter of mathematics, we shall point out two examples.

Let us consider a curve L drawn in a vertical plane and a particle moving from a point A to a point B along the curve under the influence of gravity. What should be the shape of the curve L in order for the descent of the particle to take the minimum time? This problem leads to the determination of a function y of x such that a certain integral

$$I = \int_a^b f(x, y, y') dx$$

will be a minimum, where y' is the derivative of y with respect to x. This problem is called the problem of the brachistochrones; the curve L must be a cycloid.

Here is the second example. If we immerse a wire contour in a soapy liquid, we find on withdrawing it that a thin film of liquid remains stretched taut over the interior of the contour. Plateau, who made this experiment, showed that the area of the surface, of which this film is the representation, must be a minimum. Plateau's problem consists in the determination of surfaces of minimum area bounded by a given contour, it comes down to the investigation of the minimum of a double integral and is very difficult. Generally, the calculus of variations sets itself the problem of determining a function of one or several variables furnishing an extremum of an integral involving an unknown function of the independent variables of the unknown function and certain derivatives of the function.

In the integration of differential equations and of partial differential equations, and in the calculus of variations, it is a question of determining functions of one or several variables satisfying fixed conditions. These are problems which come under what has been termed functional analysis which, under the influence of Volterra and Hadamard, has expanded greatly. There exist numerous kinds of functional equations; we shall restrict ourselves to giving one example of them. It has been experimentally proven that the state of a spring does not depend only upon the tension which it undergoes, but also upon the various tensions it has undergone previously; there is in some way a kind of heredity which must be taken into account. Problems of this nature lead to integral equations (or integro-differential equations) of the form

$$\phi(x) + \int_a^x K(x, s)\phi(s)\, ds = F(x),$$

where it is a question of determining the function $\phi(x)$, the functions $K(x, s)$ and $F(x)$ being given.

Concluding Remarks

We have tried to make comprehensible the nature of the problems engrossing mathematicians; our account is very incomplete and we should have made some mention of numerous problems, of important subjects such as the theory of numbers,[5] topology,[6] etc. We could not hope to be complete. Before concluding, permit us to add a few words on a present trend in mathematical research. When geometers attack a problem for the first time, they introduce hypotheses which let them achieve the desired end with the means at their disposal. Later, the question can be reopened and certain hypotheses seen as unnecessary—they are in a certain sense hypotheses of convenience. We are thus led, given a theorem concerning a certain mathematical entity, to look for the minimum number of hypotheses to be introduced. We shall explain by an example.

It is known that we call a point where the tangent to a plane curve crosses the curve a point of inflexion of the curve. Let us consider a real plane algebraic curve of the third order, that is, the locus of real points x, y in the plane satisfying an equation $f(x, y) = 0$, where f is a polynomial of the third degree, with real coefficients. A line of the plane cuts this curve at three real points at most. The curve has a tangent at each point and possesses nine points of inflexion, only three of which are real. Restricting ourselves to real points, we can therefore say that the curve possesses three points of inflection.

Let us now suppose a continuous curve drawn in the plane, having a tangent at each point, intersected at three points at most by a line of the plane. The Danish geometer Juel has proven that this curve possesses three points of inflection. This curve is not algebraic and it is seen that the hypothesis of its algebraic character was superfluous; it was simply a hypothesis of convenience. What is of importance is the fact that the curve is intersected at three points at most by a line.[7]

Analogous examples are numerous; it is the improvement of the methods of investigation which allows us to uncover them.

---

[5] The latter has recently taken on a new impetus with the Russian school of Vinogradov, who has particularly studied the Goldbach conjecture.

[6] Under the impetus of Fréchet, the idea of abstract space has recently experienced a great development. (See the article by Fréchet: "From Three-Dimensional Space to the Abstract Spaces." Part I, Book Two, pp. 115–124. (Note by F. LL.)

[7] Such propositions turn up in a chapter of mathematics of recent creation, called *finite geometry*. It has been expounded in France by Bouligand, Marchaud, Montel and others. (Note by F. LL.)

HENRI LEBESGUE: RENEWER OF MODERN ANALYSIS

by Louis Perrin
PROFESSOR AT THE COLLÈGE MODERNE ET TECHNIQUE OF COLMAR

THE object of this article is to pay a pious homage to the memory of one of the most eminent mathematicians of our time, the renowned creator of the modern theory of the integral. Henri Lebesgue, so modest, has never been regarded in his proper importance even by the scientific public. He must, however, be considered as one of the great renewers of mathematical analysis; we shall try to prove this by analyzing his most important works. His creations—so original and so daring—have invaded the many regions of analysis where they immediately permitted considerable progress. While recounting the life of this late mathematician, we shall underscore his strong personality and his great goodness.

Born at Beauvais on June 28, 1875, Henri Lebesgue came from a very modest background. His father, a typographer, had acquired an austere library, finding some pleasure in instructive reading. This intelligent father passed on, alas, too soon. But, as early as primary school, the child gave proof of remarkable aptitude for computation, and thanks to an instructor who guessed the exceptional gifts of the future mathematician, young Lebesgue was able to pursue his secondary studies at first in his native city, then in the Parisian *lycées*. A headstrong worker, he had but one object, the Ecole Normale. He was admitted there in 1894. Already his acute critical sense was revealed when, a crumpled paper in his hand, he sought to convince his fellow students of the falsity of an important geometrical theorem. The rumpled sheet of paper, the very image of irregularity, can be mapped on the plane, whereas classical theory teaches us that this is possible only with cones, cylinders and surfaces generated by the tangents to a skew curve. The result of classical geometry came from hypotheses of regularity, *convenient* but *irrelevant* to the problem in question. In his work, Lebesgue endeavored to solve problems in a "natural" manner, that is to say, by avoiding superfluous hypotheses.[1]

Having passed his competitive examination in 1897, Henri Lebesgue was appointed professor at the Nancy Lycée. Despite the schedule of courses, very heavy at this time, he prepared his doctoral thesis entitled *Integral—Length—Area*, in which he expounded a new theory of integration of functions of real variables; he also showed how to define correctly the area of a curved surface. The great pundits of the day (about the year 1900) were not at all favorable to this work of audacious, indeed revolutionary, appearance, and the author was able to sustain his thesis, in 1902, only after having surmounted some difficulties.

It must nevertheless be acknowledged that this paper, received with some mistrust, is a sheer masterpiece, whose reverberation became prodigious in a few years. The new integral quickly dethroned Riemann's; it seemed more flexible, more easily handled, and it was not long before it was used by a large number of analysts the world over. To work out his theory, Lebesgue had to introduce the idea of the *measure of a set* of points. He gives a *descriptive* definition for this measure, that is, one in terms of compatible axioms, while Emile Borel, in 1894, had given a *constructive* definition, in certain respects less easy to work with than Lebesgue's. For forty years now, mathematicians have constantly spoken, in analysis, of (L) measurable or (L) integrable sets or functions. Very often, Lebesgue and his followers encountered a property satisfied by all the points of a set, except, at most, by those of a subset of measure zero; we then say that the property is true *almost everywhere*. This convenient expression has enjoyed an extraordinary success. It has become quite ordinary in modern mathematical literature.

What is the *Lebesgue integral*? Let us try to give a summary. We suppose that the reader is familiar with rectangular coordinate graphs. Let us consider a function $f(x)$ represented by a curve AMB; let Aa and Bb be parallels to $0y$ drawn through A and B, the points a and b

---

[1] This tendency should be compared to that which Godeaux brings out in the last paragraphs of his article. (Note by F. LL.)

being on the x-axis, $0x$. The problem of integrating $f(x)$ consists in determining the curvilinear area AMBba. The quotient of the area AMBba by ab gives the *mean ordinate* of the curve AMB. If the function is an elementary one (polynomial, sine curve), integration is then the inverse operation of differentiation, and is often resolved without difficulty. Thus it is that the derivative of sin x is cos x and, inversely, the integral (called in this case the primitive) of cos x is sin x, except for an additive constant. In the case where the function $f(x)$ is merely required to be *continuous* and *bounded*, the French mathematician Cauchy has proven that the investigation of the area AMBba by decomposition into slices by parallels to the $0y$ axis, always gives a finite number. Let us attack more complicated cases.

Let us suppose that we wish to look for the average atmospheric temperature during the month. We represent this temperature graphically by marking off the time as abscissas and the temperature as ordinates. This function of the time is, generally, continuous, but we can imagine another magnitude whose representative curve has numerous discontinuities. Riemann, the great German geometer, divided the time axis into days and multiplied the fraction $\frac{1}{30}$ (a 30-day month) by one of the values taken on by the temperature during a day, then found the sum of the results obtained for each of the 30 days of the month. The same procedure, applied to a division of the month into hours, gives another number. Continuing indefinitely the subdivision of the time axis into smaller and smaller segments, and approaching zero, it may happen that the results obtained approach a *finite limit*. We then say that the function under consideration is integrable in the Riemann sense. Otherwise, we say that it is not integrable. Lebesgue's method is based on a different idea: Instead of subdividing the time axis, he subdivides the temperature axis, between the maximum and minimum ordinates of the function. For definiteness, let the subdivision be in degrees; then, the time values for which the function is contained between 15° and 16° form a set of points having a *measure m*. Lebesgue takes the product of any ordinate whatever, between 15° and 16°, by the number m/T where T is the total time of the experiment—30 days in this case. He next adds all these products relative to each of the intervals of the subdivision into degrees and obtains a number M_1. Repeating the process for smaller and smaller subdivisions, the lengths of whose intervals approaches zero, he obtains a sequence of numbers $M_1, M_2, M_3, \ldots$ which may have a *finite limit*. The function being considered is then said to be integrable in the Lebesgue sense, or *L*-integrable, or "summable." If not, the function is not *L*-integrable. The improvement made is considerable, for the interpolation of the measure of a set permits easy demonstration of the properties of the new integral, and the class of *L*-integrable functions is very much wider than that of *R* (Riemann)-integrable functions. It may be asked why one tries to integrate complicated functions; it is not just to extend gratuitously the field of analysis, but because these functions, whether we like it or not, do appear in the "natural" problems of classic theories.

Let us leave this major part of Henri Lebesgue's work to set forth the content of his great paper, published in 1905, *On Analytically Representable Functions*. A work of the very first order, it is a veritable mine of extraordinary riches and the point of departure for important works of the Polish and Russian schools of mathematics on the sets called *analytic*. In this wonderful publication, too little known, Lebesgue, following the eminent French mathematician René Baire, treats the problem of the representation of discontinuous functions by other simpler or even continuous functions. Baire had divided functions into *classes* of which the first, class zero, is composed of continuous functions—class one, by discontinuous limits of continuous functions, and so on. Lebesgue gave various new proofs of Baire's fundamental theorem; he also succeeded in proving the existence of functions of every class. Certain parts of the paper are on the borderline of mathematics and philosophy. Here Lebesgue takes, with his idea of "nommable," a position intermediate between the two extreme views: that of the *pure idealist* for whom a mathematical entity exists if it is possible to discourse upon it without running into any contradiction; and that of the pure *empiricist* who admits the existence of a mathematical entity only if it can be constructed by means of a finite number of well-defined operations. The idea of "Nommable" is one of Lebesgue's most original creations: "An object," writes this great scholar, "is defined or given when a finite number of words applying to this object and this alone have been uttered; that is, when a characteristic property of the object has been *named*." In this great work of 1905, the illustrious author *names* a function incapable of any mode of analytic representation: an example of great daring for the time, the source of many and often fruitful discussions. Directly drawing inspiration from Lebesgue's paper, the mathematicians Lusin and Sierpinski introduced into the subject very general classes of *analytic* sets and *projective* sets. We can be proud of the essentially

French origin of these theories, three of our great pioneers in analysis being the renewers there: René Baire, Emile Borel and Henri Lebesgue.

His works, beyond comparison, had earned for Lebesgue a lectureship at the Faculty of Sciences at Rennes (1902–1906), then a chair at the University of Poitiers (1906–1910). Finally, in 1910, he was appointed lecturer at the Sorbonne until 1919. In 1920 and 1921, Lebesgue exercised the functions of titulary professor at the Sorbonne, which he soon left to occupy one of the chairs in mathematics at the Collège de France.

We cannot pass over in silence a very important application of the Lebesgue integral, the calculation of the coefficients of a *trigonometric series* converging everywhere to a given function $f(x)$. Such a series is the sum of an infinite number of terms of the form $a_n \cos nx + b_n \sin nx$, where n varies from 1 to infinity. It is a question of calculating the coefficients a_n and b_n, given the function $f(x)$, the sum of the series (assumed convergent for every value of x). If the function $f(x)$ is assumed bounded or L-integrable, the solution is given by the Lebesgue integral. If we take account of the importance of trigonometric series in analysis and in mathematical physics, we can only feel a profound admiration for Lebesgue's method, so marvelously adapted to these branches of mathematics.

In many of his works, the famous author started from a very simple and intuitive, but very profound, observation which no one before him had even foreseen. In the words of one of his emulators, "He looked at old things with new eyes." On this subject, it is of interest to mention that it was the careful observation of brick walls which led to an important lemma proved by Lebesgue. This lemma considerably simplifies the proof of the theorem on the invariance of the number of dimensions under *topological* transformations (biunique and bicontinuous). This fruitful remark nevertheless seems trivial: if the bricks are small enough, there exists at least one point common to four bricks, and it can be arranged so that no point is common to more than four bricks.

In 1922, Henri Lebesgue was elected a member of the Academy of Sciences, geometry section, following the death of the great geometer Camille Jordan. Honors came by degrees in great numbers: He was welcomed by the Royal Society of London, by the celebrated Accademia dei Lincei in Rome, by the Royal Academies of Denmark, Belgium and Romania, by the Polish Academy of Sciences and

Letters. Finally, several universities conferred upon him the title of doctor *honoris causa*.

For about fifteen years, he was above all concerned with elementary questions and in 1937 his very interesting work *Sur la mesure des grandeurs* (On the Measure of Magnitudes) appeared—a work which should be read and thought about by all professors of mathematics. Shortly before his death, he still had the fortitude to dictate the closing chapters of a book on conics. Always deeply interested in pedagogical problems, Lebesgue taught for a long time at the Ecole Normale Supérieure at Sèvres and was a brilliant contributor to the Swiss review *L'Enseignement mathématique* (Mathematical Teaching).

He scarcely spoke of his works and led a very simple life. Courteous and very gracious, he liked to converse in a friendly fashion with his students. A man of quiet generosity and rare goodness, he was always profoundly humane and very much in love with justice. After four months of illness which left him lucid, although his strength declined rapidly, he died on July 26, 1941.

But his work—so beautiful, so original and so profound—will survive him for a long time. The ideas of this great analyst, there can be no doubt, will abide, as have those of Gauss, Cauchy, Riemann, giants of mathematical creation. A renewer whose works reflect French thought so well—moderation, balance, good sense—Lebesgue, daring pioneer of modern analysis, is reckoned in the rank of the greatest mathematicians of the beginning of the 20th century.

26

DAVID HILBERT (1862-1943)

by Jean Dieudonné
PROFESSOR AT THE FACULTÉ DES SCIENCES OF NANCY

It is a somewhat childish pastime to attempt to draw up a list of honors of the most eminent scholars of a single period. Is it not the characteristic of genius to be free of all rules as well as all comparison? And what common measure can be established between two important results, which are found situated at antipodes from one another in the vast universe which a science such as mathematics represents? Nevertheless, from time to time there appear men in whom profundity of thought is combined with a matchless universality; it seems that they cannot attack a problem without immediately putting it in a new light, and the brilliance of their discoveries quickly confers on them an intellectual primacy acknowledged by all. After Gauss, the *princeps mathematicorum* [or the prince of mathematicians, Tr.] par excellence, Riemann, then Poincaré, had known this supremacy. Since the death of Poincaré, it is to David Hilbert that the sceptre of mathematics has been handed down, by almost unanimous consent.

The life of Hilbert is quite simple, as is very often that of a scholar busy above all with his work, especially if he has the luck to live and to produce during a period free from great political and social upheavals. Born at Königsberg in 1862, of a middle-class family, Hilbert did the greatest part of his studies at the University of that city, where he wrote his inaugural dissertation in 1885, then became Privatdocent from 1886 to 1892, and titular professor from 1893 to 1895. He did not

---

In this article we have made a great use of important commentaries on Hilbert's work, inserted in the edition of his works (*Gesammelte Abhandlungen*, 3 vols., Berlin, Springer, 1935) as well as the excellent biographical study by Blumenthal which appears there (in Vol. III).

at first attract attention by great precocity, and it was only in 1888 that he came to the notice of the scientific world through his first great work on the theory of invariants. From that time on, discoveries of the first order were to follow one another in rapid succession for more than twenty years, bringing to him worldwide reknown, which the International Congress of Mathematicians in 1900 recognized, then the arrival of numerous pupils, Germans as well as foreigners, who crowded around him at Göttingen, where he had been a professor since 1895. He brought his career to a close in this university which he had made into one of the leading mathematical centers of Germany and the world, and which he never wanted to leave; and it was at Göttingen that he died peacefully on February 14, 1943.

What strikes one from the very first in Hilbert's works is the pure beauty of their imposing architecture; they do not give the impression of superficial "elegance," resulting from calculations cleverly carried out, but a much more profound esthetic satisfaction, which flows from the perfect harmony between the end pursued and the means employed to achieve it. The latter are most often of a disconcerting simplicity; it is usually not a perfecting, however ingenious, of the methods of his predecessors that leads Hilbert to his great discoveries, but, on the contrary, a voluntary return to the beginning of the problem under consideration, and a freeing from the raw material, where no one had been able to see them, the fundamental principles which allowed him to take the "royal road" to the solution, vainly sought for until then.

This trait was already manifest in the works of Hilbert on the theory of invariants.[1] Some forty years old, this was at the time one of the branches of mathematics most "in vogue." But the copious literature devoted to it was characterized above all by an amorphous mass of endless calculations, from which some general ideas emerged with great difficulty, too inadequate, it seemed, to permit the estab-

---

[1] A number associated with a geometric configuration formed by a certain number of points and algebraic curves is an *invariant* of this configuration with respect to certain geometric transformations when it is the same for the configuration considered and for all those which are derived from it by applying to it the transformations of the type considered. For example, the distance from a point to a line is an invariant of the configuration formed by the point and the line, for every *displacement*; the ratio of two segments of a line is an invariant of the configuration formed by their extremities, for every *similitude*; the anharmonic ratio of four concurrent lines is an invariant of these lines for every *projective* transformation.

lishing of universal laws[2] whose existence was strongly suspected, but whose validity could be proved only by very laborious arguments in each particular case. We therefore understand the surprise created in the mathematical world when, in 1888, Hilbert proved these general theorems in a few pages and almost without calculation. "This is no longer mathematics, it is theology!" exclaimed Gordan, one of those who had contributed the most to erecting the theory. An unexpected consequence was that almost from one day to the next, the theory of invariants found itself virtually abandoned; in resolving the principal problems it raised, Hilbert had dealt it a mortal blow. But, while doing this, he had at the same time laid the foundations for a new branch of algebra, the theory of *polynomial ideals*, which was, from the beginning of the 20th century, to revive the old algebraic geometry, and to form one of the pillars of the modern algebra erected by Emmy Noether and Emil Artin.

The situation was entirely different in the theory of algebraic numbers when Hilbert began his important investigations in this area around 1895. After Gauss, the founder of the theory, the entire effort of the 19th century had been directed toward the study of the laws of divisibility of algebraic integers of the same given field of numbers;[3] the persevering and profound works of Dirichlet, Kummer, Kronecker, Dedekind and Minkowski had made possible the disentanglement of the essential idea giving the key to the problem, that of "ideal," and the obtainment finally of a perfect generalization of the properties of divisibility of the rational integers. But it was still necessary to elucidate the phenomena which arise when passing from a given field to a greater "extension field"—that is to say, to determine how the "prime ideals" of the initial field are decomposed in the extension field. At the time of which we are speaking, it was known how to solve this problem only in very particular cases, and it did not seem possible to deduce any general law from the known facts. It was the great

---

[2] The most important of these laws is that every invariant of a given configuration can be expressed by rational combinations of a *finite* number of them.

[3] A field of numbers is a set of numbers (real or complex) such that every rational operation (addition, subtraction, multiplication, division) carried out on numbers of the set gives another number of the set. For example, all the numbers $a + b\sqrt{2}$, where a and b are *rational*, form a field. In such a field we are led to distinguish numbers called "algebraic integers," which have properties analogous to those of ordinary integers; for example, the numbers $a + b\sqrt{2}$, where a and b are *integers* (ordinary) are the algebraic integers of the preceding field.

attainment of Hilbert, and an astonishing proof of his remarkable wisdom, to be the first to know how to formulate these general laws. Previously, in a volume remaining a classic, he had synthesized into a homogeneous whole the totality of results achieved up to that time, using the theory of algebraic numbers; and he had introduced on this occasion many new ideas whose full fecundity was to be demonstrated in later works. He had, however, been able to verify the correctness of his general statements only in some special cases, which today seem to us a narrow base indeed to justify so powerful an induction; and it was only twenty-five years later that the last bastions of the fortress fell, and all the results stated by Hilbert were established in the most general case. But the impetus had been given, and it can be said that during this whole period, the theory of algebraic numbers was entirely imbued with Hilbert's methods and ideas, and tended toward the goal which he had been the first to perceive and to assign to it.

Up to about 1900, the works of Hilbert of which we have just been speaking had revealed him as a peerless algebraist and arithmetician; in the early years of the 20th century, he was to prove himself no less remarkable as an analyst. It was first of all in the calculus of variations that he opened an entirely new approach, since called the "direct method"; by applying it to Dirichlet's problem, he was the first to succeed in making rigorous the minimum method sketched by Riemann fifty years earlier;[4] numerous researchers have since then successfully employed the same principles in analogous problems.

But the most striking of Hilbert's studies in analysis are devoted to the theory of integral equations. Following the decisive works of Volterra, Poincaré and Fredholm, this entirely new theory was then in the foreground of mathematical current events. Here again, Hilbert is the lawgiver par excellence; beginning again with the general idea of "passage from the finite to the infinite," already employed by his predecessors, he decided to found, in complete analogy with linear and elementary quadratic algebra and its translation into geometric language in spaces of a finite number of dimensions, a new "functional algebra" in which the elements coming into play are no longer points, as in ordinary analytic geometry, but *functions*. It was not, however, a question here of a purely formal and almost mechanical analogy, and the essential point consists of intro-

---

[4] See the allusions to this question in the articles of Valiron, Part I, Book Two, p. 161, and Montel, Part I, Book Two, p. 174. (Note by F. LL.)

ducing in a suitable way what we today call topological ideas adapted to the theory. That was the most important progress realized by Hilbert—being the first to define and study the space of an infinite number of dimensions which now bears his name and which permits a transposition to functional algebra of the geometric expression of the theorems of elementary algebra. Fortified by this leading clue, Hilbert realized in the papers which he published between 1904 and 1920 the essentials of his program, which were to be completed by the works of his numerous pupils, among whom we must especially mention Schmidt, Weyl, Hellinger and Toeplitz. But Hilbert did not stop here; an entire part of his papers—and not the least important—is devoted to a multitude of applications of integral equations, abounding in methods and results as new as they are ingenious. Here again, it is the path traced out by Hilbert that mathematicians have followed for thirty years in the development of the branch of mathematics which he had founded, and which remains one of the most important chapters of modern functional analysis.

We have not yet said anything of the works of Hilbert which, perhaps more than all the previous admirable papers, have contributed to make his name known to all people concerned with mathematics, even in its most elementary aspect; I mean the famous *Grundlagen der Geometrie* (Foundations of Geometry), published in 1899, with which Hilbert became in one stroke the most noted representative of the trend called "axiomatic." Without doubt, he was not without precursors in this path; the discovery of non-Euclidean geometries, almost a century old, had taught mathematicians to consider only as relative the "truths" of classic elementary geometry; and in the last 20 years of the 19th century, many felt the need to throw full light on the logical bases of geometric proofs, freeing them from all recourse to intuition. But before Hilbert, nobody had yet known how to realize this program with as much determination and clarity, and no one had brought out so well the fundamental principle that in mathematics, the precise *nature* of the entities studied does not matter; it is the *relations* between these entities which alone are of importance. "Instead of the words 'points,' 'line,' 'plane' in geometry, we ought to be able everywhere to say without ill consequence 'table,' 'chair' and 'glass of beer,'" he said as early as 1891 in a sally almost reproduced in the famous opening (then considered so revolutionary) of the *Grundlagen*: "Let us think of three systems of things which we shall call points, lines and planes."

But the *Grundlagen* was not limited to a rigorous account of the origins of Euclidean geometry; further extending the analysis of the logical machinery, Hilbert began here a whole series of investigations into the mutual independence of the various axioms of geometry, and the delimitation of the parts of geometry in which each axiom plays an essential part; studies where free play is given to the inventive genius of their author, who knew how to discover even in the tritest subjects new properties, as remarkable as they were unsuspected.

Toward the end of his career, Hilbert was to return to mathematical logic, which he had put aside for more than fifteen years. This time, he attacked a vaster problem, that of the logical foundations of *all* mathematical theories—in particular those of arithmetic—and of the "non-contradiction" of these theories. It seems that one of the incentives which led Hilbert to resume studies of this kind (to which, moreover, his earlier work naturally led him) was a reaction against the movement of the mathematicians called "intuitionists," who, in order to avoid the logical difficulties raised by the "paradoxes" of set theory, did not hesitate proposing to sacrifice not only the essential part of Cantor's work, but a whole part of classic mathematical analysis. Against so radical a position, Hilbert revolted with all the force of his vigorous optimism, of his unshakable confidence in the power of human thought. As early as his lecture at the Congress of 1900, he did not hesitate to affirm his conviction that for every mathematical problem, we should succeed one day or another either in solving it or in showing that it had no solution; and we find again a similar act of faith in a lecture given in 1930 (one of his last publications), which he ended with these words: "We must know and we shall know"—a philosophic testament worthy of the man who had brought to the science such an ample harvest of new results. It is in this spirit that Hilbert, in his works on mathematical logic, proposed to show how, without giving up any earlier conquests, it was possible to give incontestable foundations to the structure of mathematics, and establish rigorously that it is impossible for an argument resting on these foundations ever to lead to a contradiction. On this last point, it seems that Hilbert's intuition involved him, for once, in hopes that were a little exaggerated, and at the present time there are good reasons for doubting the possibility of such "proofs." Nonetheless, through the interest and the discussions which his works on logic and those of his school have given rise to, Hilbert has contributed greatly to the clarification of the delicate question of the foundations of mathematics and the nature of logical reasoning, particularly by showing that thinking is clear only when it can be expressed completely by explicit symbols (therefore *finite* in number), and that, despite this restriction, it is perfectly allowable for the mathematician to argue correctly about "infinity."

What we have said thus far naturally leads us to see in Hilbert a constructor of vast syntheses, interesting himself in particular problems only to the degree that they are capable of being dominated by a general idea. Such a view would not be entirely inaccurate, since it indeed presents one of the fundamental aspects of Hilbert's mind, but it would give only a distorted and quite incomplete image of his richness and his astonishing variety. If Hilbert is one of the masters of axiomatics, he never suffered from the failings of certain of his followers, who created a grand theory for the sake of a few meagre applications and generalized for the sake of generalizing; and few mathematicians have had as much as he the fondness, so characteristic of the true mathematician, for the special problem, exact and "concrete," if one may say that, whether such a problem is related or not to a well-developed general theory. Without speaking of his famous lecture at the Congress of 1900 on "Mathematical Problems," where a good part of those enumerated consisted precisely of "isolated" problems, questions of this nature which Hilbert attacked with success are numerous; we shall limit ourselves to mentioning three of these results, which are worthy of his stature, and each of which would have been sufficient to establish the fame of an investigator: the first proof of Waring's hypothesis,[5] the theorem of irreducibility[6] and the proof of the existence of singularities on every surface of constant negative curvature immersed in three-dimensional space.

It is not to be wondered at that the influence of a man like Hilbert on the mathematical thought of his time has been considerable and lasting; it has without doubt not yet reached its zenith, and the numerous pupils whom he trained—and many of whom have become

---

[5] The question is about the following theorem: for every integer $s > 0$, there exists a number $g(s)$, *depending only upon* s, such that *every* integer can be expressed as a sum of at most $g(s)$ numbers, each the sth power of an integer. Stated by Waring at the end of the 18th century, this result had been proven, before Hilbert, for only a very small number of particular values of the exponent s.

[6] If $P(x, t)$ is a polynomial in two variables, with *rational* coefficients, and irreducible in the field of rational numbers, there exists an infinite number of rational values t_0 of t, such that the polynomial in one variable $P(x, t_0)$ is irreducible over the field of rational numbers.

eminent teachers—have made no small contribution to the increase of its wide circulation. We have seen above that entire branches of mathematics owe to him their orientation or their methods; but his activity is perceptible even in theories on which he never worked directly himself, as in general topology, for example. More than by his ingenious discoveries, it is perhaps indeed by the cast of his mind that Hilbert has had the most profound effect on the mathematical world; he has taught mathematicians to *think axiomatically*, that is to say, to strive to reduce each theory to its strictest logical schema, disentangled from contingencies of calculation. It is impossible to count the new and important results to which the application of this doctrine has led, and which have assured its triumph; but, more than by its immediate usefulness, it can be said that it is by its esthetic and, even in some way, moral attraction that it has won over most young mathematicians; by his intense need to *understand*, his more and more exacting intellectual integrity, by his untiring aspiration toward a science more and more unified, pure and stripped-down, Hilbert truly personified, for the generation between the two World Wars, the ideal of the mathematician.

27

THE INTERNATIONAL CONGRESSES OF MATHEMATICIANS

by Rolin Wavre
PROFESSOR AT THE UNIVERSITÉ DE GENÈVE

We apologize for whatever may be grim in the enumeration of the various international meetings in the area of the mathematical sciences. To be brief and to avoid troublesome repetitions, we shall designate the international congresses by the letters I.C. and the international congresses of mathematicians by the letters I.C.M.

If the 12th and 13th centuries saw the establishment of the first universities and the 16th and 17th the first societies of learned men, it was necessary to wait for the 19th for the appearance of scientific congresses in the form they have today. In the first half of the century the national associations for the advancement of the sciences were established; in the second half, the mathematical societies. It is no lack of idealism to say that the development of means of transportation was at the root of international meetings. The mathematicians were not the first to arrive on the scene. As early as 1863, anthropologists, astronomers, botanists, doctors, geographers, geologists began to gather together. It was toward the end of the century that the need for international collaboration made itself felt in the area of the mathematical sciences.

In 1894 the *Intermédiaire des Mathématiciens* was established; here everyone asked the questions troubling him or supplied the answers; on page 113 of the *Intermédiaire* we read: "Several of our correspondents . . . have spontaneously submitted to us the idea of the organization of great international mathematical congresses." This review was directed by Lemoine and by Laisant, who wrote later: "If the first idea came to us from the outside, if Georg Cantor told us of it, Lemoine thought of it at exactly the same time. He applied himself to it without sparing time or trouble, with the ardor of an apostle, busying himself with correspondence and personal meetings." Shortly thereafter, there gathered together at Paris a congress of "Mathematical Bibliography," to which many foreigners were invited.

In August 1897, there gathered at Zurich the first I.C.M., 209 participants attending. The following resolutions were adopted: "In the future, I.C.M.'s will succeed each other at intervals of from three to five years. Account will be taken, in the choice of location, of the justifiable wishes of the various countries. At the end of each congress, the date and location of the following congress will be chosen, as well as the agencies or associations in charge of organizing and preparing it." Finally, "the congress has for its purpose the encouragement of personal relations among mathematicians of different countries ... and the presentation of a survey of the present state of the various branches of mathematics." Several interesting projects for study were made much of: the establishment of annual publications, such as an address book of mathematicians, a biographical dictionary of contemporary mathematicians with their portraits; finally, a bibliographic headquarters operating rapidly and continually.

From the very beginning, the I.C.M. included general lectures and more particular communications presented at sessions of sections which met simultaneously: arithmetic and algebra, analysis and the theory of functions, geometry, mechanics and mathematical physics, history and philosophy.

Permit me to recall here the words of Poincaré read at the time of the first general conference: "You have doubtless often been asked what is the good of mathematics and whether its delicate structures drawn entirely from our minds are not artificial and brought forth by our caprice. I must make a distinction between the people who ask this question: practical men demand of us only a means of making money. These do not deserve any answer from us; rather, they are the ones who should be asked what is the good of accumulating so much wealth and whether, in order to have the time to acquire it, it is necessary to neglect art and science which alone make us beings able to enjoy it, *et propter vitam vivendi perdere causas*."

In 1899, the review *L'Enseignement mathématique* was founded, based in Geneva; in addition to its program of pedagogic material, it has contained since then the news of the mathematical life of the whole world.

The second I.C.M. met in Paris in 1900 (262 attending). It took up again the bibliographic and biographic projects of the previous congress (one and only one directory appeared in 1901 in the shape of *L'Annuaire des Mathématiciens*). In a general conference, Hilbert spoke of future problems in mathematics and concluded with these words: "The foregoing problems show us the increasing variety of mathematics. Is it not to be feared that our science will split up into several branches with hardly any relation to one another? We do not think so nor do we want to hope so. Moreover we see that as mathematics develops, far from losing its nature as a single science, it shows it more clearly day by day."

In 1900, also, there met at Paris the first session of the I.C. of philosophy which we also mention, for mathematical philosophy still occupies an important place there; the same is true of the I.C. of the history of sciences (Rome, 1902) and of the I.C. of the sciences at St. Louis, Mo.

The third I.C.M. met at Heidelberg in 1904. The centenary of the birth of Jacobi was celebrated at it. Various desires were expressed: the creation of university chairs in the history of the mathematical sciences; introduction of the elements of the history of sciences into higher secondary instruction; introduction at the university level of courses in elementary mathematics.

The second session of the I.C. of philosophy held its meetings in Geneva in 1904 and consisted of sections in logic, philosophy of science and history of science. It formulated a motion for the adoption of an international auxiliary language: "Men do not understand each other because they do not all speak the same language."

The fourth I.C.M. assembled at Rome in 1908, with almost 500 in attendance. There were established: the international commission of mathematical education of which Fehr is the secretary and which finds its natural spokesman in the review *L'Enseignement mathématique*; and the commission for the standardization of vectorial notation, which, since then, has endeavored to curb our diagrammatic fantasies.

At Heidelberg, in 1908, the congress of philosophy had its third session. At Milan, in 1911, the I.C. of mathematical education gathered.

The fifth I.C.M. got together at Cambridge in 1912 with some 500 participants; the scientific communications testified to the intense activity of the research men of that time; they were even too numerous according to some and too often overlapped. Mittag-Leffler invited the next congress to meet at Stockholm in 1916.

The first I.C. of mathematical philosophy met at Paris in 1914.

Then came the war with all its consequences. That of 1870 is hardly noticeable in the reports of the Academies of Paris and Berlin. But it was impossible after 1918 to revive the congress in its old form. Here are several quotations which will show how gloomy the skies had become:

"Earlier wars did not destroy the mutual respect of the scholars of the belligerent countries for one another; peace was able to expunge after a few years the trace of past struggles. Today (1918) conditions are quite otherwise.... So we must give up the old international associations and establish new ones among allies, with the possible participation of neutrals." The Inter-allied Conference of the Academies of Sciences preoccupied itself with "the conditions which international scientific collaboration will have to submit to after the war." The International Council of Research was founded at Brussels in 1919, to which were attached through adherence to certain general ideas ... international unions corresponding to different divisions of science. The International Union of Mathematicians was then born. The admission of a country depended upon conditions set by the regulations of the International Council of Research. The task of organizing the international congresses was incumbent upon the Union. In 1920, Strasbourg received the sixth I.C.M. under the presidency of Emile Picard. Some 200 participants attended.

On the initiative of Dutch professors, the I.C. of applied mechanics met at Delft, in April 1924. On this subject, let us mention the I.C. of technical mechanics at Leyden (1922), Zurich (1926) and Stockholm (1930).

The seventh I.C.M. gathered at Toronto in 1924, with some 400 persons attending. Résumés of the papers contributed were printed in advance. A happy innovation! The International Union of Mathematicians held a session there and appointed a commission on bibliography.

The eighth I.C.M. met at Bologna in 1928. Pincherle was pleased at having been able to put an end to an uncomfortable situation; certain countries which had refused two years before to adhere to the International Council of Research were represented at Bologna. Some personal abstentions were still greatly regretted. The international mathematics union maintained its position, but referred the matter to its committee to study the situation after Bologna.

From this time on congresses of Scandinavian and Balkan mathematicians were held as well as those of the U.S.S.R., then those of the Slavic countries. Since 1932 the International Council of Research has called itself the International Council of Scientific Unions.

In 1932, back to Zurich, where the ninth I.C.M. was held under the presidency of Fueter. It joined 667 participants and 40 countries were represented. A commission was appointed to study once more international cooperation in the field of mathematics, and the "liquidation" of the Union was decided upon. The Congress accepted with gratitude a gift of the late Professor Fields enabling it to award two gold medals to young mathematicians every four years.

In 1933, there was established at Geneva, with the name of the International Conferences of Mathematical Sciences, a series of conferences and of *international colloquia* devoted to the study of particular fields; a colloquium on quantum theory, October 1933; on mathematical logic, June 1934; on the theory of electrons in metals, October 1934; on topology, October 1935; on partial differential equations, June 1935; on theoretical probability, October 1937; on applied probability, July 1939; this last colloquium was organized with the cooperation of the Commission for Intellectual Cooperation which had, in 1938, organized at Zurich the "Conferences on the Fundamentals of Mathematics." In May 1934, a conference on tensor differential geometry met at Moscow. The conference resolved to form groups of scholars for a planned work. The members of each group would maintain a scientific correspondence and would communicate interesting problems with one another. In August 1935, there met at Brussels the first I.C. of mathematical recreations. In September 1935 at Moscow was held the first international conference on topology and at about the same time in Paris the first I.C. of scientific philosophy. The necessity of uniting the specialists of the whole world on particular problems then made itself imperiously felt. We must hope that a fragmentation of mathematics does not result but that, on the contrary, following Hilbert's views, the study of a particular subject will only be able to strengthen the whole.

The tenth I.C.M. met at Oslo in 1936. The absence of the Russians and the Italians was grievously felt. The commission appointed by the I.C. of Zurich was not able, for various reasons, to arrive at a unanimous decision on the question of an international organization of mathematics. It wished that in the future the question raised might receive a solution. Let us mention finally the congress of applied mechanics at Paris, 1946.

Harvard University was to receive the great family of mathematics in 1940; the I.C. was postponed until 1950. Its organization was entrusted to the American Mathematical Society and Prof. Marston Morse of Princeton was to preside over the organization committee.

Let us now take a look at the present time and if possible at the future.

It was in 1919 that the International Council of Research as well as the International Union of Mathematics were founded. In 1931 the International Council of Research changed its rules and became the International Council of Scientific Unions. But the International Union of Mathematics was in principle dissolved at Zurich in September 1932 after criticisms were directed against it. Nevertheless efforts were made at the last minute to arrange a program of possible activity. The Union had no secretariat in 1931 and 1932 and was unable to perform the service mathematicians expected of it.

The International Council of Scientific Unions, which grouped together seven unions ranging from astronomy through geodesy, chemistry, pure physics and the biological sciences to geography, became connected by an agreement with the International Organization for Intellectual Cooperation of the League of Nations. The latter is today replaced by UNESCO (the United Nations Educational, Scientific and Cultural Organization) which apparently will possess more powerful financial means than the earlier Organization. The secretariat of the International Council of Scientific Unions signed an agreement on December 16, 1946, with UNESCO, for mutual recognition and consultation, reciprocal representation and exchange of information, and article eight concerns financial aid to be furnished by UNESCO to the Council and to the Unions to allow them to carry out their program successfully. In particular, this agreement provides a contribution to the financing of international projects, either proposed or in the course of execution, on the condition that the scope and international character of these projects justify such aid.

But, as we have said, the International Union of Mathematics, in fact, no longer exists. There would be an advantage, however, in reorganizing it. It could make contact with the agencies of the United Nations and reply to all consultations, whether on behalf of the United Nations or other international organizations. It is needless to insist further on the value of this preliminary understanding between mathematicians and within the framework of the International Council of Scientific Unions, in order that the latter can present our demands and our desiderata to UNESCO. Already in March, a meeting took place with the aim of reviving the International Union of Mathematics; at the end of spring, in Paris, on the occasion of two international colloquia, subsidized by the Rockefeller Institute, the question will be taken up again, in a session which will bring together mathematicians of different countries.

UNESCO possesses a section on the sciences, one of whose principal aims is to "facilitate," by its secretariat and its capital, certain international organizations already existing. Along with the great I.C.M., UNESCO seems disposed to organize symposia or international colloquia of the same type as those already organized by the International Institute for Intellectual Cooperation.

Let us mention, to bring this account to an end, the brochure entitled *UNESCO and Science* (International Scientific Cooperation, tasks and functions of the Section on Sciences of the Secretariat of the United Nations Organization for Education, Science and Culture). It sets forth the aims and methods of this new agency of the United Nations. Let us notice simply the following ideas: the creation of a joint committee on scientific publications to distribute all papers in the form of off-prints and forward them to the specialists in the subject treated.

UNESCO is considering supporting certain scientific journals of a well defined international character and underwriting the translation of scientific works written in languages not widely prevalent. It would be able to promote the adoption of a worldwide language of communication among scholars, a role which Latin played in the 16th century. There is even a mention in this brochure of the establishment of an international university, a very important task which would require the expenditure of considerable capital.

All these projects are splendid. If the United Nations (UN) succeeds in being worldwide, if its Section on Sciences can work in an atmosphere of political understanding, everything will be for the best, and we can expect from it unlimited advantages of international cooperation. But if political understanding does not exist, the pursuit of scientific progress in an unselfish purpose, or toward a humanitarian end, would gain great advantages from setting itself apart as much as possible from the strife and wars in which, alas! politics too often involves the man of science.

Book Three

THE FUTURE

The future of mathematics is a function of so many variables (many of which are extraneous to mathematics) that an attempt at long-range predictions would be somewhat presumptuous. We can remain on solid ground only by limiting our objective to the examination of a "circle of convergence" of sufficiently small radius, in which we must be content to develop those current ideas which seem filled with promise. That is what we have aspired to in this work. It is, in short, much more a question of the directions of research of contemporary mathematicians, of their preoccupations, of their ambitions and of their hopes than of a vain effort with a view to defining a more remote future which will doubtless bring about staggering surprises to future investigators. Even in this limited sense, in order to master this formidable subject, there was needed an intimacy with mathematics and an uncommon gift of acuteness, of which hardly any convincing examples are known.[1]

André Weil, who has taken the leading role in the creation and direction of the Bourbaki school, is one of the most outstanding among the mathematicians who are continuing the work of the great preceding generation. His works, which have dealt with the theory of numbers as well as with algebraic geometry and topology, have exerted a great influence in the years preceding the Second World War. Thus it is not just a compilation that Weil offers us, but a vigorous syn-

---

[1] It is known that aside from rather unevenly successful attempts, such as those of Gaston Darboux (congress at St. Louis, 1904) and of Henri Poincaré (congress at Rome, 1908), few probing thrusts into the future proved themselves so astonishingly accurate as the famous "Hilbert Program" (congress at Paris, 1900). On this subject, see the articles by Dieudonné, Part II, Book Two, pp. 304–311, and Wavre, pp. 312–318. We intend to return in the second volume [never published] to an examination of these 23 problems of Hilbert and to state the current position for each of them.

thesis aiming at encompassing the problem of the future of mathematics. In its consideration of the connections of mathematics with the development of society, as in its technical and internal point of view, his article reflects the very plan of this work. It is on the one hand a genuine ethic, individual and social, which the author is led to develop from his investigation of the relations of mathematics and the human environment in which it is situated. On the other hand, the immense mathematical culture of Weil was required to unravel the most promising directions from the latest attainments of current work. On the strength of its very contents, this article included statements which required the support of a technical argumentation, which the author has given, as briefly and as concisely as possible. The reader not fortified with considerable mathematical knowledge will not need to retain from this part of the article any more than the vitality of a science in which so many problems arise at each step.

The author of a remarkable thesis on Hilbert's spaces, Roger Godement is one of the hopes of the young French school of mathematics. The ingeniously reasoned explanation which he gives us completes that of Weil in stressing the relations of the abstract and concrete, a precise understanding of which will determine to a large degree the future of mathematics. For Godement the abstract and concrete are not in opposition. It is necessary to advance unceasingly toward greater generality without, however, wanting to cut the bonds which unite abstraction to concrete reality. Here, indeed, is the right path. Icarus and Antaeus are but two faces of Prometheus.

F. LL.

28

THE FUTURE OF MATHEMATICS*

by André Weil

"At one time," says Poincaré in his Rome conference on the future of mathematics, "there were prophets of misfortune; they reiterated that all the problems had been solved, that after them there would be nothing but gleanings left...." "But," he added, "the pessimists have always been compelled to retreat... so that I believe there are none left to-day."

Our faith in progress, our belief in the future of our civilization are no longer as strong; they have been too rudely shaken by brutal shocks. To us, it hardly seems legitimate to "extrapolate" from the past and present to the future, as Poincaré did not hesitate to do. If the mathematician is asked to express himself as to the future of his science, he has a right to raise the preliminary question: what kind of future is mankind preparing for itself? Are our modes of thought, fruits of the sustained efforts of the last four or five millennia, anything more than a vanishing flash? If, unwilling to stumble into metaphysics, one should prefer to remain on the hardly more solid ground of history, the same questions reappear, although in different guise: are we witnessing the beginning of a new eclipse of civilization? Rather than to abandon ourselves to the selfish joys of creative work, is it not our duty to put the essential elements of our culture in order, for the mere purpose of preserving it, so that at the dawn of a new Renaissance, our descendants may one day find them intact?

These questions are not purely rhetorical; upon each man's answer, or rather (for such questions do not have answers), upon the

---

* Authorized translation by Arnold Dresden reproduced by permission of The Mathematical Association of America.

attitude which he takes in front of them, depends in large measure the trend of his intellectual efforts. It was necessary, before writing about the future of mathematics, to formulate these questions, just as the faithful cleansed themselves before consulting the oracle. Let us now interrogate destiny.

Mathematics, as we know it, appears to us as one of the necessary forms of our thought. The archaeologist and the historian have shown us civilizations from which mathematics were absent. It is indeed doubtful whether they would ever have become more than a technique, at the service of technologies, if it had not been for the Greeks; and it is possible that, under our very eyes, a type of human society is being evolved in which they will be nothing but that. But for us, whose shoulders sag under the weight of the heritage of Greek thought and who walk in the paths traced out by the heroes of the Renaissance, a civilization without mathematics is unthinkable. Like the parallel postulate, the postulate that mathematics will survive has been stripped of its "evidence"; but, while the former is no longer necessary, we would not be able to get on without the latter.

The clinical student of ideas who limits his prognosis to the immediate future, and does not risk long-range prophecies, certainly observes more than one favorable symptom in contemporary mathematics. To begin with, while some sciences, conferring, as they now do, an almost unlimited power upon a ruthless possessor of their results, tend to become caste monopolies, treasures jealously guarded under a seal of secrecy which must of necessity become fatal to any genuine scientific activity, the real mathematician does not seem to be exposed to the temptations of power nor to the strait-jacket of state secrecy. "Mathematics" said G. H. Hardy in substance in a famous inaugural lecture, "is a useless science. By this I mean that it can contribute directly neither to the exploitation of our fellowmen, nor to their extermination."

It is certain that few men of our times are as completely free as the mathematician in the exercise of their intellectual activity. Even if some State ideologies sometimes attack his person, they have never yet presumed to judge his theorems. Every time that so-called mathematicians, to please the powers that be, have tried to subject their colleagues to the yoke of some orthodoxy, their only reward has been contempt. Let others besiege the offices of the mighty in the hope of getting the expensive apparatus, without which no Nobel prize comes within reach. Pencil and paper is all the mathematician needs; he can even sometimes get along without these. Neither are there Nobel prizes to tempt him away from slowly maturing work, towards a brilliant but ephemeral result. Mathematics is taught the world over, well here, badly there; the exiled mathematician—and who among us can to-day feel free from the danger of exile—can find everywhere the modest livelihood which allows him to pursue his work to some extent. Even in jail one can do good mathematics if one's courage fail him not.

To these "objective conditions," or rather, as the physician would say, to these external symptoms, must be added others revealed by a more penetrating clinical examination. In recent times mathematics has demonstrated its vitality by passing through one of these periods of growing pains, to which it has been accustomed for a long time, and which are designated by the strange name of "foundation crises." It has come through it, not only without damage, but with great gain. Whenever wide domains have been added to the field of mathematical reasoning, it is necessary to inquire what techniques are allowed in the exploration of the new territory. One wants certain objects to have certain properties, one wants certain modes of reasoning to be admissible and one behaves as if they were. But the pioneer who proceeds in this way knows very well that some day the police will come to put an end to the disorder and to bring everything under the control of the general law. Thus, when the Greeks defined the ratio of two magnitudes for the first time with enough precision to raise the problem of the existence of incommensurable magnitudes, they seem to have believed and to have wanted all ratios to be rational and to have based the first sketch of their geometrical reasonings on this provisional hypothesis; some of the greatest advances in Greek mathematics are connected with the discovery of their initial error at this point. In the same way, at the beginning of the era of the theory of functions and of the infinitesimal calculus, one wished every analytical expression to define a function, and every function to have a derivative; we know to-day that these requirements were incompatible. The last crisis, which grew out of the sophistries, for which the "naive" theory of sets opened a way in its early stages, has led for us to a no less happy result, which can now be considered as permanently established. We have learned to trace our entire science back to a single source, constituted by a few signs and by a few rules for their use; this is unquestionably an unassailable stronghold, inside which we could scarcely confine ourselves without risk of famine, but to which we are always free to retire in case of uncertainty or of external danger. Only a few backward spirits still maintain the position that the mathematician must forever draw on his "intuition" for new, alogical or "prelogical" elements of reasoning. If certain branches of mathematics have not yet been axiomatized, *i.e.*, reduced to a form of exposition in which all terms are defined, and all axioms made explicit, in terms of the basic notions of set theory, this is simply because there has not yet been the time to do it. It is of course possible that some day our successors will want to introduce into set-theory modes of reasoning which we do not admit. It is even possible that the germ of a contradiction, which we do not perceive to-day, may later be discovered in the modes of reasoning we now use, although the work of the modern logicians makes this very unlikely. A general revision will then become necessary; one can feel certain even now that this will not affect the essential elements of our science.

But, if logic is the hygiene of the mathematician, it is not his source of food; the great problems furnish the daily bread on which he thrives. "A branch of science is full of life," said Hilbert, "as long as it offers an abundance of problems; a lack of problems is a sign of death." They are certainly not lacking in our mathematics; and the present time might not be ill chosen for drawing up a list, as Hilbert did in the famous lecture from which we have just quoted. Even among those of Hilbert, there are still several which stand out as distant, although not inaccessible, goals which will continue to suggest research for perhaps more than a generation; an example is furnished by his fifth problem, on Lie groups. The Riemann hypothesis, after the attempts to prove it by function-theoretic methods had been given up, appears to-day in a new light, which shows it to be closely connected with the conjecture of Artin on the L-functions, thus making these two problems two aspects of the same arithmetico-algebraic question, in which the simultaneous study of all the cyclotomic extensions of a given number field will undoubtedly play a decisive role. Gaussian arithmetic was centered around the law of quadratic reciprocity; we know now that this law is only a first example, we might better say the pattern of, the laws of "class fields," which control the abelian extensions of algebraic number-fields; we know how to formulate these laws so as to make them look like a coherent set. But, pleasant as this façade may be to the eye, we do not know whether it might not hide deeper lying symmetries. The automorphisms induced in the class groups by the automorphisms of the field, the properties of the norm-residues in the non-cyclic cases, the passage to the limit (inductive or projective) when the base field is replaced by extensions, for example, cyclotomic extensions, of indefinitely increasing degree, all these are questions on which our ignorance is almost complete and in whose study the key to the Riemann hypothesis is perhaps to be found. Closely connected with these questions is the study of Artin's conductor and, in particular, in the local case, the search for the representation, whose trace can be expressed by means of simple characters with coefficients equal to the exponents of their conductors. These are some of the directions which can and must be followed up in order to penetrate the mystery of non-abelian extensions; it is not impossible that we are here close to principles of extraordinary fertility and that, once the first decisive step on this road will have been taken, we shall gain access to vast domains whose existence is hardly suspected. For, however wide our generalizations of Gauss' results may be, we can hardly claim to have as yet really moved beyond them.

Even in the realm of abelian extensions, we have not made any progress towards the generalization of the theorems of "Kronecker's youth dream," the generation of class fields, whose existence is known, by means of values of analytic functions. While it has been possible, without serious difficulties, to complete Kronecker's unfinished work and to obtain the solution of this problem, in the case of imaginary quadratic fields, by means of complex multiplication, the key to the general problem, considered by Hilbert as one of the most important of modern mathematics, still escapes us, in spite of the conjectures of Hilbert himself and the efforts of his pupils. Must we look for it perhaps in the new automorphic functions of Siegle, in his modular functions of several variables? Or can the theory of the endomorphisms of abelian varieties, which has now made considerable progress, be of some help here? It is too early to risk acceptable conjectures on these questions; but their closer examination is bound to produce interesting results, even though they should be negative in character.

The foregoing discussion shows clearly not only the vitality of modern arithmetic, but also the close ties which connect it, to-day as in the days of Euler and the days of Jacobi, with the most deep-lying parts of the theory of groups and of the theory of functions. This essential unity, which appears in so many and in such diverse ways, is also found in many other places. The introduction by Hermite of continuous variables into the theory of numbers has led to the systematic study of discontinuous groups of arithmetical nature by means of the continuous groups in which they can be imbedded, of the symmetric Riemannian spaces associated with these groups, of the differential and topological properties of their fundamental domains (or rather, in modern terminology, of their quotient spaces), and of the automorphic functions which belong to them. The work of Siegel, continuing the great tradition of Dirichlet, of Hermite and of Minkowski, has opened entirely new paths here. On the one hand, we connect with Fermat, Lagrange and Gauss, the representation of numbers by forms, and the genera of quadratic forms. At the same time, we begin to see in outline the fertile principle, according to which the global aspect of an arithmetical problem can, under certain circumstances, be reconstructed from its local aspects. For instance, in the work of Siegel we see repeatedly that the number of solutions of some arithmetical problem in the field of rational numbers is expressed by means of numbers defined by the corresponding local problems, density of solutions in the real field and in the p-adic fields for all values of the prime p. This is a principle, analogous to Cauchy's theorem for the Riemann surface of an algebraic curve, with which one may connect also the famous "singular series" which appear in the application of the method of Hardy-Littlewood to problems in the analytic theory of numbers. Is it possible to formulate this principle in a general statement, which would allow us to obtain at one stroke all results of this character, just as the discovery of Cauchy's theorem made it possible to calculate by a single method a number of integrals and of series which were formerly treated by special distinct processes? It looks as if this were not yet a problem for the immediate future; so much the more reason to prepare for its solution by the study of well-chosen particular cases. It may be that this same principle will one day reveal the deep reason for the existence of Eulerian products, of which the extreme importance for the theory of numbers and the theory of functions has only become clear through the work of Hecke. Here we deal with the classes of quadratic forms, and not merely, as in the work of Siegel, with their genera; at the same time, we find ourselves at the core of the theory of modular functions, which has been infused with new life by these studies, and of the theory of theta functions. This domain is still so full of mystery, the questions which it raises so numerous and fascinating, that it would be premature to try to arrange them in order of importance.

At the same time, Siegel has taught us to construct discontinuous groups and automorphic functions by arithmetical methods; in this field the theory of functions, by its own efforts, had been unable to move forward since Poincaré. Indeed it is very likely, that, just as in the case of functions of a single variable, the thorough study of special functions of several complex variables will have to prepare the ground for an attack on the general theory. In the work of Siegel, the local and global geometrical study of fundamental domains, in effect of manifolds with a complex-analytic structure, tends to occupy a dominating role. Along this road, connection is made with the immense work accomplished by E. Cartan and its various extensions; at the same time one gets into the center of modern topology, the theory of fibre-spaces; and the invariants of Sitefel-Whitney appear, along with their generalizations. The intimate connections between these two domains had been suspected for some time, but their actual merging was made possible only through the recent discoveries of Chern, stimulated in their turn, at least in part, by considerations of algebraic geometry. Indeed, algebraic varieties, at least varieties without singularities in the complex field, are nothing but a special, and particulary interesting, class of manifolds with a complex-analytic structure; more precisely, they are manifolds on which, at least in all known cases, one can define one of these remarkable Hermitian metrics which were introduced by Kähler in connection with functions of several complex variables, and of which results of S. Bergmann, not yet fully clarified, furnish other examples. By a systematic, although not explicit use of these metrics, Hodge has recently obtained the first existence theorems for this type of manifold, generalizing the classical results of Riemann. While it may be too much to hope that such methods may one day lead to the uniformization of algebraic varieties (which, contrary to what happens in the case of curves, can not be done in general by means of unramified functions), there is little doubt that they can be extended to integrals of the third kind. The analogous generalization of the methods of Hodge to differential forms with singularities in the real domain raises still more important problems. It appears to be connected, on the one hand, with local properties of the equations of elliptic type which harmonic forms satisfy; on the other hand, it seems to be linked with an extension of de Rham's theory which would make it possible to obtain the homologic torsion of a manifold by means of differential forms with singularities. De Rham's results have, as a matter of fact, definitely clarified a certain aspect of the relation between homology groups and multiple integrals, and this accounts for the fundamental role they play in the work of Hodge and of Chern; but until now they have only made the homology groups with real coefficients accessible to differential methods; moreover, the striking and fertile analogy between chains and differential forms, which is expressed in these results, remains a mere heuristic principle, until we succeed in finding a common basis for these two concepts. To convert this principle into a method of proof has thus far succeeded only in a few special cases, for example, in some of the beautiful papers with which Ahlfors has in recent years given fresh life to the theory of analytic functions.

But, while algebraic geometry, as we have just seen, receives a fresh stimulus from the most recent developments in topology and in differential geometry, this field does not lack purely algebraic problems; and, thanks to the methods of modern algebra, our understanding of them no longer need depend upon flashes of intuition of a few privileged mortals. At present, the theory of surfaces, brilliantly but too rapidly developed by the Italian school, must yield place to a general theory of algebraic varieties, freed from restrictive assumptions as to the nature of the base field and as to the absence of singularities. The structure of the groups of divisor classes, with respect to the different known concepts of equivalence (linear, continuous, numerical), the study of unramified extensions of a field of algebraic functions, both abelian and non-abelian, these are the questions that call for solution first of all. Thanks to the results obtained, or at least made plausible, by the Italian geometers, we can more or less guess the answers; and their solution, perhaps already within our reach, must open the road for important advances. On the other hand, the study of algebraic geometry over various special fields of constants is still gropingly taking its first steps. In view of the fact that algebraic geometry over the complex field, studied for almost a century, has arrived by its own methods (topological and transcendental) at well-known important results, it is probable that other fields, finite fields, p-adic fields, fields of algebraic numbers, deserve to be studied, each by itself, by methods suited to their purpose. From this point of view, geometry over a finite field appears somewhat like a turntable, from which one may at will direct one's further progress either towards algebraic geometry proper, with the powerful tools already at its disposal, or towards the theory of numbers; it is precisely in that way that we are beginning to get a better insight into the nature of the Zeta-function and into the true nature of the Riemann hypothesis. In the same way, before undertaking the determination of the extensions of a field of algebraic numbers by means of their local properties, it might be indicated to solve the analogous problem, already difficult enough, concerning algebraic functions of one variable over a finite field, *i.e.*, to extend Riemann's existence theorem to such functions. To mention merely a particular case, one might ask whether the modular group, whose structure determines the fields of functions of a complex variable with only three points of ramification, plays the same role, at least with respect to the extensions of degree prime to the characteristic, when the field of constants is finite. It is not impossible that all questions of this kind can be treated by a uniform method, which would make it possible to deduce, from a result established (for instance by topological methods) for characteristic 0, the corresponding result for characteristic p; the discovery of such a principle would constitute an advance of the greatest importance. Of the same character, but still more difficult, are the problems arising in the modern study of finite groups. Is the theory of finite simple groups an anologue of the theory of simple Lie groups? It would probably be premature to make a frontal attack on this question at the present stage; by means of indirect procedures, in particular the study of p-groups, some progress has been made in this direction in recent years. As in many other questions of algebra and of the theory of numbers, so also here a new element has recently been introduced by the definition of the homology groups of an abstract group. The discovery of this concept, which generalizes the fruitful concepts of character and factor-set, is due to Eilenberg and MacLane, who introduced it in connection with H. Hopf's studies in pure combinatorial topology; it will have to be subjected for some time to systematic study, before its scope and possibilities of application can be estimated.

While arithmetic, in the widest sense, is for its devotees always the queen of mathematics, and while for that reason we have allowed ourselves to dwell on it with predilection, this is not to say that other branches of mathematics do not offer as many problems worthy of sustained attention. The work of a Cartan alone contains enough material to keep busy several generations of geometers. His general theory of systems in involution has not been carried to its conclusion by its author, who appears not to have been able to overcome all the difficulties of algebraic character which it involves. Concerning the theory of "infinite Lie groups," undoubtedly very important, but for us very obscure, we know nothing beyond what is found in the memoirs of Cartan, a first exploration into an almost impenetrable jungle; this jungle threatens to overgrow the paths which already have been marked out, if the indispensable task of clearing is not undertaken very soon. The modern theory of Lie groups proper, studied by a combination of Cartan's methods with those of modern topology, is far from complete; even in the theory of semi-simple groups and in the theory of the symmetric Riemannian spaces associated with them, a good many results are attainable only by *a posteriori* verification, making use of our knowledge (also due to Cartan) of all simple groups. But, as has already been suggested, it is principally in the topological theory of fibre spaces, in the theorems of de Rham and in the notion of homotopy group, that we now find the tools best suited to the global study of the generalized geometries of Cartan. To give but a single example, the classical Gauss-Bonnet formula, until recently the only result in which a topological invariant was expressed by means of the integral of an invariant differential form, appears to us now as but the first term of a whole sequence of formulas, to which Chern's methods give access and whose systematic study has barely been started.

But, even though involutory systems should, in principle, enable us to obtain everything which can be reduced to the local problem of Cauchy-Kowalewski in the theory of partial differential equations, this is merely one aspect of the existence problem for solutions of these equations; and, from several points of view, it is not its most interesting aspect. Beyond this, we find important results concerning equations of very special types, chiefly elliptic and hyperbolic, some of which are of very recent date; but, although the study of these types, to which our predecessors were led by mathematical physics more than a century ago, is far from complete, it will not do to stop indefinitely at this point. The system which is satisfied by the real part of an analytic function of several complex variables does not belong to any of these simple types; however, function-theory has taught us, for example, that the most general singularities which they can have are, in a sense which is still not easy to specify, made up out of elementary singularities which are characteristic varieties; at any rate, one can interpret the theorems of Hartogs and of E. E. Levi in this manner. In this form, they present an obvious analogy to known results concerning hyperbolic equations; it is this analogy which suggests that we look for the germ of a general theory in a fuller development of the concepts of characteristic variety and of elementary solution. On the other hand, in the work of Delsarte, and in that of S. Bergmann and of his pupils, we find the first examples of the transformation of differential equations by means of integral or of integro-differential operators. It looks as if we have here the germ of entirely new developments and of a classification of systems of partial differential equations, which falls entirely outside the framework of classical methods. In particular, as was shown by Delsarte, the series of orthogonal functions to which elliptic problems lead naturally, are found to be transformed into series of much more general types; some isolated examples of these are found in classical analysis, but their general study presents problems of the greatest interest. The mathematician can no longer be satisfied here with Hilbert space, with which he has become as familiar as with the Taylor series or the Lebesgue integral; must he look for the most appropriate tool in the theory of Banach spaces, or must he have recourse to more general spaces? It must be admitted that Banach spaces, interesting and useful as they have already proved to be, have not yet brought about the revolution in analysis which some people expected of them; but it would be a counsel of despair to abandon their study now before the various possibilities of application have been more fully explored. However it is possible that they are both too general to be suited to as exact a theory as that of Hilbert spaces, and too special to lend themselves to the study of the most significant operators. They do not include, for instance, the space of indefinitely differentiable functions; and it is only in that space that the operators of L. Schwartz can be defined, which represent formally the derivatives of all orders of arbitrary functions. Perhaps the basis for a new calculus, founded on the generalized Stokes theorem, is to be looked for here; and this may give access to the relations between differential operators and integral operators. Ideas of this character have already proved very useful in special problems, for instance in the calculus of variations under the name of Haar's lemma as well as in certain papers of Freidrichs. Similarly, the well-known theorem, which asserts that the mean of a harmonic function on a circle is equal to its value at the center, expresses the fact that a certain operator, defined by a mass distribution in the plane, is, in a certain sense, a linear combination of the values of the Laplacian in the closed domain bounded by the circle. Also connected with these questions is the problem, mentioned above, of the representation of differential forms as sums of chains, which arises from the theory of de Rham. It is possible that we have in these researches the dim outlines of an operational calculus, destined to become in one or two centuries as powerful an instrument as the differential calculus has been for our predecessors and for ourselves.

All of this has to do only with the local or semi-local study of partial differential equations; indeed, apart from the simple cases which can be treated by means of the theory of Hilbert spaces or by the direct methods of the calculus of variations, the study of partial differential equations in the large, for instance on a compact analytic manifold, appears too difficult to justify any hope that it can be attacked for a long time to come. On the other hand, the study in the large of ordinary differential equations raises a large number of interesting problems; they are difficult, but within our reach. It will suffice to mention as an example the recent beautiful proof, by E. Hopf, of the ergodic character of the geodesics on every compact Riemannian manifold of everywhere negative curvature. Related to this subject is also the study of van der Pol's equation and of relaxation oscillations, one of the few interesting problems which contemporary physics has suggested to mathematics; for the study of nature, which was formerly one of the main sources of great mathematical problems, seems in recent years to have borrowed from us more than it has given us.

But, incomplete as the foregoing enumeration cannot fail to seem to our colleagues, it has undoubtedly exhausted the attention of more than one reader; and yet, for lack of space and for lack of necessary competence, we have not spoken of the geometry of numbers, nor of diophantine approximations, of the calculus of variations, of the calculus of probabilities, or of hydrodynamics, neither have we mentioned at all several problems, to-day in the background of interest, which could be reactivated by a new idea and restored to the vital stream of mathematics. As a matter of fact, we neither can nor want to lay out a route for the future development of our science; this would be a futile task, indeed it would be a ridiculous enterprise, for the great mathematicians of the future, like those of the past, will flee from the beaten track. They will solve the great problems which we shall bequeath to them, through unexpected connections, which our imagination will not have succeeded in discovering, and by looking at them in a new light. It was our purpose, in passing some of the principal branches of our mathematics in review, to draw attention to their robust vitality and to their fundamental unity. We believe to have shown, not only that there are large numbers of problems, but also that there are very few really important problems which are not intimately related to others which, at first sight, seem to be far removed from them. When a branch of mathematics ceases to interest any but the specialists, it is very near to its death, or at any rate dangerously close to a paralysis, from which it can be rescued only by being plunged back into the vivifying sources of the science. "Mathematics," said Hilbert at the end of his 1900 lecture (and it would be quite in order to quote the conclusion of this lecture in full), "is an organism for whose vital strength the indissoluble union of the parts is a necessary condition."

Does this mean that mathematics is becoming a science for erudites, and that it will no longer be possible to do creative work in mathematics until one has grown gray in the harness, and exhausted from burning the midnight oil for many years in the company of dusty tomes? This would at the same time be a sign of its decline; for be it strength or weakness, mathematics is not a science that prospers on details, painstakingly collected in the course of a long career, on patient reading, on observations or on filing cards, amassed one by one so as to form a bundle from which an idea will ultimately come forth. Perhaps it is more true in mathematics than in any other branch of knowledge that the idea comes forth in full armor from the brain of the creator. Moreover, mathematical talent usually shows itself at an early age; and the workers of the second rank play a smaller role in it than elsewhere, the role of a sounding board for sounds in whose production they had no part. There are examples to show that in mathematics an old person can do useful work, even inspired work; but they are rare, and each case fills us with wonder and admiration. Therefore, if mathematics is to continue to exist in the way in which it has manifested itself to its votaries until now, the technical complications with which more than one of its subjects is now studded, must be superficial or of only temporary character; in the future, as in the past, the great ideas must be simplifying ideas, the creator must always be one who clarifies, for himself and for others, the most complicated tissues of formulas and concepts. Hilbert indeed asked himself: "Is it not going to become impossible for the individual worker to embrace all the branches of our science?" and he justified his negative answer, not only by his example, but by remarking that every important advance in mathematics is related to the simplification of methods, to the disappearance of old procedures which have lost their usefulness, and to the unification of branches which were until then foreign to each other. It is quite likely that the contemporaries of Apollonius for example, or those of Lagrange, were familiar with the same feeling of growing complexity which tends to overwhelm us to-day. It is undoubtedly true that the modern mathematician does not know certain details of the theory of conic sections as well as Apollonius did, or as a candidate for a French competitive examination, but this does not lead any one to think that the theory of conic sections should form an autonomous science. Perhaps the same fate is in store for some of the theories of which we are proudest. The unity of mathematics would not be threatened by such an occurrence.

The danger lies elsewhere. Although it is more contingent in character, it does not strike us as less serious; and it seems to us that we cannot bring our reflections on the future of mathematics to a conclusion without saying something of it. We have already said that our civilization itself seems to be under attack from all sides; but this remark was couched in too general terms. "Ne sutor ultra crepidam": it is as mathematicians that we must look at the contemporary world. Our tradition is healthy; are we assured of transmitting it undamaged? In some European countries, particulary in Germany until the start of the Hitler regime, there existed, still a short time ago, university instruction, based on a solid secondary education, which made sure that the mathematical apprentice acquired specific subject-matter knowledge and also the general culture without which nothing of importance can be accomplished. What do we see to-day? In France, none of the essential parts of modern mathematics is taught in our universities, except by a lucky chance. One looks in vain for a university course which puts the advanced student in contact with any one of the great problems which we have listed.* Even the elements of the science are too frequently taught in

---

* (Author's footnote) This was written in 1946; the same would not be unqualifiedly true to-day.

such a way that the student has to learn everything over again if he wants to push on; the extreme rigidity of a mandarin-caste founded on obsolete academic institutions is the cause of the fact that every attempt at modernization is doomed to failure, unless it remains restricted to verbal changes. Italy, which had formerly a flourishing mathematical school, seems to have fallen into a state of sclerosis analogous to that with which France is threatened, but which has had there still more immediate and destructive effects. We do not know what principles guide, at present, secondary and higher education in the U.S.S.R. There are in that country a number of first-rate mathematicians, but they seem to be absolutely prohibited from crossing the frontiers; and, if such practice should persist, it can hardly in the long run have any other result than the slow choking off of all scientific life. The most remote, as well as the most recent history of our science, shows sufficiently to what extent the contacts between one country and another, prolonged sojourns of students and of teachers at foreign universities, not official sessions at which one drinks toasts while waiting for the next airplane, are an indispensable condition for all progress. We believe that more favorable conditions are found in England and in some of those nations of western Europe which are small only in military statistics. As to Germany, only the future can show whether she will find within herself the necessary elements for linking up with the brilliant tradition interrupted by fifteen years of organized stupidity. Beyond the Atlantic finally, we find a large country, which counts its universities by the hundreds, its students by the hundred thousands, and where, in the words of H. Morrison, the great American specialist in educational problems, "one wanted the education of the masses, one has mass production in education." Thorstein Veblen once sketched in a small book, too little read, the plan of higher education in the United States, and he has done it in a masterly manner; let us merely indicate how the future mathematician is formed in this country which produces more "mathematicians" than perhaps all the rest of the world. In the most favorable cases, one sees a student who, towards the end of his stay at the University, has three or four years at his disposal in which to acquire at the same time the knowledge, the method of work and the elementary intellectual apprenticeship, for which nothing that he has experienced before has in the least prepared him. His only way out under these circumstances is to seek his salvation in the most narrow specialization; in this way, he can often, if he is intelligent and has good guidance, do useful work. Beyond this, he runs the great risk of not being able to survive the stupefying effects of the purely mechanical teaching which he will have to inflict on others, in order to earn his living, after having undergone it himself for too long a time. Whether, in other fields, mass production, thus understood, may produce good results, we are not qualified to determine; we hope to have made it clear that this can not be the case in mathematics. If, unfortunately, the plausible doctrine of making education available to all has had such consequences in a country which lacks, it is true, a strong intellectual tradition, do we not have reason to fear the spread of the contagion to a Europe enfeebled by a catastrophe without precedent?

But if, as Panurge, we ask the oracle questions which are too indiscreet, the oracle will answer us as it did Panurge: *Trinq*! This advice the mathematician follows gladly, pleased as he is to believe that he will be able to slake his thirst at the very sources of knowledge, convinced as he is that they will always continue to pour forth, pure and abundant, while others have to have recourse to the muddy streams of a sordid reality. If he be reproached with the haughtiness of his attitude, if he be summoned to do his part, if he be asked why he persists on the high glaciers whither no one but his own kind can follow him, he will answer, with Jacobi: For the honor of the human spirit.

29

MODERN METHODS AND THE FUTURE OF APPLIED MATHEMATICS

by Roger Godement
DEPUTY LECTURER AT THE FACULTÉ DES SCIENCES OF NANCY

To speak of the future of mathematics is a capricious undertaking which cannot be too strongly advised against; strictly speaking, it is absurd. It will therefore be understood that I acknowledge frankly—at the same time while regretting it—my inability to draw up an exact balance sheet of the state of mathematics in the year 2000; a state which, there is no doubt, will be marvelous if the atomic physicists or some peace conferences do not happen to break the course of progress. To outline the *future* of mathematics is simply to endeavor to discern in the *present* the principal currents of thought and to draw in some way the "tangents" to them.

The period from 1919 to 1940 saw the accomplishment of a veritable revolution in mathematics. Principally under the impetus of the German, Polish and Russian schools—of course, there are exceptions, represented in France by Elie Cartan—theories of an essentially new character were born: modern algebra, spaces of an infinite number of dimensions, topology, non-Euclidean spaces. Toward the end of this period the French school—which until then had remained almost entirely apart from these studies—participated itself; and it is, doubtless, to the young and multiple genius of Bourbaki, that the future will owe the systematization of a multitude of results from which the legendary French clarity was until then unfortunately missing. The recent political upheavals have, moreover, not slowed down the mathematicians' labors; they have simply altered their geographical distribution so that nowadays the United States, along with the Soviet

Union, are the two principal sources of progress (from the point of view of our present topic, at least).

Today, the structure of "modern" theories is sufficiently advanced for a new phase in the evolution of mathematics to be envisaged: that of the application to "applied" problems of the new tools possessed (which naturally does not exclude the pursuit of investigations already begun). In order to make this point better, it will not perhaps be superfluous to make precise the meaning of certain terms—at least the one which the author gives to them.

Mathematics can be classified in two ways: the first distinguishes *abstract* theories from *concrete* theories; the second, *modern* theories from *classic* theories.[1] An abstract theory ignores the nature of its objects and concerns itself only with the logical relations which they have among themselves; on the contrary, a concrete theory is applied to entities of a completely determined nature. For example, arithmetic such as it is taught in secondary classes is a concrete theory; on the other hand, if we substitute for the system of integers a set of "entities" on which we assume defined a priori certain operations satisfying fixed formal laws:

$$x + y = y + x, \qquad x(y + z) = xy + xz, \qquad \ldots,$$

there can be deduced from these postulates, by the laws of logic alone, a large number of properties—which the integers in particular possess—and which, without involving the nature of the objects to which they apply, constitute an abstract theory.

The advantage of an abstract theory is that, assuming nothing about the nature of its objects, it admits precisely for that reason of applications to different branches of mathematics—namely: to all those where the postulates which are its foundation have a meaning. Conversely, every concrete theory gives rise to an abstract theory which constitutes a sort of logical skeleton of it. But here two cases can occur: Either the several concrete domains of application of an abstract theory are "isomorphic" to one another (that is to say, any two whatever of these domains differ *only* in the *nature* of their elements, every property which is true in one being true, with suitable changes in interpretation and notation, in all the others); or else *essentially* different domains of

---

[1] Let us make clear that the qualifiers "modern" and "classic" are not to be taken in their historical sense; classic mathematics is always being practiced, and will be again in the future. We could also speak of "multivalent" theories instead of modern, and of "univalent" instead of classic.

validity can be found. For example, it is well known (since Descartes) that to speak of points in a plane or of pairs of algebraic numbers (x, y) is one and the same thing; a line is an equation

$$ax + by + c = 0, \text{etc.} \ldots;$$

the language is different, but the logical structure of the theories is the same; they are thus two isomorphic domains. On the other hand, the set of integers and that of polynomials, although both contained in the general theory of "rings" (existence of the following operations: addition, subtraction, multiplication), are essentially distinct; even setting aside the clearly different natures of integers and polynomials, we can distinguish these two domains by purely logical means. We thus arrive at the stated distinction: A modern theory has essentially distinct "interpretations"—while those of a classic theory are, in short, only expressions in different language of one and the same doctrine. Naturally, just as an abstract theory can be modern or classic, a concrete theory can also belong to one of these two types.

At the point where mathematical research finds itself at the present time, there is no doubt that the future will see more and more the development of the influence of modern methods on concrete theories. First of all, a concrete theory, reasoning about well-determined objects, inevitably has a tendency to lay stress on *all* the properties of these objects; but very often (or more precisely when this concrete theory is of the modern type) it is found out, after a more or less concise analysis, that certain of these properties are perfectly unnecessary and in fact play no part in the construction of the contemplated doctrine. The role of modern methods is thus to rid concrete theories of all superfluous hypotheses which encumber them and which generally conceal their true meaning; in some way, to rethink them completely, extracting from them their major structural lines and, automatically, increasing considerably their domain of validity. For example, until recently, Fourier series and integrals were studied separately and by different methods; to construct these theories, appeal was made to "special functions" which, by a mysterious and providential mechanism, enabled us to obtain the desired properties. Today it is known that almost all the results discovered in this category of ideas come from the fact that the real numbers, for example, form an "Abelian group" (existence of a commutative addition: $x + y = y + x$, and of a subtraction; the existence of a multiplicative operation xy plays no role in this question). In other words, we have succeeded in integrating several concrete theories—which seem to present only an apparent analogy in their results—into a same pattern, which moreover surpasses each by a great deal. We could say as much for many other branches of mathematics: algebraic equations, integral equations, the theory of surfaces, the geometry of sets, integration—they have totally changed in appearance. The theory of functions of complex variables itself, which up to now had resisted this synthesis, is beginning to deliver up its secrets. Let us make clear besides that even in branches where modern methods have been able to penetrate with success, important progress remains to be made; for example, can we extend the Fourier transformation to non-commutative groups? Moreover, it is generally true that, as soon as one studies algebraic systems where the hypothesis of the commutativity of multiplication ($xy = yx$) is abandoned, we find ourselves facing problems infinitely more difficult, which constitute one of the most exciting aspects of mathematical progress.

This simplifying activity of modern methods is not, however, the ultimate aim of mathematics. Until now, progress has consisted almost exclusively of an *extension of the domains of validity* of concrete results, and not of a true *fathoming* of the latter; it is also to this last task that future mathematicians should devote themselves. The following example will perhaps make clear the possibility of such an action. Let us consider functions $f(x)$ integrable over $(-\infty, +\infty)$; we can define in this set an addition and subtraction in an obvious way, and a multiplication ("product of composition") by the formula

$$f \times g(x) = \int_{-\infty}^{\infty} f(x - y) g(y) \, dy.$$

We have therefore in this manner a "ring." The difference—essential—which it presents from the rings studied until now is its infinity of dimensions, and the necessity of including in it, along with purely algebraic methods, topological considerations which make its study much more difficult: Every extension to our ring of already known properties for rings of finite dimensions will lead us, so to speak, automatically, to essentially new progress concerning the functions in question. For example, the fact that every "closed ideal" is contained in a "closed maximal deal," properly interpreted, leads to the following theorem (proven in 1945 by Beurling): Let $f(x)$ be a function continuous and bounded on $(-\infty, +\infty)$; let us consider the "translations" $f(x + h)$ of $f(x)$, the linear combinations of the former, and functions which can be approached, uniformly on every bounded segment, by such combinations; then the set of functions thus obtained contains at least one exponential function e^{ix}. This result, which without any doubt will constitute the point of departure for a "spectral theory" whose possibilities can hardly be foreseen, is thus reduced to a theorem of algebra which, in the ordinary cases, is almost trivial. This example—and there already exist others—shows us also that, contrary to the opinion of certain persons, modern mathematics is not a mere juggling destined to occupy the leisure time of seekers without imagination; it is often forgotten that new methods date from 1920 and not from Pythagoras; it is now that their influence can make itself felt with profit.

Another aspect, quite fascinating, of the recent evolution of mathematics and which, to tell the truth, is not really contained in the general scope of this account—but mathematics has many aspects—is the systematic study of "mixed" theories, that is, ones which call upon different structures: algebraic, topological, differential, etc. From this point of view, it is the quite recent works of Hopf and his pupils which are the most representative: How do the algebraic and differential properties of Lie groups influence their topological behavior? It is noteworthy that by taking this latter point of view—topological—it has been possible to regain almost entirely the classification of a great part of these groups—a classification which Cartan had obtained, a long time before, by totally different methods. Perhaps the origin of these phenomena should be sought for in a topological theory of differential forms which is beginning to be glimpsed. Here, as in so many other places, Henri Poincaré will have been the initiator, and his genius will soar above the works of his successors for a long time.

I could not bring to an end this account—unfortunately a little abstract, but it is difficult to speak of future mathematics while remaining in the domain of elementary geometry—without noting the increasing role played by the new theories of physics in the development of mathematics. It does not appear doubtful that, without relativity and quantum mechanics, several extremely important branches of mathematics would have remained somewhat in the shadows. It is not without significance to note that one of the first papers of Elie Cartan on differential geometry was entitled "On Einstein's equations of Gravitation"; one of the best known works on Hilbert space was called *Mathematische Grundlagen der Quantenmechanik*

(Von Neumann); another on the theory of groups was called *Gruppentheorie und Quantenmechanik* (Weyl). It is essential to observe that the mathematical apparatus of modern physics is, almost completely, of recent creation and belongs to what we have styled "modern"; that at the present time the development of physics is still strictly parallel to that of mathematics; Hilbert spaces with an undefined metric were introduced in 1941 by the physicist Dirac, and studied two years later by the mathematician Pontriagine; the unitary representations of infinite dimension of any groups whatever, studied systematically in 1943 by the mathematicians Gelfand and Raïkov, are met again (with supplementary properties) in 1945 by the same Dirac in the case of the Lorentz group. Dirac's opinion—and it can be believed that it is authorized—according to which every new mathematical concept has its interpretation in nature, however paradoxical it may appear at first, comes very close to the truth. That this invasion of mathematics by physics seems regrettable to certain "rationalistic" philosophers—Julien Benda, for example—is an opinion that can be sustained from the esthetic point of view (but, from this point of view, the contrary opinion is no less justified; it could as well be deplored that man is not a quadruped!). As for us, we believe that the constant interaction of these two disciplines (pure reason and the material universe) is a fact at which there is no cause to be astonished, and that it is vain to want to express value judgments about them. The unity of the human mind and nature seems to us a hypothesis superabundantly justified by its consequences, and infinitely more acceptable than that of some place of spirits, of a somewhat undetermined appearance, whence Minerva, carrying the principle of the excluded middle slung across her back, comes down periodically to bestow the atomic bomb upon stupefied humanity. We believe on the contrary that the human mind is a "meteor" of the same quality as the rainbow—a phenomenon of nature; and that Hilbert, achieving the "spectral decomposition" of linear operators, Perrin analyzing the blue of the sky, Monet, Debussy and Proust recreating for our astonishment the sparkling of the light on the sea, all were working toward the same end, which will also be that of the future: knowledge of the entire universe.

A CATALOGUE OF SELECTED DOVER BOOKS
IN ALL FIELDS OF INTEREST

A CATALOGUE OF SELECTED DOVER BOOKS
IN ALL FIELDS OF INTEREST

AMERICA'S OLD MASTERS, James T. Flexner. Four men emerged unexpectedly from provincial 18th century America to leadership in European art: Benjamin West, J. S. Copley, C. R. Peale, Gilbert Stuart. Brilliant coverage of lives and contributions. Revised, 1967 edition. 69 plates. 365pp. of text.
21806-6 Paperbound $2.75

FIRST FLOWERS OF OUR WILDERNESS: AMERICAN PAINTING, THE COLONIAL PERIOD, James T. Flexner. Painters, and regional painting traditions from earliest Colonial times up to the emergence of Copley, West and Peale Sr., Foster, Gustavus Hesselius, Feke, John Smibert and many anonymous painters in the primitive manner. Engaging presentation, with 162 illustrations. xxii + 368pp.
22180-6 Paperbound $3.50

THE LIGHT OF DISTANT SKIES: AMERICAN PAINTING, 1760-1835, James T. Flexner. The great generation of early American painters goes to Europe to learn and to teach: West, Copley, Gilbert Stuart and others. Allston, Trumbull, Morse; also contemporary American painters—primitives, derivatives, academics—who remained in America. 102 illustrations. xiii + 306pp. 22179-2 Paperbound $3.00

A HISTORY OF THE RISE AND PROGRESS OF THE ARTS OF DESIGN IN THE UNITED STATES, William Dunlap. Much the richest mine of information on early American painters, sculptors, architects, engravers, miniaturists, etc. The only source of information for scores of artists, the major primary source for many others. Unabridged reprint of rare original 1834 edition, with new introduction by James T. Flexner, and 394 new illustrations. Edited by Rita Weiss. 6⅝ x 9⅝.
21695-0, 21696-9, 21697-7 Three volumes, Paperbound $13.50

EPOCHS OF CHINESE AND JAPANESE ART, Ernest F. Fenollosa. From primitive Chinese art to the 20th century, thorough history, explanation of every important art period and form, including Japanese woodcuts; main stress on China and Japan, but Tibet, Korea also included. Still unexcelled for its detailed, rich coverage of cultural background, aesthetic elements, diffusion studies, particularly of the historical period. 2nd, 1913 edition. 242 illustrations. lii + 439pp. of text.
20364-6, 20365-4 Two volumes, Paperbound $5.00

THE GENTLE ART OF MAKING ENEMIES, James A. M. Whistler. Greatest wit of his day deflates Oscar Wilde, Ruskin, Swinburne; strikes back at inane critics, exhibitions, art journalism; aesthetics of impressionist revolution in most striking form. Highly readable classic by great painter. Reproduction of edition designed by Whistler. Introduction by Alfred Werner. xxxvi + 334pp.
21875-9 Paperbound $2.25

CATALOGUE OF DOVER BOOKS

VISUAL ILLUSIONS: THEIR CAUSES, CHARACTERISTICS, AND APPLICATIONS, Matthew Luckiesh. Thorough description and discussion of optical illusion, geometric and perspective, particularly; size and shape distortions, illusions of color, of motion; natural illusions; use of illusion in art and magic, industry, etc. Most useful today with op art, also for classical art. Scores of effects illustrated. Introduction by William H. Ittleson. 100 illustrations. xxi + 252pp.
21530-X Paperbound $1.50

A HANDBOOK OF ANATOMY FOR ART STUDENTS, Arthur Thomson. Thorough, virtually exhaustive coverage of skeletal structure, musculature, etc. Full text, supplemented by anatomical diagrams and drawings and by photographs of undraped figures. Unique in its comparison of male and female forms, pointing out differences of contour, texture, form. 211 figures, 40 drawings, 86 photographs. xx + 459pp. 5⅜ x 8⅜.
21163-0 Paperbound $3.00

150 MASTERPIECES OF DRAWING, Selected by Anthony Toney. Full page reproductions of drawings from the early 16th to the end of the 18th century, all beautifully reproduced: Rembrandt, Michelangelo, Dürer, Fragonard, Urs, Graf, Wouwerman, many others. First-rate browsing book, model book for artists. xviii + 150pp. 8⅜ x 11¼.
21032-4 Paperbound $2.00

THE LATER WORK OF AUBREY BEARDSLEY, Aubrey Beardsley. Exotic, erotic, ironic masterpieces in full maturity: Comedy Ballet, Venus and Tannhauser, Pierrot, Lysistrata, Rape of the Lock, Savoy material, Ali Baba, Volpone, etc. This material revolutionized the art world, and is still powerful, fresh, brilliant. With *The Early Work*, all Beardsley's finest work. 174 plates, 2 in color. xiv + 176pp. 8⅛ x 11.
21817-1 Paperbound $2.75

DRAWINGS OF REMBRANDT, Rembrandt van Rijn. Complete reproduction of fabulously rare edition by Lippmann and Hofstede de Groot, completely reedited, updated, improved by Prof. Seymour Slive, Fogg Museum. Portraits, Biblical sketches, landscapes, Oriental types, nudes, episodes from classical mythology—All Rembrandt's fertile genius. Also selection of drawings by his pupils and followers. "Stunning volumes," *Saturday Review*. 550 illustrations. lxxviii + 552pp. 9⅛ x 12¼.
21485-0, 21486-9 Two volumes, Paperbound $6.50

THE DISASTERS OF WAR, Francisco Goya. One of the masterpieces of Western civilization—83 etchings that record Goya's shattering, bitter reaction to the Napoleonic war that swept through Spain after the insurrection of 1808 and to war in general. Reprint of the first edition, with three additional plates from Boston's Museum of Fine Arts. All plates facsimile size. Introduction by Philip Hofer, Fogg Museum. v + 97pp. 9⅜ x 8¼.
21872-4 Paperbound $1.75

GRAPHIC WORKS OF ODILON REDON. Largest collection of Redon's graphic works ever assembled: 172 lithographs, 28 etchings and engravings, 9 drawings. These include some of his most famous works. All the plates from *Odilon Redon: oeuvre graphique complet*, plus additional plates. New introduction and caption translations by Alfred Werner. 209 illustrations. xxvii + 209pp. 9⅛ x 12¼.
21966-8 Paperbound $4.00

CATALOGUE OF DOVER BOOKS

DESIGN BY ACCIDENT; A BOOK OF "ACCIDENTAL EFFECTS" FOR ARTISTS AND DESIGNERS, James F. O'Brien. Create your own unique, striking, imaginative effects by "controlled accident" interaction of materials: paints and lacquers, oil and water based paints, splatter, crackling materials, shatter, similar items. Everything you do will be different; first book on this limitless art, so useful to both fine artist and commercial artist. Full instructions. 192 plates showing "accidents," 8 in color. viii + 215pp. 8⅜ x 11¼. 21942-9 Paperbound $3.50

THE BOOK OF SIGNS, Rudolf Koch. Famed German type designer draws 493 beautiful symbols: religious, mystical, alchemical, imperial, property marks, ines, etc. Remarkable fusion of traditional and modern. Good for suggestions of timelessness, smartness, modernity. Text. vi + 104pp. 6⅛ x 9¼. 20162-7 Paperbound $1.25

HISTORY OF INDIAN AND INDONESIAN ART, Ananda K. Coomaraswamy. An unabridged republication of one of the finest books by a great scholar in Eastern art. Rich in descriptive material, history, social backgrounds; Sunga reliefs, Rajput paintings, Gupta temples, Burmese frescoes, textiles, jewelry, sculpture, etc. 400 photos. viii + 423pp. 6⅜ x 9¾. 21436-2 Paperbound $3.50

PRIMITIVE ART, Franz Boas. America's foremost anthropologist surveys textiles, ceramics, woodcarving, basketry, metalwork, etc.; patterns, technology, creation of symbols, style origins. All areas of world, but very full on Northwest Coast Indians. More than 350 illustrations of baskets, boxes, totem poles, weapons, etc. 378 pp. 20025-6 Paperbound $2.50

THE GENTLEMAN AND CABINET MAKER'S DIRECTOR, Thomas Chippendale. Full reprint (third edition, 1762) of most influential furniture book of all time, by master cabinetmaker. 200 plates, illustrating chairs, sofas, mirrors, tables, cabinets, plus 24 photographs of surviving pieces. Biographical introduction by N. Bienenstock. vi + 249pp. 9⅞ x 12¾. 21601-2 Paperbound $3.50

AMERICAN ANTIQUE FURNITURE, Edgar G. Miller, Jr. The basic coverage of all American furniture before 1840. Individual chapters cover type of furniture—clocks, tables, sideboards, etc.—chronologically, with inexhaustible wealth of data. More than 2100 photographs, all identified, commented on. Essential to all early American collectors. Introduction by H. E. Keyes. vi + 1106pp. 7⅞ x 10¾. 21599-7, 21600-4 Two volumes, Paperbound $7.50

PENNSYLVANIA DUTCH AMERICAN FOLK ART, Henry J. Kauffman. 279 photos, 28 drawings of tulipware, Fraktur script, painted tinware, toys, flowered furniture, quilts, samplers, hex signs, house interiors, etc. Full descriptive text. Excellent for tourist, rewarding for designer, collector. Map. 146pp. 7⅞ x 10¾. 21205-X Paperbound $2.00

EARLY NEW ENGLAND GRAVESTONE RUBBINGS, Edmund V. Gillon, Jr. 43 photographs, 226 carefully reproduced rubbings show heavily symbolic, sometimes macabre early gravestones, up to early 19th century. Remarkable early American primitive art, occasionally strikingly beautiful; always powerful. Text. xxvi + 207pp. 8⅜ x 11¼. 21380-3 Paperbound $3.00

CATALOGUE OF DOVER BOOKS

ALPHABETS AND ORNAMENTS, Ernst Lehner. Well-known pictorial source for decorative alphabets, script examples, cartouches, frames, decorative title pages, calligraphic initials, borders, similar material. 14th to 19th century, mostly European. Useful in almost any graphic arts designing, varied styles. 750 illustrations. 256pp. 7 x 10. 21905-4 Paperbound $3.50

PAINTING: A CREATIVE APPROACH, Norman Colquhoun. For the beginner simple guide provides an instructive approach to painting: major stumbling blocks for beginner; overcoming them, technical points; paints and pigments; oil painting; watercolor and other media and color. New section on "plastic" paints. Glossary. Formerly *Paint Your Own Pictures*. 221pp. 22000-1 Paperbound $1.75

THE ENJOYMENT AND USE OF COLOR, Walter Sargent. Explanation of the relations between colors themselves and between colors in nature and art, including hundreds of little-known facts about color values, intensities, effects of high and low illumination, complementary colors. Many practical hints for painters, references to great masters. 7 color plates, 29 illustrations. x + 274pp.
20944-X Paperbound $2.50

THE NOTEBOOKS OF LEONARDO DA VINCI, compiled and edited by Jean Paul Richter. 1566 extracts from original manuscripts reveal the full range of Leonardo's versatile genius: all his writings on painting, sculpture, architecture, anatomy, astronomy, geography, topography, physiology, mining, music, etc., in both Italian and English, with 186 plates of manuscript pages and more than 500 additional drawings. Includes studies for the Last Supper, the lost Sforza monument, and other works. Total of xlvii + 866pp. $7\frac{7}{8}$ x $10\frac{3}{4}$.
22572-0, 22573-9 Two volumes, Paperbound $10.00

MONTGOMERY WARD CATALOGUE OF 1895. Tea gowns, yards of flannel and pillow-case lace, stereoscopes, books of gospel hymns, the New Improved Singer Sewing Machine, side saddles, milk skimmers, straight-edged razors, high-button shoes, spittoons, and on and on . . . listing some 25,000 items, practically all illustrated. Essential to the shoppers of the 1890's, it is our truest record of the spirit of the period. Unaltered reprint of Issue No. 57, Spring and Summer 1895. Introduction by Boris Emmet. Innumerable illustrations. xiii + 624pp. $8\frac{1}{2}$ x $11\frac{5}{8}$.
22377-9 Paperbound $6.95

THE CRYSTAL PALACE EXHIBITION ILLUSTRATED CATALOGUE (LONDON, 1851). One of the wonders of the modern world—the Crystal Palace Exhibition in which all the nations of the civilized world exhibited their achievements in the arts and sciences—presented in an equally important illustrated catalogue. More than 1700 items pictured with accompanying text—ceramics, textiles, cast-iron work, carpets, pianos, sleds, razors, wall-papers, billiard tables, beehives, silverware and hundreds of other artifacts—represent the focal point of Victorian culture in the Western World. Probably the largest collection of Victorian decorative art ever assembled—indispensable for antiquarians and designers. Unabridged republication of the Art-Journal Catalogue of the Great Exhibition of 1851, with all terminal essays. New introduction by John Gloag, F.S.A. xxxiv + 426pp. 9 x 12.
22503-8 Paperbound $4.50

CATALOGUE OF DOVER BOOKS

A HISTORY OF COSTUME, Carl Köhler. Definitive history, based on surviving pieces of clothing primarily, and paintings, statues, etc. secondarily. Highly readable text, supplemented by 594 illustrations of costumes of the ancient Mediterranean peoples, Greece and Rome, the Teutonic prehistoric period; costumes of the Middle Ages, Renaissance, Baroque, 18th and 19th centuries. Clear, measured patterns are provided for many clothing articles. Approach is practical throughout. Enlarged by Emma von Sichart. 464pp. 21030-8 Paperbound $3.00

ORIENTAL RUGS, ANTIQUE AND MODERN, Walter A. Hawley. A complete and authoritative treatise on the Oriental rug—where they are made, by whom and how, designs and symbols, characteristics in detail of the six major groups, how to distinguish them and how to buy them. Detailed technical data is provided on periods, weaves, warps, wefts, textures, sides, ends and knots, although no technical background is required for an understanding. 11 color plates, 80 halftones, 4 maps. vi + 320pp. $6\frac{1}{8}$ x $9\frac{1}{8}$. 22366-3 Paperbound $5.00

TEN BOOKS ON ARCHITECTURE, Vitruvius. By any standards the most important book on architecture ever written. Early Roman discussion of aesthetics of building, construction methods, orders, sites, and every other aspect of architecture has inspired, instructed architecture for about 2,000 years. Stands behind Palladio, Michelangelo, Bramante, Wren, countless others. Definitive Morris H. Morgan translation. 68 illustrations. xii + 331pp. 20645-9 Paperbound $2.50

THE FOUR BOOKS OF ARCHITECTURE, Andrea Palladio. Translated into every major Western European language in the two centuries following its publication in 1570, this has been one of the most influential books in the history of architecture. Complete reprint of the 1738 Isaac Ware edition. New introduction by Adolf Placzek, Columbia Univ. 216 plates. xxii + 110pp. of text. $9\frac{1}{2}$ x $12\frac{3}{4}$.
21308-0 Clothbound $10.00

STICKS AND STONES: A STUDY OF AMERICAN ARCHITECTURE AND CIVILIZATION, Lewis Mumford. One of the great classics of American cultural history. American architecture from the medieval-inspired earliest forms to the early 20th century; evolution of structure and style, and reciprocal influences on environment. 21 photographic illustrations. 238pp. 20202-X Paperbound $2.00

THE AMERICAN BUILDER'S COMPANION, Asher Benjamin. The most widely used early 19th century architectural style and source book, for colonial up into Greek Revival periods. Extensive development of geometry of carpentering, construction of sashes, frames, doors, stairs; plans and elevations of domestic and other buildings. Hundreds of thousands of houses were built according to this book, now invaluable to historians, architects, restorers, etc. 1827 edition. 59 plates. 114pp. $7\frac{7}{8}$ x $10\frac{3}{4}$.
22236-5 Paperbound $3.00

DUTCH HOUSES IN THE HUDSON VALLEY BEFORE 1776, Helen Wilkinson Reynolds. The standard survey of the Dutch colonial house and outbuildings, with constructional features, decoration, and local history associated with individual homesteads. Introduction by Franklin D. Roosevelt. Map. 150 illustrations. 469pp. $6\frac{5}{8}$ x $9\frac{1}{4}$. 21469-9 Paperbound $3.50

CATALOGUE OF DOVER BOOKS

THE ARCHITECTURE OF COUNTRY HOUSES, Andrew J. Downing. Together with Vaux's *Villas and Cottages* this is the basic book for Hudson River Gothic architecture of the middle Victorian period. Full, sound discussions of general aspects of housing, architecture, style, decoration, furnishing, together with scores of detailed house plans, illustrations of specific buildings, accompanied by full text. Perhaps the most influential single American architectural book. 1850 edition. Introduction by J. Stewart Johnson. 321 figures, 34 architectural designs. xvi + 560pp.
22003-6 Paperbound $3.50

LOST EXAMPLES OF COLONIAL ARCHITECTURE, John Mead Howells. Full-page photographs of buildings that have disappeared or been so altered as to be denatured, including many designed by major early American architects. 245 plates. xvii + 248pp. 7⅞ x 10¾. 21143-6 Paperbound $3.00

DOMESTIC ARCHITECTURE OF THE AMERICAN COLONIES AND OF THE EARLY REPUBLIC, Fiske Kimball. Foremost architect and restorer of Williamsburg and Monticello covers nearly 200 homes between 1620-1825. Architectural details, construction, style features, special fixtures, floor plans, etc. Generally considered finest work in its area. 219 illustrations of houses, doorways, windows, capital mantels. xx + 314pp. 7⅞ x 10¾. 21743-4 Paperbound $3.50

EARLY AMERICAN ROOMS: 1650-1858, edited by Russell Hawes Kettell. Tour of 12 rooms, each representative of a different era in American history and each furnished, decorated, designed and occupied in the style of the era. 72 plans and elevations, 8-page color section, etc., show fabrics, wall papers, arrangements, etc. Full descriptive text. xvii + 200pp. of text. 8⅜ x 11¼.
21633-0 Paperbound $4.00

THE FITZWILLIAM VIRGINAL BOOK, edited by J. Fuller Maitland and W. B. Squire. Full modern printing of famous early 17th-century ms. volume of 300 works by Morley, Byrd, Bull, Gibbons, etc. For piano or other modern keyboard instrument; easy to read format. xxxvi + 938pp. 8⅜ x 11.
21068-5, 21069-3 Two volumes, Paperbound $8.00

HARPSICHORD MUSIC, Johann Sebastian Bach. Bach Gesellschaft edition. A rich selection of Bach's masterpieces for the harpsichord: the six English Suites, six French Suites, the six Partitas (Clavierübung part I), the Goldberg Variations (Clavierübung part IV), the fifteen Two-Part Inventions and the fifteen Three-Part Sinfonias. Clearly reproduced on large sheets with ample margins; eminently playable. vi + 312pp. 8⅛ x 11. 22360-4 Paperbound $5.00

THE MUSIC OF BACH: AN INTRODUCTION, Charles Sanford Terry. A fine, non-technical introduction to Bach's music, both instrumental and vocal. Covers organ music, chamber music, passion music, other types. Analyzes themes, developments, innovations. x + 114pp. 21075-8 Paperbound $1.25

BEETHOVEN AND HIS NINE SYMPHONIES, Sir George Grove. Noted British musicologist provides best history, analysis, commentary on symphonies. Very thorough, rigorously accurate; necessary to both advanced student and amateur music lover. 436 musical passages. vii + 407 pp. 20334-4 Paperbound $2.25

CATALOGUE OF DOVER BOOKS

JOHANN SEBASTIAN BACH, Philipp Spitta. One of the great classics of musicology, this definitive analysis of Bach's music (and life) has never been surpassed. Lucid, nontechnical analyses of hundreds of pieces (30 pages devoted to St. Matthew Passion, 26 to B Minor Mass). Also includes major analysis of 18th-century music. 450 musical examples. 40-page musical supplement. Total of xx + 1799pp.
(EUK) 22278-0, 22279-9 Two volumes, Clothbound $15.00

MOZART AND HIS PIANO CONCERTOS, Cuthbert Girdlestone. The only full-length study of an important area of Mozart's creativity. Provides detailed analyses of all 23 concertos, traces inspirational sources. 417 musical examples. Second edition. 509pp. (USO) 21271-8 Paperbound $2.50

THE PERFECT WAGNERITE: A COMMENTARY ON THE NIBLUNG'S RING, George Bernard Shaw. Brilliant and still relevant criticism in remarkable essays on Wagner's Ring cycle, Shaw's ideas on political and social ideology behind the plots, role of Leitmotifs, vocal requisites, etc. Prefaces. xxi + 136pp.
21707-8 Paperbound $1.50

DON GIOVANNI, W. A. Mozart. Complete libretto, modern English translation; biographies of composer and librettist; accounts of early performances and critical reaction. Lavishly illustrated. All the material you need to understand and appreciate this great work. Dover Opera Guide and Libretto Series; translated and introduced by Ellen Bleiler. 92 illustrations. 209pp.
21134-7 Paperbound $1.50

HIGH FIDELITY SYSTEMS: A LAYMAN'S GUIDE, Roy F. Allison. All the basic information you need for setting up your own audio system: high fidelity and stereo record players, tape records, F.M. Connections, adjusting tone arm, cartridge, checking needle alignment, positioning speakers, phasing speakers, adjusting hums, trouble-shooting, maintenance, and similar topics. Enlarged 1965 edition. More than 50 charts, diagrams, photos. iv + 91pp. 21514-8 Paperbound $1.25

REPRODUCTION OF SOUND, Edgar Villchur. Thorough coverage for laymen of high fidelity systems, reproducing systems in general, needles, amplifiers, preamps, loudspeakers, feedback, explaining physical background. "A rare talent for making technicalities vividly comprehensible," R. Darrell, *High Fidelity*. 69 figures. iv + 92pp. 21515-6 Paperbound $1.00

HEAR ME TALKIN' TO YA: THE STORY OF JAZZ AS TOLD BY THE MEN WHO MADE IT, Nat Shapiro and Nat Hentoff. Louis Armstrong, Fats Waller, Jo Jones, Clarence Williams, Billy Holiday, Duke Ellington, Jelly Roll Morton and dozens of other jazz greats tell how it was in Chicago's South Side, New Orleans, depression Harlem and the modern West Coast as jazz was born and grew. xvi + 429pp.
21726-4 Paperbound $2.00

FABLES OF AESOP, translated by Sir Roger L'Estrange. A reproduction of the very rare 1931 Paris edition; a selection of the most interesting fables, together with 50 imaginative drawings by Alexander Calder. v + 128pp. 6½x9¼.
21780-9 Paperbound $1.25

CATALOGUE OF DOVER BOOKS

AGAINST THE GRAIN (A REBOURS), Joris K. Huysmans. Filled with weird images, evidences of a bizarre imagination, exotic experiments with hallucinatory drugs, rich tastes and smells and the diversions of its sybarite hero Duc Jean des Esseintes, this classic novel pushed 19th-century literary decadence to its limits. Full unabridged edition. Do not confuse this with abridged editions generally sold. Introduction by Havelock Ellis. xlix + 206pp. 22190-3 Paperbound $2.00

VARIORUM SHAKESPEARE: HAMLET. Edited by Horace H. Furness; a landmark of American scholarship. Exhaustive footnotes and appendices treat all doubtful words and phrases, as well as suggested critical emendations throughout the play's history. First volume contains editor's own text, collated with all Quartos and Folios. Second volume contains full first Quarto, translations of Shakespeare's sources (Belleforest, and Saxo Grammaticus), Der Bestrafte Brudermord, and many essays on critical and historical points of interest by major authorities of past and present. Includes details of staging and costuming over the years. By far the best edition available for serious students of Shakespeare. Total of xx + 905pp.
21004-9, 21005-7, 2 volumes, Paperbound $5.25

A LIFE OF WILLIAM SHAKESPEARE, Sir Sidney Lee. This is the standard life of Shakespeare, summarizing everything known about Shakespeare and his plays. Incredibly rich in material, broad in coverage, clear and judicious, it has served thousands as the best introduction to Shakespeare. 1931 edition. 9 plates. xxix + 792pp. (USO) 21967-4 Paperbound $3.75

MASTERS OF THE DRAMA, John Gassner. Most comprehensive history of the drama in print, covering every tradition from Greeks to modern Europe and America, including India, Far East, etc. Covers more than 800 dramatists, 2000 plays, with biographical material, plot summaries, theatre history, criticism, etc. "Best of its kind in English," *New Republic*. 77 illustrations. xxii + 890pp.
20100-7 Clothbound $7.50

THE EVOLUTION OF THE ENGLISH LANGUAGE, George McKnight. The growth of English, from the 14th century to the present. Unusual, non-technical account presents basic information in very interesting form: sound shifts, change in grammar and syntax, vocabulary growth, similar topics. Abundantly illustrated with quotations. Formerly *Modern English in the Making*. xii + 590pp.
21932-1 Paperbound $3.50

AN ETYMOLOGICAL DICTIONARY OF MODERN ENGLISH, Ernest Weekley. Fullest, richest work of its sort, by foremost British lexicographer. Detailed word histories, including many colloquial and archaic words; extensive quotations. Do not confuse this with the Concise Etymological Dictionary, which is much abridged. Total of xxvii + 830pp. 6½ x 9¼.
21873-2, 21874-0 Two volumes, Paperbound $5.50

FLATLAND: A ROMANCE OF MANY DIMENSIONS, E. A. Abbott. Classic of science-fiction explores ramifications of life in a two-dimensional world, and what happens when a three-dimensional being intrudes. Amusing reading, but also useful as introduction to thought about hyperspace. Introduction by Banesh Hoffmann. 16 illustrations. xx + 103pp. 20001-9 Paperbound $1.00

CATALOGUE OF DOVER BOOKS

POEMS OF ANNE BRADSTREET, edited with an introduction by Robert Hutchinson. A new selection of poems by America's first poet and perhaps the first significant woman poet in the English language. 48 poems display her development in works of considerable variety—love poems, domestic poems, religious meditations, formal elegies, "quaternions," etc. Notes, bibliography. viii + 222pp.

22160-1 Paperbound $2.00

THREE GOTHIC NOVELS: THE CASTLE OF OTRANTO BY HORACE WALPOLE; VATHEK BY WILLIAM BECKFORD; THE VAMPYRE BY JOHN POLIDORI, WITH FRAGMENT OF A NOVEL BY LORD BYRON, edited by E. F. Bleiler. The first Gothic novel, by Walpole; the finest Oriental tale in English, by Beckford; powerful Romantic supernatural story in versions by Polidori and Byron. All extremely important in history of literature; all still exciting, packed with supernatural thrills, ghosts, haunted castles, magic, etc. xl + 291pp.

21232-7 Paperbound $2.00

THE BEST TALES OF HOFFMANN, E. T. A. Hoffmann. 10 of Hoffmann's most important stories, in modern re-editings of standard translations: Nutcracker and the King of Mice, Signor Formica, Automata, The Sandman, Rath Krespel, The Golden Flowerpot, Master Martin the Cooper, The Mines of Falun, The King's Betrothed, A New Year's Eve Adventure. 7 illustrations by Hoffmann. Edited by E. F. Bleiler. xxxix + 419pp. 21793-0 Paperbound $2.25

GHOST AND HORROR STORIES OF AMBROSE BIERCE, Ambrose Bierce. 23 strikingly modern stories of the horrors latent in the human mind: The Eyes of the Panther, The Damned Thing, An Occurrence at Owl Creek Bridge, An Inhabitant of Carcosa, etc., plus the dream-essay, Visions of the Night. Edited by E. F. Bleiler. xxii + 199pp. 20767-6 Paperbound $1.50

BEST GHOST STORIES OF J. S. LEFANU, J. Sheridan LeFanu. Finest stories by Victorian master often considered greatest supernatural writer of all. Carmilla, Green Tea, The Haunted Baronet, The Familiar, and 12 others. Most never before available in the U. S. A. Edited by E. F. Bleiler. 8 illustrations from Victorian publications. xvii + 467pp. 20415-4 Paperbound $2.50

THE TIME STREAM, THE GREATEST ADVENTURE, AND THE PURPLE SAPPHIRE—THREE SCIENCE FICTION NOVELS, John Taine (Eric Temple Bell). Great American mathematician was also foremost science fiction novelist of the 1920's. *The Time Stream,* one of all-time classics, uses concepts of circular time; *The Greatest Adventure,* incredibly ancient biological experiments from Antarctica threaten to escape; The *Purple Sapphire,* superscience, lost races in Central Tibet, survivors of the Great Race. 4 illustrations by Frank R. Paul. v + 532pp.

21180-0 Paperbound $2.50

SEVEN SCIENCE FICTION NOVELS, H. G. Wells. The standard collection of the great novels. Complete, unabridged. *First Men in the Moon, Island of Dr. Moreau, War of the Worlds, Food of the Gods, Invisible Man, Time Machine, In the Days of the Comet.* Not only science fiction fans, but every educated person owes it to himself to read these novels. 1015pp. 20264-X Clothbound $5.00

CATALOGUE OF DOVER BOOKS

LAST AND FIRST MEN AND STAR MAKER, TWO SCIENCE FICTION NOVELS, Olaf Stapledon. Greatest future histories in science fiction. In the first, human intelligence is the "hero," through strange paths of evolution, interplanetary invasions, incredible technologies, near extinctions and reemergences. Star Maker describes the quest of a band of star rovers for intelligence itself, through time and space: weird inhuman civilizations, crustacean minds, symbiotic worlds, etc. Complete, unabridged. v + 438pp. 21962-3 Paperbound $2.00

THREE PROPHETIC NOVELS, H. G. WELLS. Stages of a consistently planned future for mankind. *When the Sleeper Wakes*, and *A Story of the Days to Come*, anticipate *Brave New World* and *1984*, in the 21st Century; *The Time Machine*, only complete version in print, shows farther future and the end of mankind. All show Wells's greatest gifts as storyteller and novelist. Edited by E. F. Bleiler. x + 335pp. (USO) 20605-X Paperbound $2.00

THE DEVIL'S DICTIONARY, Ambrose Bierce. America's own Oscar Wilde—Ambrose Bierce—offers his barbed iconoclastic wisdom in over 1,000 definitions hailed by H. L. Mencken as "some of the most gorgeous witticisms in the English language." 145pp. 20487-1 Paperbound $1.25

MAX AND MORITZ, Wilhelm Busch. Great children's classic, father of comic strip, of two bad boys, Max and Moritz. Also Ker and Plunk (Plisch und Plumm), Cat and Mouse, Deceitful Henry, Ice-Peter, The Boy and the Pipe, and five other pieces. Original German, with English translation. Edited by H. Arthur Klein; translations by various hands and H. Arthur Klein. vi + 216pp.
20181-3 Paperbound $1.50

PIGS IS PIGS AND OTHER FAVORITES, Ellis Parker Butler. The title story is one of the best humor short stories, as Mike Flannery obfuscates biology and English. Also included, That Pup of Murchison's, The Great American Pie Company, and Perkins of Portland. 14 illustrations. v + 109pp. 21532-6 Paperbound $1.00

THE PETERKIN PAPERS, Lucretia P. Hale. It takes genius to be as stupidly mad as the Peterkins, as they decide to become wise, celebrate the "Fourth," keep a cow, and otherwise strain the resources of the Lady from Philadelphia. Basic book of American humor. 153 illustrations. 219pp. 20794-3 Paperbound $1.25

PERRAULT'S FAIRY TALES, translated by A. E. Johnson and S. R. Littlewood, with 34 full-page illustrations by Gustave Doré. All the original Perrault stories—Cinderella, Sleeping Beauty, Bluebeard, Little Red Riding Hood, Puss in Boots, Tom Thumb, etc.—with their witty verse morals and the magnificent illustrations of Doré. One of the five or six great books of European fairy tales. viii + 117pp. 8⅛ x 11. 22311-6 Paperbound $2.00

OLD HUNGARIAN FAIRY TALES, Baroness Orczy. Favorites translated and adapted by author of the *Scarlet Pimpernel*. Eight fairy tales include "The Suitors of Princess Fire-Fly," "The Twin Hunchbacks," "Mr. Cuttlefish's Love Story," and "The Enchanted Cat." This little volume of magic and adventure will captivate children as it has for generations. 90 drawings by Montagu Barstow. 96pp.
(USO) 22293-4 Paperbound $1.95

CATALOGUE OF DOVER BOOKS

THE RED FAIRY BOOK, Andrew Lang. Lang's color fairy books have long been children's favorites. This volume includes Rapunzel, Jack and the Bean-stalk and 35 other stories, familiar and unfamiliar. 4 plates, 93 illustrations x + 367pp.
21673-X Paperbound $1.95

THE BLUE FAIRY BOOK, Andrew Lang. Lang's tales come from all countries and all times. Here are 37 tales from Grimm, the Arabian Nights, Greek Mythology, and other fascinating sources. 8 plates, 130 illustrations. xi + 390pp.
21437-0 Paperbound $1.95

HOUSEHOLD STORIES BY THE BROTHERS GRIMM. Classic English-language edition of the well-known tales — Rumpelstiltskin, Snow White, Hansel and Gretel, The Twelve Brothers, Faithful John, Rapunzel, Tom Thumb (52 stories in all). Translated into simple, straightforward English by Lucy Crane. Ornamented with headpieces, vignettes, elaborate decorative initials and a dozen full-page illustrations by Walter Crane. x + 269pp.
21080-4 Paperbound $1.75

THE MERRY ADVENTURES OF ROBIN HOOD, Howard Pyle. The finest modern versions of the traditional ballads and tales about the great English outlaw. Howard Pyle's complete prose version, with every word, every illustration of the first edition. Do not confuse this facsimile of the original (1883) with modern editions that change text or illustrations. 23 plates plus many page decorations. xxii + 296pp.
22043-5 Paperbound $2.00

THE STORY OF KING ARTHUR AND HIS KNIGHTS, Howard Pyle. The finest children's version of the life of King Arthur; brilliantly retold by Pyle, with 48 of his most imaginative illustrations. xviii + 313pp. 6⅛ x 9¼.
21445-1 Paperbound $2.00

THE WONDERFUL WIZARD OF OZ, L. Frank Baum. America's finest children's book in facsimile of first edition with all Denslow illustrations in full color. The edition a child should have. Introduction by Martin Gardner. 23 color plates, scores of drawings. iv + 267pp.
20691-2 Paperbound $1.95

THE MARVELOUS LAND OF OZ, L. Frank Baum. The second Oz book, every bit as imaginative as the Wizard. The hero is a boy named Tip, but the Scarecrow and the Tin Woodman are back, as is the Oz magic. 16 color plates, 120 drawings by John R. Neill. 287pp.
20692-0 Paperbound $1.75

THE MAGICAL MONARCH OF MO, L. Frank Baum. Remarkable adventures in a land even stranger than Oz. The best of Baum's books not in the Oz series. 15 color plates and dozens of drawings by Frank Verbeck. xviii + 237pp.
21892-9 Paperbound $2.00

THE BAD CHILD'S BOOK OF BEASTS, MORE BEASTS FOR WORSE CHILDREN, A MORAL ALPHABET, Hilaire Belloc. Three complete humor classics in one volume. Be kind to the frog, and do not call him names ... and 28 other whimsical animals. Familiar favorites and some not so well known. Illustrated by Basil Blackwell. 156pp.
(USO) 20749-8 Paperbound $1.25

CATALOGUE OF DOVER BOOKS

EAST O' THE SUN AND WEST O' THE MOON, George W. Dasent. Considered the best of all translations of these Norwegian folk tales, this collection has been enjoyed by generations of children (and folklorists too). Includes True and Untrue, Why the Sea is Salt, East O' the Sun and West O' the Moon, Why the Bear is Stumpy-Tailed, Boots and the Troll, The Cock and the Hen, Rich Peter the Pedlar, and 52 more. The only edition with all 59 tales. 77 illustrations by Erik Werenskiold and Theodor Kittelsen. xv + 418pp. 22521-6 Paperbound $3.00

GOOPS AND HOW TO BE THEM, Gelett Burgess. Classic of tongue-in-cheek humor, masquerading as etiquette book. 87 verses, twice as many cartoons, show mischievous Goops as they demonstrate to children virtues of table manners, neatness, courtesy, etc. Favorite for generations. viii + 88pp. $6\frac{1}{2}$ x $9\frac{1}{4}$.
22233-0 Paperbound $1.25

ALICE'S ADVENTURES UNDER GROUND, Lewis Carroll. The first version, quite different from the final *Alice in Wonderland,* printed out by Carroll himself with his own illustrations. Complete facsimile of the "million dollar" manuscript Carroll gave to Alice Liddell in 1864. Introduction by Martin Gardner. viii + 96pp. Title and dedication pages in color. 21482-6 Paperbound $1.00

THE BROWNIES, THEIR BOOK, Palmer Cox. Small as mice, cunning as foxes, exuberant and full of mischief, the Brownies go to the zoo, toy shop, seashore, circus, etc., in 24 verse adventures and 266 illustrations. Long a favorite, since their first appearance in St. Nicholas Magazine. xi + 144pp. $6\frac{5}{8}$ x $9\frac{1}{4}$.
21265-3 Paperbound $1.50

SONGS OF CHILDHOOD, Walter De La Mare. Published (under the pseudonym Walter Ramal) when De La Mare was only 29, this charming collection has long been a favorite children's book. A facsimile of the first edition in paper, the 47 poems capture the simplicity of the nursery rhyme and the ballad, including such lyrics as I Met Eve, Tartary, The Silver Penny. vii + 106pp. 21972-0 Paperbound $1.25

THE COMPLETE NONSENSE OF EDWARD LEAR, Edward Lear. The finest 19th-century humorist-cartoonist in full: all nonsense limericks, zany alphabets, Owl and Pussycat, songs, nonsense botany, and more than 500 illustrations by Lear himself. Edited by Holbrook Jackson. xxix + 287pp. (USO) 20167-8 Paperbound $1.75

BILLY WHISKERS: THE AUTOBIOGRAPHY OF A GOAT, Frances Trego Montgomery. A favorite of children since the early 20th century, here are the escapades of that rambunctious, irresistible and mischievous goat—Billy Whiskers. Much in the spirit of *Peck's Bad Boy,* this is a book that children never tire of reading or hearing. All the original familiar illustrations by W. H. Fry are included: 6 color plates, 18 black and white drawings. 159pp. 22345-0 Paperbound $2.00

MOTHER GOOSE MELODIES. Faithful republication of the fabulously rare Munroe and Francis "copyright 1833" Boston edition—the most important Mother Goose collection, usually referred to as the "original." Familiar rhymes plus many rare ones, with wonderful old woodcut illustrations. Edited by E. F. Bleiler. 128pp. $4\frac{1}{2}$ x $6\frac{3}{8}$. 22577-1 Paperbound $1.25

CATALOGUE OF DOVER BOOKS

TWO LITTLE SAVAGES; BEING THE ADVENTURES OF TWO BOYS WHO LIVED AS INDIANS AND WHAT THEY LEARNED, Ernest Thompson Seton. Great classic of nature and boyhood provides a vast range of woodlore in most palatable form, a genuinely entertaining story. Two farm boys build a teepee in woods and live in it for a month, working out Indian solutions to living problems, star lore, birds and animals, plants, etc. 293 illustrations. vii + 286pp.
20985-7 Paperbound $1.95

PETER PIPER'S PRACTICAL PRINCIPLES OF PLAIN & PERFECT PRONUNCIATION. Alliterative jingles and tongue-twisters of surprising charm, that made their first appearance in America about 1830. Republished in full with the spirited woodcut illustrations from this earliest American edition. 32pp. 4½ x 6⅜.
22560-7 Paperbound $1.00

SCIENCE EXPERIMENTS AND AMUSEMENTS FOR CHILDREN, Charles Vivian. 73 easy experiments, requiring only materials found at home or easily available, such as candles, coins, steel wool, etc.; illustrate basic phenomena like vacuum, simple chemical reaction, etc. All safe. Modern, well-planned. Formerly *Science Games for Children*. 102 photos, numerous drawings. 96pp. 6⅛ x 9¼.
21856-2 Paperbound $1.25

AN INTRODUCTION TO CHESS MOVES AND TACTICS SIMPLY EXPLAINED, Leonard Barden. Informal intermediate introduction, quite strong in explaining reasons for moves. Covers basic material, tactics, important openings, traps, positional play in middle game, end game. Attempts to isolate patterns and recurrent configurations. Formerly *Chess*. 58 figures. 102pp. (USO) 21210-6 Paperbound $1.25

LASKER'S MANUAL OF CHESS, Dr. Emanuel Lasker. Lasker was not only one of the five great World Champions, he was also one of the ablest expositors, theorists, and analysts. In many ways, his Manual, permeated with his philosophy of battle, filled with keen insights, is one of the greatest works ever written on chess. Filled with analyzed games by the great players. A single-volume library that will profit almost any chess player, beginner or master. 308 diagrams. xli x 349pp.
20640-8 Paperbound $2.50

THE MASTER BOOK OF MATHEMATICAL RECREATIONS, Fred Schuh. In opinion of many the finest work ever prepared on mathematical puzzles, stunts, recreations; exhaustively thorough explanations of mathematics involved, analysis of effects, citation of puzzles and games. Mathematics involved is elementary. Translated by F. Göbel. 194 figures. xxiv + 430pp. 22134-2 Paperbound $3.00

MATHEMATICS, MAGIC AND MYSTERY, Martin Gardner. Puzzle editor for Scientific American explains mathematics behind various mystifying tricks: card tricks, stage "mind reading," coin and match tricks, counting out games, geometric dissections, etc. Probability sets, theory of numbers clearly explained. Also provides more than 400 tricks, guaranteed to work, that you can do. 135 illustrations. xii + 176pp.
20338-2 Paperbound $1.50

CATALOGUE OF DOVER BOOKS

"ESSENTIAL GRAMMAR" SERIES

All you really need to know about modern, colloquial grammar. Many educational shortcuts help you learn faster, understand better. Detailed cognate lists teach you to recognize similarities between English and foreign words and roots—make learning vocabulary easy and interesting. Excellent for independent study or as a supplement to record courses.

ESSENTIAL FRENCH GRAMMAR, Seymour Resnick. 2500-item cognate list. 159pp.
(EBE) 20419-7 Paperbound $1.25

ESSENTIAL GERMAN GRAMMAR, Guy Stern and Everett F. Bleiler. Unusual shortcuts on noun declension, word order, compound verbs. 124pp.
(EBE) 20422-7 Paperbound $1.25

ESSENTIAL ITALIAN GRAMMAR, Olga Ragusa. 111pp.
(EBE) 20779-X Paperbound $1.25

ESSENTIAL JAPANESE GRAMMAR, Everett F. Bleiler. In Romaji transcription; no characters needed. Japanese grammar is regular and simple. 156pp.
21027-8 Paperbound $1.25

ESSENTIAL PORTUGUESE GRAMMAR, Alexander da R. Prista. vi + 114pp.
21650-0 Paperbound $1.25

ESSENTIAL SPANISH GRAMMAR, Seymour Resnick. 2500 word cognate list. 115pp.
(EBE) 20780-3 Paperbound $1.25

ESSENTIAL ENGLISH GRAMMAR, Philip Gucker. Combines best features of modern, functional and traditional approaches. For refresher, class use, home study. x + 177pp.
21649-7 Paperbound $1.25

A PHRASE AND SENTENCE DICTIONARY OF SPOKEN SPANISH. Prepared for U. S. War Department by U. S. linguists. As above, unit is idiom, phrase or sentence rather than word. English-Spanish and Spanish-English sections contain modern equivalents of over 18,000 sentences. Introduction and appendix as above. iv + 513pp.
20495-2 Paperbound $2.00

A PHRASE AND SENTENCE DICTIONARY OF SPOKEN RUSSIAN. Dictionary prepared for U. S. War Department by U. S. linguists. Basic unit is not the word, but the idiom, phrase or sentence. English-Russian and Russian-English sections contain modern equivalents for over 30,000 phrases. Grammatical introduction covers phonetics, writing, syntax. Appendix of word lists for food, numbers, geographical names, etc. vi + 573 pp. $6\frac{1}{8}$ x $9\frac{1}{4}$.
20496-0 Paperbound $3.00

CONVERSATIONAL CHINESE FOR BEGINNERS, Morris Swadesh. Phonetic system, beginner's course in Pai Hua Mandarin Chinese covering most important, most useful speech patterns. Emphasis on modern colloquial usage. Formerly *Chinese in Your Pocket*. xvi + 158pp.
21123-1 Paperbound $1.50

CATALOGUE OF DOVER BOOKS

HOW TO KNOW THE WILD FLOWERS, Mrs. William Starr Dana. This is the classical book of American wildflowers (of the Eastern and Central United States), used by hundreds of thousands. Covers over 500 species, arranged in extremely easy to use color and season groups. Full descriptions, much plant lore. This Dover edition is the fullest ever compiled, with tables of nomenclature changes. 174 full-page plates by M. Satterlee. xii + 418pp. 20332-8 Paperbound $2.50

OUR PLANT FRIENDS AND FOES, William Atherton DuPuy. History, economic importance, essential botanical information and peculiarities of 25 common forms of plant life are provided in this book in an entertaining and charming style. Covers food plants (potatoes, apples, beans, wheat, almonds, bananas, etc.), flowers (lily, tulip, etc.), trees (pine, oak, elm, etc.), weeds, poisonous mushrooms and vines, gourds, citrus fruits, cotton, the cactus family, and much more. 108 illustrations. xiv + 290pp. 22272-1 Paperbound $2.00

HOW TO KNOW THE FERNS, Frances T. Parsons. Classic survey of Eastern and Central ferns, arranged according to clear, simple identification key. Excellent introduction to greatly neglected nature area. 57 illustrations and 42 plates. xvi + 215pp. 20740-4 Paperbound $1.75

MANUAL OF THE TREES OF NORTH AMERICA, Charles S. Sargent. America's foremost dendrologist provides the definitive coverage of North American trees and tree-like shrubs. 717 species fully described and illustrated: exact distribution, down to township; full botanical description; economic importance; description of subspecies and races; habitat, growth data; similar material. Necessary to every serious student of tree-life. Nomenclature revised to present. Over 100 locating keys. 783 illustrations. lii + 934pp. 20277-1, 20278-X Two volumes, Paperbound $6.00

OUR NORTHERN SHRUBS, Harriet L. Keeler. Fine non-technical reference work identifying more than 225 important shrubs of Eastern and Central United States and Canada. Full text covering botanical description, habitat, plant lore, is paralleled with 205 full-page photographs of flowering or fruiting plants. Nomenclature revised by Edward G. Voss. One of few works concerned with shrubs. 205 plates, 35 drawings. xxviii + 521pp. 21989-5 Paperbound $3.75

THE MUSHROOM HANDBOOK, Louis C. C. Krieger. Still the best popular handbook: full descriptions of 259 species, cross references to another 200. Extremely thorough text enables you to identify, know all about any mushroom you are likely to meet in eastern and central U. S. A.: habitat, luminescence, poisonous qualities, use, folklore, etc. 32 color plates show over 50 mushrooms, also 126 other illustrations. Finding keys. vii + 560pp. 21861-9 Paperbound $3.95

HANDBOOK OF BIRDS OF EASTERN NORTH AMERICA, Frank M. Chapman. Still much the best single-volume guide to the birds of Eastern and Central United States. Very full coverage of 675 species, with descriptions, life habits, distribution, similar data. All descriptions keyed to two-page color chart. With this single volume the average birdwatcher needs no other books. 1931 revised edition. 195 illustrations. xxxvi + 581pp. 21489-3 Paperbound $3.25

CATALOGUE OF DOVER BOOKS

AMERICAN FOOD AND GAME FISHES, David S. Jordan and Barton W. Evermann. Definitive source of information, detailed and accurate enough to enable the sportsman and nature lover to identify conclusively some 1,000 species and sub-species of North American fish, sought for food or sport. Coverage of range, physiology, habits, life history, food value. Best methods of capture, interest to the angler, advice on bait, fly-fishing, etc. 338 drawings and photographs. 1 + 574pp. 6⅝ x 9⅜.
22383-1 Paperbound $4.50

THE FROG BOOK, Mary C. Dickerson. Complete with extensive finding keys, over 300 photographs, and an introduction to the general biology of frogs and toads, this is the classic non-technical study of Northeastern and Central species. 58 species; 290 photographs and 16 color plates. xvii + 253pp.
21973-9 Paperbound $4.00

THE MOTH BOOK: A GUIDE TO THE MOTHS OF NORTH AMERICA, William J. Holland. Classical study, eagerly sought after and used for the past 60 years. Clear identification manual to more than 2,000 different moths, largest manual in existence. General information about moths, capturing, mounting, classifying, etc., followed by species by species descriptions. 263 illustrations plus 48 color plates show almost every species, full size. 1968 edition, preface, nomenclature changes by A. E. Brower. xxiv + 479pp. of text. 6½ x 9¼.
21948-8 Paperbound $5.00

THE SEA-BEACH AT EBB-TIDE, Augusta Foote Arnold. Interested amateur can identify hundreds of marine plants and animals on coasts of North America; marine algae; seaweeds; squids; hermit crabs; horse shoe crabs; shrimps; corals; sea anemones; etc. Species descriptions cover: structure; food; reproductive cycle; size; shape; color; habitat; etc. Over 600 drawings. 85 plates. xii + 490pp.
21949-6 Paperbound $3.50

COMMON BIRD SONGS, Donald J. Borror. 33⅓ 12-inch record presents songs of 60 important birds of the eastern United States. A thorough, serious record which provides several examples for each bird, showing different types of song, individual variations, etc. Inestimable identification aid for birdwatcher. 32-page booklet gives text about birds and songs, with illustration for each bird.
21829-5 Record, book, album. Monaural. $2.75

FADS AND FALLACIES IN THE NAME OF SCIENCE, Martin Gardner. Fair, witty appraisal of cranks and quacks of science: Atlantis, Lemuria, hollow earth, flat earth, Velikovsky, orgone energy, Dianetics, flying saucers, Bridey Murphy, food fads, medical fads, perpetual motion, etc. Formerly "In the Name of Science." x + 363pp.
20394-8 Paperbound $2.00

HOAXES, Curtis D. MacDougall. Exhaustive, unbelievably rich account of great hoaxes: Locke's moon hoax, Shakespearean forgeries, sea serpents, Loch Ness monster, Cardiff giant, John Wilkes Booth's mummy, Disumbrationist school of art, dozens more; also journalism, psychology of hoaxing. 54 illustrations. xi + 338pp.
20465-0 Paperbound $2.75

CATALOGUE OF DOVER BOOKS

THE PRINCIPLES OF PSYCHOLOGY, William James. The famous long course, complete and unabridged. Stream of thought, time perception, memory, experimental methods—these are only some of the concerns of a work that was years ahead of its time and still valid, interesting, useful. 94 figures. Total of xviii + 1391pp.
20381-6, 20382-4 Two volumes, Paperbound $6.00

THE STRANGE STORY OF THE QUANTUM, Banesh Hoffmann. Non-mathematical but thorough explanation of work of Planck, Einstein, Bohr, Pauli, de Broglie, Schrödinger, Heisenberg, Dirac, Feynman, etc. No technical background needed. "Of books attempting such an account, this is the best," Henry Margenau, Yale. 40-page "Postscript 1959." xii + 285pp. 20518-5 Paperbound $2.00

THE RISE OF THE NEW PHYSICS, A. d'Abro. Most thorough explanation in print of central core of mathematical physics, both classical and modern; from Newton to Dirac and Heisenberg. Both history and exposition; philosophy of science, causality, explanations of higher mathematics, analytical mechanics, electromagnetism, thermodynamics, phase rule, special and general relativity, matrices. No higher mathematics needed to follow exposition, though treatment is elementary to intermediate in level. Recommended to serious student who wishes verbal understanding. 97 illustrations. xvii + 982pp. 20003-5, 20004-3 Two volumes, Paperbound $5.50

GREAT IDEAS OF OPERATIONS RESEARCH, Jagjit Singh. Easily followed non-technical explanation of mathematical tools, aims, results: statistics, linear programming, game theory, queueing theory, Monte Carlo simulation, etc. Uses only elementary mathematics. Many case studies, several analyzed in detail. Clarity, breadth make this excellent for specialist in another field who wishes background. 41 figures. x + 228pp. 21886-4 Paperbound $2.25

GREAT IDEAS OF MODERN MATHEMATICS: THEIR NATURE AND USE, Jagjit Singh. Internationally famous expositor, winner of Unesco's Kalinga Award for science popularization explains verbally such topics as differential equations, matrices, groups, sets, transformations, mathematical logic and other important modern mathematics, as well as use in physics, astrophysics, and similar fields. Superb exposition for layman, scientist in other areas. viii + 312pp.
20587-8 Paperbound $2.25

GREAT IDEAS IN INFORMATION THEORY, LANGUAGE AND CYBERNETICS, Jagjit Singh. The analog and digital computers, how they work, how they are like and unlike the human brain, the men who developed them, their future applications, computer terminology. An essential book for today, even for readers with little math. Some mathematical demonstrations included for more advanced readers. 118 figures. Tables. ix + 338pp. 21694-2 Paperbound $2.25

CHANCE, LUCK AND STATISTICS, Horace C. Levinson. Non-mathematical presentation of fundamentals of probability theory and science of statistics and their applications. Games of chance, betting odds, misuse of statistics, normal and skew distributions, birth rates, stock speculation, insurance. Enlarged edition. Formerly "The Science of Chance." xiii + 357pp. 21007-3 Paperbound $2.00

CATALOGUE OF DOVER BOOKS

MATHEMATICAL PUZZLES FOR BEGINNERS AND ENTHUSIASTS, Geoffrey Mott-Smith. 189 puzzles from easy to difficult—involving arithmetic, logic, algebra, properties of digits, probability, etc.—for enjoyment and mental stimulus. Explanation of mathematical principles behind the puzzles. 135 illustrations. viii + 248pp.
20198-8 Paperbound $1.25

PAPER FOLDING FOR BEGINNERS, William D. Murray and Francis J. Rigney. Easiest book on the market, clearest instructions on making interesting, beautiful origami. Sail boats, cups, roosters, frogs that move legs, bonbon boxes, standing birds, etc. 40 projects; more than 275 diagrams and photographs. 94pp.
20713-7 Paperbound $1.00

TRICKS AND GAMES ON THE POOL TABLE, Fred Herrmann. 79 tricks and games—some solitaires, some for two or more players, some competitive games—to entertain you between formal games. Mystifying shots and throws, unusual caroms, tricks involving such props as cork, coins, a hat, etc. Formerly *Fun on the Pool Table*. 77 figures. 95pp.
21814-7 Paperbound $1.00

HAND SHADOWS TO BE THROWN UPON THE WALL: A SERIES OF NOVEL AND AMUSING FIGURES FORMED BY THE HAND, Henry Bursill. Delightful picturebook from great-grandfather's day shows how to make 18 different hand shadows: a bird that flies, duck that quacks, dog that wags his tail, camel, goose, deer, boy, turtle, etc. Only book of its sort. vi + 33pp. $6\frac{1}{2}$ x $9\frac{1}{4}$. 21779-5 Paperbound $1.00

WHITTLING AND WOODCARVING, E. J. Tangerman. 18th printing of best book on market. "If you can cut a potato you can carve" toys and puzzles, chains, chessmen, caricatures, masks, frames, woodcut blocks, surface patterns, much more. Information on tools, woods, techniques. Also goes into serious wood sculpture from Middle Ages to present, East and West. 464 photos, figures. x + 293pp.
20965-2 Paperbound $2.00

HISTORY OF PHILOSOPHY, Julián Marias. Possibly the clearest, most easily followed, best planned, most useful one-volume history of philosophy on the market; neither skimpy nor overfull. Full details on system of every major philosopher and dozens of less important thinkers from pre-Socratics up to Existentialism and later. Strong on many European figures usually omitted. Has gone through dozens of editions in Europe. 1966 edition, translated by Stanley Appelbaum and Clarence Strowbridge. xviii + 505pp.
21739-6 Paperbound $2.75

YOGA: A SCIENTIFIC EVALUATION, Kovoor T. Behanan. Scientific but non-technical study of physiological results of yoga exercises; done under auspices of Yale U. Relations to Indian thought, to psychoanalysis, etc. 16 photos. xxiii + 270pp.
20505-3 Paperbound $2.50

Prices subject to change without notice.
Available at your book dealer or write for free catalogue to Dept. GI, Dover Publications, Inc., 180 Varick St., N. Y., N. Y. 10014. Dover publishes more than 150 books each year on science, elementary and advanced mathematics, biology, music, art, literary history, social sciences and other areas.

LIBRARY OF DAVIDSON COLLEGE